Lens Design Fundamentals

Second Edition

RUDOLF KINGSLAKE
R. BARRY JOHNSON

AMSTERDAM • BOSTON • HEIDELBERG • LONDON
NEW YORK • OXFORD • PARIS • SAN DIEGO
SAN FRANCISCO • SINGAPORE • SYDNEY • TOKYO
Academic Press is an imprint of Elsevier

Academic Press is an imprint of Elsevier
30 Corporate Drive, Suite 400
Burlington, MA 01803, USA

The Boulevard, Langford Lane
Kidlington, Oxford, OX5 1 GB, UK

Co-published by SPIE
P.O. Box 10
Bellingham, Washington 98227-0010 USA
Tel.: +1 360-676-3290 / Fax: +1 360-647-1445
Email: Books@SPIE.org
SPIE ISBN: 9780819479396
SPIE Vol: PM195

Library of Congress Cataloging-in-Publication Data
Application submitted.

ISBN: 978-0-12-374301-5

British Library Cataloguing-in-Publication Data
A catalogue record for this book is available from the British Library.

For information on all Academic Press publications
visit our Web site at *www.elsevierdirect.com*

Printed in the United States
10 11 12 13   10 9 8 7 6 5 4 3 2

Contents

Preface to the Second Edition

Inasmuch as the first edition of this book could be regarded as an extension and modernization of Professor Alexander Eugen Conrady's *Applied Optics and Optical Design*, this second edition can be viewed as a further extension and modernization of Conrady's 80-year-old treatise.[1] As was stated in the preface to the first edition, referring to Conrady's book, "This was the first practical text to be written in English for serious students of lens design, and it received a worldwide welcome." Until then, optical design was generally in a disorganized state and design procedures were often considered rather mysterious by many.

In 1917, the Department of Technical Optics at the Imperial College of Science and Technology in London was founded. Conrady was invited to the principal teaching position as a result of his two decades of success in designing new types of telescopic, microscopic, and photographic lens systems, and for his work during WWI in designing most of the new forms of submarine periscopes and some other military instruments. Arguably, his greatest achievement was to establish systematic and instructive methods for teaching practical optical design techniques to students and practitioners alike. Without question, Conrady is the father of practical lens design.[2,3]

Rudolf Kingslake (1903–2003) earned an MSc. degree under Professor Conrady, earning himself a commendable reputation while a student and during his early career. Soon after The Institute of Optics was founded in 1929 at the University of Rochester in New York, Kingslake was appointed an Assistant Professor of Geometrical Optics and Optical Design. His contributions to the fields of lens design and optical engineering are legendary. Most lens designers can trace the roots of their education back to Kingslake. Following in Conrady's footsteps, Kingslake is certainly the father of lens design in the United States.

[1] A. E. Conrady, *Applied Optics and Optical Design*, Part I, Oxford Univ. Press, London (1929); also Dover, New York (1957); Part II, Dover, New York (1960).

[2] R. Kingslake and H. G. Kingslake, "Alexander Eugen Conrady, 1866–1944," *Applied Optics*, 5(1):176–178 (1966).

[3] Conrady commented that he limited the content of his book to what the great English electrical engineer Silvamus P. Thomson called "real optics" and excluded purely mathematical acrobatics, which Thomson called "examination optics" (see Ref. 1).

Kingslake published numerous technical papers, was awarded an array of patents, wrote a variety of books, and taught classes in lens design for nearly half a century.[4] Collectively these have had a major impact on practicing lens designers and optical engineers. Perhaps his most important contribution was the first edition of *Lens Design Fundamentals* in 1978, followed in 1983 by *Optical System Design*. In the years since the first edition was published, spectacular advances in optical technology have occurred.

The pervasive infusion of optics into seemingly all areas of our lives, perhaps only dreams in 1978, has resulted in significant developments in optical theory, software, and manufacturing technology. As a consequence, a revised and expanded edition has been produced primarily to address the needs of the lens design beginner, just as was the first edition. Nevertheless, those practitioners desiring to obtain an orderly background in the subject should find this second edition an appropriate book to study because it contains about 50 percent more pages and figures than the first edition by Kingslake.

Revising this book without the participation of its first author presented somewhat of a challenge. The issues of what to retain, change, add, and so on, were given significant consideration. Having taught a number of classes in lens design and optical engineering myself during the past 35 years, often using *Lens Design Fundamentals* as the textbook, the importance of the student mastering the fundamental elements of practical lens design, rather than simply relying on a lens design program, cannot be overemphasized.

Notation and sign conventions used in lens design have varied over the years, but currently almost everyone is using a right-handed Cartesian coordinate system. In preparing this edition, figures, tables, and equations were changed from a left-handed Cartesian coordinate system with the reversed slope angles used by Conrady and Kingslake into a right-handed Cartesian coordinate system. The student may wonder why different coordinate systems have been used over the years. Minimization of manual computation effort is the answer. Elimination of as many minus signs as possible was the objective to both increase computational speed and reduce errors. Today, manual ray tracing is rarely done, so it makes good sense to use a right-handed Cartesian coordinate system, which also makes interfacing with other modeling, CAD, and manufacturing programs easier.

Since the first edition, a number of books have been published on the topic of aberration theory. Some authors of these books tend to suggest that wavefront aberrations are preferable to longitudinal or transverse ray aberrations. In reality, these aberration forms are directly related (see Chapter 4). The approach used by Conrady and Kingslake to study aberrations was to use real ray errors,

---

[4]A selected bibliography of the writings of Rudolf Kingslake is provided in the Appendix of this book.

optical path differences (OPD), and $(D - d)$ for chromatic correction, in contrast to wavefront aberrations expressed by a polynomial or Zernike expansion. In this second edition, the same approach is continued for various reasons, but primarily because experience has shown that beginning lens design students more intuitively comprehend ray aberrations.

The content here has been revised and expanded to reflect the general changes that have occurred since the first edition. Chapter titles remain the same except that a new overview chapter about aberrations has been added. All the chapters have been revised to some extent, often including new examples, significantly more literature references, and additional subject content. The final chapter, discussing automatic lens design, was completely rewritten. Although the types of optical systems had been limited to rotationally symmetric systems, the chapter on mirrors and catadioptric systems was expanded to include a variety of newer systems with some having eccentric pupils. Some material from *Optical System Design* has been incorporated without attribution. The reader will notice that trigonometric ray tracing is still discussed in this edition. The reason is that many concepts are profitably discussed using ray trace information. These discussions and examples contain the ray trace data for students to consider without having to generate it themselves.

The lack of explanations about how to use any particular computer-based lens design program was intentional because such a program is not required to learn the fundamentals; however, the student will find significant benefit in exploring many of the examples using a lens design program to replicate what is shown and perhaps to improve on or change the design. Much can be learned from such experimentation by the student. Following the philosophy of Conrady and Kingslake, this book contains essentially no problems for the student to work since there are numerous fully worked examples of the principles for students to follow and expand on themselves. Instructors can develop their own problems to supplement their teaching style, computational resources, and course objectives.

Lens design is based not only on scientific principles, but also on the talent of the designer. Shannon appropriately titled his book *The Art and Science of Optical Design*.[5] A new feature in this edition is the occasional insertion of a *Designer Note*; these provide the student with additional relevant information that is somewhat out of the flow of the basic text. Reasonable effort has been given to making this edition have improved clarity and to being more comprehensive.

Although many new technologies have become available for lens designers to employ, such as diffractive surfaces, free-form surfaces, systems without

[5]Robert R. Shannon, *The Art and Science of Optical Design*, Cambridge University Press, Cambridge (1997).

symmetry, holographic lenses, polarization, Fresnel surfaces, gradient index lenses, birefringent materials, superconic surfaces, Zernike surfaces, and so on, they intentionally have not been included. Once students and self-taught practitioners have mastered the fundamentals taught in this edition, they should be able to quickly develop the ability to use these other technologies, surfaces, and materials through study of the literature and/or the manual for the lens design program of their choice.

Acknowledgments

In 1968, it was my good fortune to meet Professor Kingslake when he gave a series of lectures on lens design at Texas Instruments and, with his encouragement, I soon went to The Institute of Optics for graduate studies. Not only was he my teacher, but he also became a good friend and mentor for decades. Without question, his teaching style and willingness to share his extraordinary knowledge positively impacted my career in optical design as it did for the multitude of others who had the occasion to study under Kingslake. I am humbled and appreciative to have had the opportunity to prepare this second edition of his book and hope that he would have approved of my revisions.

My sincere gratitude is given to Dr. Jean Michel Taguenang and Mr. Allen Mann whose careful reading of, and comments about, the manuscript resulted in a better book; to Professor Brian Thompson and Mr. Martin Scott for providing access to early documents containing Kingslake's work; and to Thompson for writing "A Special Tribute to Rudolf Kingslake." I acknowledge, with thanks, Professor Jose Sasian, who suggested that I undertake the project of preparing this second edition, and Dr. William Swantner for many constructive discussions on practical optical design. The tireless efforts and professionalism of Marilyn E. Rash, an Elsevier Inc. Project Manager, during the editing, proofreading, and production stages of this book are sincerely appreciated.

R. Barry Johnson
Huntsville, Alabama

Preface to the First Edition

This book can be regarded as an extension and modernization of Conrady's 50-year-old treatise, *Applied Optics and Optical Design*, Part I of which was published in 1929.* This was the first practical text to be written in English for serious students of lens design, and it received a worldwide welcome.

It is obvious, of course, that in these days of rapid progress any scientific book written before 1929 is likely to be out of date in 1977. In the early years of this century all lens calculations were performed slowly and laboriously by means of logarithms, the tracing of one ray through one surface taking at least five minutes. Conrady, therefore, spent much time and thought on the development of ways by which a maximum of information could be extracted from the tracing of a very few rays.

Today, when this can be performed in a matter of seconds or less on a small computer—or even on a programmable pocket calculator—the need for Conrady's somewhat complicated formulas has passed, but they remain valid and can be used profitably by any designer who takes the trouble to become familiar with them. In the same way, the third-order or Seidel aberrations have lost much of their importance in lens design. Even so, in some instances such as the predesign of a triplet photographic objective, third-order calculations still save an enormous amount of time.

Since Conrady's day, a great deal of new information has appeared, and new procedures have been developed, so that a successor to Conrady's book is seriously overdue. Many young optical engineers today are designing lenses with the aid of an optimization program on a large computer, but they have little appreciation of the how and why of lens behavior, particularly as these computer programs tend to ignore many of the classical lens types that have been found satisfactory for almost a century. Anyone who has had the experience of designing lenses by hand is able to make much better use of an optimization program than someone who has just entered the field, even though that newcomer may have an excellent academic background and be an expert in computer operation.

For this reason an up-to-date text dealing with the classical processes of lens design will always be of value. The best that a computer can do is to optimize

---

*A. E. Conrady, *Applied Optics and Optical Design, Part I*, Oxford University Press, London (1929); also Dover (1957); *Part II*, Dover, New York (1960).

the system given to it, so the more understanding and competent the designer, the better the starting system he will be able to give the computer. A perceptive preliminary study of a system will often indicate how many solutions exist in theory and which one is likely to yield the best final form.

A large part of this book is devoted to a study of possible design procedures for various types of lens or mirror systems, with fully worked examples of each. The reader is urged to follow the logic of these examples and be sure that he understands what is happening, noticing particularly how each available degree of freedom is used to control one aberration. Not every type of lens has been considered, of course, but the design techniques illustrated here can be readily applied to the design of other, more complex systems. It is assumed that the reader has access to a small computer to help with the ray tracing; otherwise, he may find the computations so time-consuming that he is liable to lose track of what he is trying to accomplish.

Conrady's notation and sign conventions have been retained, except that the signs of the aberrations have been reversed in accordance with current practice. Frequent references to Conrady's book have been given in footnotes as "Conrady, p. ...."; and as the derivations of many important formulas have been given by Conrady and others, it has been considered unnecessary to repeat them here. In the last chapter a few notes have been added (with the help of Donald Feder) on the structure of an optimization program. This information is for those who may be curious to know what must go into such a program and how the data are handled.

This book is the fruit of years of study of Conrady's unique teaching at the Imperial College in London, of 30 years of experience as Director of Optical Design at the Eastman Kodak Company, and of almost 45 years of teaching lens design in The Institute of Optics at the University of Rochester—all of it a most rewarding and never-ending education for me, and hopefully also for my students.

Rudolf Kingslake

A Special Tribute to Rudolf Kingslake

Rudolf Kingslake's very first paper, written when a student at Imperial College London, was coauthored by L. C. Martin, a faculty member. The paper, "The Measurement of Chromatic Aberration on the Hilger Lens Testing Interferometer," was received 14 February 1924 and read and discussed 13 March 1924. Immediately following it was a paper by Miss H. G. Conrady, listed as a research scholar since she had already graduated in 1923. Miss Conrady's paper was entitled "Study and Significance of the Foucault Knife-Edge Test When Applied to Refracting Systems" (received 21 February 1924; read and discussed 13 March 1924).

The formal degree program in optics at Imperial College was founded in the summer of 1917 and entered its first class in 1920. Hilda Conrady was a member of that class. Her father was A. E. Conrady, who had been appointed a Professor of Optical Design. Professor Conrady's work and publications were definitive in the literature and in the teaching of optical design. In 1991, Hilda wrote a fine article in *Optics and Photonics News* describing "The First Institute of Optics in the World."

Hilda and Rudolf became lifetime partners when they married on September 14, 1929, soon before they left England because Rudolf had been appointed as the first member in the newly formed Institute of Applied Optics at the University of Rochester in New York. It is interesting to note that for the academic year 1936–1937, L. C. Martin, on the faculty of the Technical Optics Department at Imperial College London, and Rudolf Kingslake exchanged faculty positions. With Rudolf's usual sense of humor, he commented that "Martin and I exchanged jobs, houses and cars . . . but not wives."

With the publication of this new edition of *Lens Design Fundamentals*, which originally appeared in 1978, Kingslake's published works cover a period of 86 years! His last major new publication was *The Photographic Manufacturing Companies of Rochester, New York,* published by The International Museum of Photography at the George Eastman House in 1997; so even using this data point his publications covered 73 years! We should also note that his extensive teaching record extended well into his 80s and touched thousands of students. His "Summer School" courses were indeed legendary.

The Early Years

Rudolf Kingslake's interest in optics started in his school days; he wrote about his "entrance into optics" and said, "father had a camera handbook issued by Beck that contained many diagrams of lens sections, which got me wondering why camera lenses had four or six even eight elements?" This interest continued and he noted, "so when I found out that lens design was taught at Imperial College in South Kensington, I was determined to go there. The college fees were not too expensive and father soon agreed to my plan." Thus Rudolf entered the program in 1921, graduated in 1924, continued on into graduate school with a two-year fellowship, and earned his M.Sc. degree in 1926. And so, a very distinguished career was launched.

His graduate work at Imperial College was very productive, and a number of significant papers were published including works such as "A New Type of Nephelometer," "The Interferometer Patterns due to Primary Aberrations," "Recent Developments of the Hartmann Test to the Measurement of Oblique Aberrations," "The Analysis of an Interferogram," "Increased Resolving Power in the Presence of Spherical Aberration," and "An Experimental Study of the Best Minimum Wavelength for Visual Achromatism."

After graduation Rudolf was appointed to a position at Sir Howard Grubb Parsons and Co. in Newcastle-upon-Tyne as an optical designer. His notes say, "designed Hartmann Plate, measuring microscopic and readers for Edinburgh 30-inch Reflector. Took many photographs, translated German papers, Canberra 18-inch Coelostat device, Mica tests, etc." In June 1928, he published a paper in *Nature* entitled "18-inch Coelostat for Canberra Observatory."

Apparently Parsons didn't have enough work for him to do, so he accepted an appointment with International Standard Electric Company in Hendon, North London. In Hendon he "worked on speech quality over telephone lines and made lab measurements of impedance using Owen's bridge at various frequencies from 50 to 800 (cps). This experience was good for me as it gave me a glance at the business of electronics, designing telephones. I was paid weekly, so gave them a week's notice when I went to America."

The Institute of Applied Optics

Once in the Institute, Kingslake quickly developed the necessary courses and laboratory work in the Eastman Building on the Prince Street Campus. Dr. A. Maurice Taylor, also from England, joined the Institute with responsibility for physical optics. The permanent home was the fourth floor of the newly constructed Bausch and Lomb Hall on the River Campus. Despite a heavy teaching and planning load, Rudolf managed to produce a number of significant publications for major journals. These included "A New Bench for Testing

Photographic Lenses," which became the standard in the United States. A joint paper with A. B. Simmons, who was an M.S. graduate student in optics, reported on "A Method of Projecting Star Images Having Coma and Astigmatism." Then followed "The Development of the Photographic Objective" and "The Measurement of the Aberrations of a Microscope Objective."

The final paper during that period (1929–1937) was a joint paper with Hilda Kingslake writing under her maiden name of H. G. Conrady entitled "A Refractometer for the Near Infrared"; she was working as an independent researcher. Rudolf reports that "in this joint paper the design of the refractometer was mine. Miss Conrady assisted with the assembly, adjustment and calibration and made many of the measurements on glass prisms."

The Kodak Years

Even though Rudolf moved in 1937 to Eastman Kodak at the request of Dr. Mees, Kodak's Director of Research, a very important arrangement was made for Kingslake to continue to teach on a half-time basis—a position that he held long after his retirement. His last Summer School in Optical Design was held in his 90th year.

Although the work at Kodak was often proprietary (and even classified during the war years), he was able to publish a continual stream of important papers in a wide range of professional refereed journals associated with major scientific and engineering societies. At the time of his move to Kodak, Rudolf commented that his "industrial experience had been lamentably brief—that more than anything else, he needed experience in industry for greater competence in teaching an applied subject." He was correct of course. In 1939, the Institute of Applied Optics had a slight name change to The Institute of Optics.

Once Kingslake joined Kodak, he quickly made significant contributions to the design and evaluation of photographic lenses for both still photography and motion picture equipment. Topics included wide-aperture photographic objectives, resolution testing on 16-mm projection lenses, lenses for aerial photography, new optical glasses, zoom lenses, and much more (see the Appendix for specifics).

Some of the summary articles give an excellent perspective of the state of the art and its impact. His paper "The Contributions of Optics to Modern Technology and a Buoyant Economy" is a good example of the results of his exposure to the industrial world. In a joint paper, "Optical Design at Kodak," with two members of his team, he summarized his work at Kodak. Finally in 1982 he produced "My Fifty Years of Lens Design." What a good summary!

Books

Kingslake had an impact on the discipline of optical science and engineering through his writings in a number of texts and contributions he made to various handbooks. His first single-author volume, *Lenses in Photography,* was published in 1951; the 1963 second edition turned out to be a classic.

In 1929, Professor Conrady had published Part I of his book, *Applied Optics and Optical Design,* but he was not able to complete Part II before his death in 1944. He did, however, leave "a well advanced manuscript in his remarkably clear handwriting." Rudolf and Hilda worked together to compile and edit the manuscript for publication in 1960. Hilda added a biography of her father that appears as an appendix in Part II. Part I and Part II were released together by Dover. The *Journal of Applied Optics* published a revised version of the Conrady biography (see *Appl. Opt.,* 5(1):176–178, 1966). Next came two chapters in the *SPSE Handbook of Photographic Science and Engineering* on "Classes of Lenses" and "Projection." "Camera Optics" appeared in the Fifteenth Edition of the *Leica Manual.*

His major work, however, was *Lens Design Fundamentals* published by Academic Press in 1978. This new edition is authored by R. Kingslake and R. Barry Johnson and is significantly revised and expanded to encompass many of the significant advances in optical design that have occurred in the past three decades. Academic Press published two more Kingslake books: *Optical System Design* (1983) and *A History of the Photographic Lens* (1989). In 1992, SPIE Optical Engineering Press published *Optics in Photography,* which was a much revised version of *Lenses in Photography.*

Kingslake's final single-author volume, mentioned earlier, *The Photographic Manufacturing Companies of Rochester, New York,* was published by The International Museum of Photography at the George Eastman House. Rudolf was a dedicated volunteer expert curator of the camera collection together with auxiliary equipment. As a result of his work, he wrote many articles in the Museum's house journal *Image.* These articles started in 1953 and continued into the 1980s; Rudolf called them notes!

Working with his publisher, Academic Press, Rudolf launched and edited the series *Applied Optics and Optical Engineering.* The first three volumes were published in 1965 and Kingslake contributed chapters to all of them. The next two volumes appeared in 1967 and 1969; they were devoted to "Optical Instruments" as a two-volume set (Part I and Part II). This writer was asked to join Rudolf as a coeditor of Volume VI (and to contribute a chapter, of course). The series continued under the editorship of Robert Shannon and James Wyant with Rudolf Kingslake as Consulting Editor.

Acknowledgments

In the later years of Rudolf Kingslake's life, he asked me if I would take charge of his affairs. After consultation with his son and with the family lawyer, I agreed to take power of attorney and subsequently be executor of his will, since his son predeceased him. At his request, I agreed to take all his professional and personal papers to create an archive in the Department of Rare Books and Special Collections of the Rush Rhees Library at the University of Rochester.

Martin Scott, a long-time colleague and friend of Rudolf's at Kodak and at the International Museum of Photography, joined me and did the major part of the work of putting the archive together. In addition, Nancy Martin, the John M. and Barbara Keil University Archivist, was invaluable, knowledgeable, and dedicated to our task. The catalogue of this archive is now available online.

This tribute to him is a selected and revised version of a Plenary address, "Life and Works of Rudolf Kingslake," presented at the conference on optical design and engineering held in St. Etienne, France, 30 September to 3 October 2003; it was published in 2004 in the proceedings (*Proc. SPIE*, 5249:1–21).

Brian J. Thompson
Provost Emeritus
University of Rochester
Rochester, New York

Chapter 1

The Work of the Lens Designer

Before a lens can be constructed it must be designed, that is to say, the radii of curvature of the surfaces, the thicknesses, the air spaces, the diameters of the various components, and the types of glass to be used must all be determined and specified.[1,2] The reason for the complexity in lenses is that in the ideal case all the rays in all wavelengths originating at a given object point should be made to pass accurately through the image of that object point, and the image of a plane object should be a plane, without any appearance of distortion (curvature) in the images of straight lines.

Scientists always try to break down a complex situation into its constituent parts, and lenses are no exception. For several hundred years various so-called aberrations have been recognized in the imperfect image formed by a lens, each of which can be varied by changing the lens structure. Typical aberrations are spherical aberration, comatic, astigmatic, and chromatic, but in any given lens all the aberrations appear mixed together, and correcting (or eliminating) one aberration will improve the resulting image only to the extent of the amount of that particular aberration in the overall mixture. Some aberrations can be easily varied by merely changing the shape of one or more of the lens elements, while others require a drastic alteration of the entire system.

The lens parameters available to the designer for change are known as "degrees of freedom." They include the radii of curvature of the surfaces, the thicknesses and airspaces, the refractive indices and dispersive powers of the glasses used for the separate lens elements, and the position of the "stop" or aperture-limiting diaphragm or lens mount. However, it is also necessary to maintain the required focal length of the lens at all times, for otherwise the relative aperture and image height would vary and the designer might end up with a good lens but not the one he set out to design. Hence each structural change that we make must be accompanied by some other change to hold the focal length constant. Also, if the lens is to be used at a fixed magnification, that magnification must be maintained throughout the design.

The word "lens" is ambiguous, since it may refer to a single element or to a complete objective such as that supplied with a camera. The term "system" is often used for an assembly of units such as lenses, mirrors, prisms, polarizers, and detectors. The name "element" always refers to a single piece of glass having polished surfaces, and a complete lens thus contains one or more elements. Sometimes a group of elements, cemented or closely airspaced, is referred to as a "component" of a lens. However, these usages are not standardized and the reader must judge what is meant when these terms appear in a book or article.

1.1 RELATIONS BETWEEN DESIGNER AND FACTORY

The lens designer must establish good relations with the factory because, after all, the lenses that he designs must eventually be made. He should be familiar with the various manufacturing processes and work closely with the optical engineers. He must always bear in mind that lens elements cost money, and he should therefore use as few of them as possible if cost is a serious factor. Sometimes, of course, image quality is the most important consideration, in which case no limit is placed on the complexity or size of a lens. Far more often the designer is urged to economize by using fewer elements, flatter lens surfaces so that more lenses can be polished on a single block, lower-priced types of glass, and thicker lens elements since they are easier to hold by the rim in the various manufacturing operations.

1.1.1 Spherical versus Aspheric Surfaces

In almost all cases the designer is restricted to the use of spherical refracting or reflecting surfaces, regarding the plane as a sphere of infinite radius. The standard lens manufacturing processes[3,4,5,6,7] generate a spherical surface with great accuracy, but attempts to broaden the designer's freedom by permitting the use of nonspherical or "aspheric" surfaces historically lead to extremely difficult manufacturing problems; consequently such surfaces were used only when no other solution could be found. The aspheric plate in the Schmidt camera is a classic example. In recent years, significant effort has been expended in developing manufacturing and testing technology to fabricate, on a commercial scale, aspheric surfaces for elements such as mirrors, infrared lenses, and glass lenses.[8,9,10,11,12] New fabrication technologies such as single-point diamond turning, reactive ion etching, and computer-controlled free-form grinding and polishing have greatly increased the design space for lens designers. Also, molded aspheric surfaces are very practical and can be used wherever the production rate is sufficiently high to justify the cost of the mold; this applies particularly to plastic lenses made by injection molding.

In addition to the problem of generating and polishing a precise aspheric surface, there is the further matter of centering. Centered lenses with spherical surfaces have an optical axis that contains the centers of curvature of all the surfaces, but an aspheric surface has its own independent axis, which must be made to coincide with the axis containing all the other centers of curvature in the system. In the first edition of this book, it was noted that most astronomical instruments and a few photographic lenses and eyepieces have been made with aspheric surfaces, but the lens designer was advised to avoid such surfaces if at all possible.

Today, the situation has changed significantly and aspheric lenses are more commonly incorporated in designs primarily because of advances in manufacturing technologies that provide quality surfaces in a reasonable time frame and at a reasonable cost. Many of the better photographic lenses now sold by companies such as Canon and Nikon, for example, incorporate one or more aspheric surfaces. The lens designer needs to be aware of which glasses can currently be molded and aspherized by grinding or other processes. As mentioned previously, maintaining good communications with the fabricator cannot be overstressed.

1.1.2 Establishment of Thicknesses

Negative-power lens elements should have a center thickness between 6 and 10% of the lens diameter,[13] but the establishment of the thickness of a positive element requires much more consideration. The glass blank from which the lens is made must have an edge thickness of at least 1 mm to enable it to be held during the grinding and polishing operations (Figure 1.1). At least 1 mm will be removed in edging the lens to its trim diameter, and we must allow at least another 1 mm in radius for support in the mount. With these allowances in mind, and knowing the surface curvatures, the minimum acceptable center thickness of a positive lens can be determined. These specific limitations refer to a lens of average size, say $\frac{1}{2}$ to 3 in. in diameter; they may be somewhat reduced for small lenses, and they must be increased for large ones. A knife-edge lens is very hard to make and handle and it should be avoided wherever possible. A discussion of these matters with the glass-shop foreman can be very profitable. Remember that the space between the clear and trim diameters shown in Figure 1.1 is where the lens is held. The lens designer needs to be sure that the mounting will not vignette any rays.

As a general rule, weak lens surfaces are cheaper to make than strong surfaces because more lenses can be polished together on a block. However,

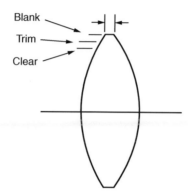

Figure 1.1 Assigning thickness to a positive element.

if only a single lens is to be made, multiple blocks will not be used, and then a strong surface is no more expensive than a weak one.

A small point but one worth noting is that a lens that is nearly equiconvex is liable to be accidentally cemented or mounted back-to-front in assembly. If possible such a lens should be made exactly equiconvex by a trifling bending, any aberrations so introduced being taken up elsewhere in the system. Another point to note is that a very small edge separation between two lenses is hard to achieve, and it is better either to let the lenses actually touch at a diameter slightly greater than the clear aperture, or to call for an edge separation of one millimeter or more, which can be achieved by a spacer ring or a rigid part of the mounting. Remember that the clearance for a shutter or an iris diaphragm must be counted from the bevel of a concave surface to the vertex of a convex surface.

Some typical forms of lens mount are shown in Figure 1.2. When designing a lens, it is wise to keep in mind what type of mounting might be employed and

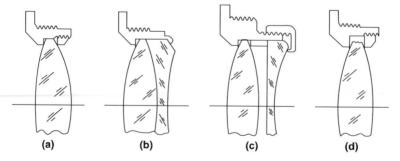

Figure 1.2 Some typical lens mounts: (a) Clamp ring, (b) spinning lip, (c) spacer and screw cap, and (d) mount centering.

any required physical adjustments for alignment. This can make the overall lens development project progress smoother. A study of optomechanics taught by Yoder can be of much benefit to the lens designer.[14,15,16] In many cases, the optomechanical structure of the lens needs to be integrated into the larger system and modeled to ensure that overall system-level performance will be realized in the actual system.[17]

1.1.3 Antireflection Coatings

Today practically all glass–air lens surfaces are given an antireflection coating to improve the light transmission and to eliminate ghost images. Since many lenses can be coated together in a large bell jar, the process is surprisingly inexpensive. However, for the most complete elimination of surface reflection over a wide wavelength range, a multilayer coating is required, and the cost then immediately rises. In the past few decades, great strides have been made in the design and production of high-efficiency antireflective coatings for optical material in both the visible and infrared spectrums.[18,19]

1.1.4 Cementing

Small lens elements are often cemented together, using either Canada balsam or some suitable organic polymer. However, in lenses of diameter over about 3 in., the differential expansion of crown and flint glasses is prone to cause warpage or even fracture if hard cement is used. Soft yielding cements or a liquid oil can be introduced between adjacent lens surfaces, but in large sizes it is more usual to separate the surfaces by small pieces of tinfoil or an actual spacer ring. The cement layer is (almost) always ignored in raytracing, the ray being refracted directly from one glass to the next.

The reasons for cementing lenses together are (a) to eliminate two-surface reflection losses, (b) to prevent total reflection at the air film, and (c) to aid in mounting by combining two strong elements into a single, much weaker cemented doublet. The relative centering of the two strong elements is accomplished during the cementing operation rather than in the lens mount, which is most generally preferred.

Cementing more than two lens elements together can be done, but it is very difficult to secure perfect centering of the entire cemented component. The designer is advised to consult with the manufacturing department before planning to use a triple or quadruple cemented component. Precise cementing of lenses is not a low-cost operation, and it is often cheaper to coat two surfaces that are airspaced in the mount rather than to cement these surfaces together.

1.1.5 Establishing Tolerances

It is essential for the lens designer to assign a tolerance to every dimension of a lens, for if he does not do so somebody else will, and that person's tolerances may be completely incorrect. If tolerances are set too loose a poor lens may result, and if too tight the cost of manufacture will be unjustifiably increased. This remark applies to radii, thicknesses, airspaces, surface quality, glass index and dispersion, lens diameters, and perfection of centering. These tolerances are generally found by applying a small error to each parameter, and tracing sufficient rays through the altered lens to determine the effects of the error.

Knowledge of the tolerances on glass index and dispersion may make the difference between being able to use a stock of glass on hand, or the necessity of ordering glass with an unusually tight tolerance, which may seriously delay production and raise the cost of the lens. When making a single high-quality lens, it is customary to design with catalog indices, then order the glass, and then redesign the lens to make use of the actual glass received from the manufacturer. On the other hand, when designing a high-production lens, it is necessary to adapt the design to the normal factory variation of about ±0.0005 in refractive index and ±0.5% in V value.[20]

Matching thicknesses in assembly is a possible though expensive way to increase the manufacturing tolerances on individual elements. For instance, in a Double-Gauss lens of the type shown in Figure 1.3, the designer may determine permissible thickness tolerances for the two cemented doublets in the following form:

each single element: ±0.2 mm
each cemented doublet: ±0.1 mm
the sum of both doublets: ±0.02 mm

Clearly such a matching scheme requires that a large number of lenses be available for assembly, with a range of thicknesses. If every lens is made on the thick side no assemblies will be possible.

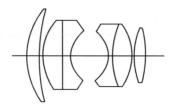

Figure 1.3   A typical Double-Gauss lens.

Very often the most important tolerances to specify are those for surface tilt and lens element decentration. A knowledge of these can have a great effect on the design of the mounting and on the manufacturability of the system. A decentered lens generally shows coma on the axis, whereas a tilted element often leads to a tilted field. Some surfaces are affected very little by a small tilt, whereas others may be extremely sensitive in this regard. A table of tilt coefficients should be in the hands of the optical engineers before they begin work on the mount design.

The subject of optical tolerancing is almost a study in itself, and the setting of realistic tolerances is far from being an obvious or simple matter. Table 1.1 presents the generally accepted tolerances for a variety of optical element attributes at three production levels, namely commercial quality, precision quality, and manufacturing limits. Tolerances for injection molded polymer optics are given in Table 1.2.[21]

Table 1.1

Optics Manufacturing Tolerances for Glass

Attribute	Commercial Quality	Precision Quality	Manufacturing Limits
Glass Quality (n_d, v_d)	±0.001, ±0.8%	±0.0005, ±0.5%	Melt controlled
Diameter (mm)	+0.00/−0.10	+0.000/−0.025	+0.000/−0.010
Center Thickness (mm)	±0.150	±0.050	±0.025
Sag (mm)	±0.050	±0.025	±0.010
Clear Aperture	80%	90%	100%
Radius	±0.2% or 5 fr	±0.1% or 3 fr	±0.0025 mm or 1 fr
Irregularity – Interferometer (fringes)	2	0.5	0.1
Irregularity – Profilometer (microns)	±10	±1	±0.1
Wedge Lens (ETD, mm)	0.050	0.010	0.002
Wedge Prism (TIA, arc min)	±5	±1	±0.1
Bevels (face width @ 45°, mm)	<1.0	<0.5	No Bevel
Scratch – Dig (MIL-PRF-13830B)	80−50	60−40	5−2
Surface Roughness (Å rms)	50	20	2
AR Coating (R_{ave})	MgF$_2$ R<1.5%	BBAR, R<0.5%	Custom Design

Source: Reprinted by permission of Optimax Systems, Inc.

Table 1.2

Optics Manufacturing Tolerances for Plastics

Attribute	Tolerances (rotationally symmetrical elements less than 75 mm in diameter)
Radius of Curvature	±0.5%
EFL	±1.0%
Center Thickness	±0.020 mm
Diameter	±0.020 mm
Wedge (TIR) in Element	<0.010 mm
S1 to S2 Displacement (across the mold parting line)	<0.020 mm
Surface Figure Error	≤2 fringes per inch (2 fringes = 1 λ)
Surface Irregularity	≤1 fringe per inch (2 fringes = 1 λ)
Scratch-Dig Specification	40–20
Surface Roughness Specification (RMS)	<50 Å
Diameter to Thickness Ratio	<4:1
Center Thickness to Edge Thickness Ratio	<3:1
Part to Part Repeatability (one cavity)	<0.50%

Source: Reprinted by permission of G-S Plastic Optics.

1.1.6 Design Tradeoffs

The lens designer is often confronted with a variety of ways to achieve a given result, and the success of a project may be greatly influenced by his choice. Some of these alternatives are as follows: Should a mirror or lens system be used? Can a strong surface be replaced by two weaker surfaces? Can a lens of high-index glass be replaced by two lenses of more common glass? Can an aspheric surface be replaced by two spherical surfaces? Can a long-focus lens working at a narrow angular field be replaced by a short-focus lens covering a wider field? Can a zoom lens be replaced by a series of normal lenses, giving a stepwise variation of magnification? If two lens systems are to be used in succession, how should the overall magnification be divided between them? Is it possible to obtain sharper definition if some unimportant aberration can be neglected?

1.2 THE DESIGN PROCEDURE

A closed mathematical solution for the constructional data of a lens in terms of its desired performance would be much too complex to be a real possibility. The best we can do is to use our knowledge of optics to set up a likely first approach to the desired lens, evaluate it, make judicious changes, reevaluate

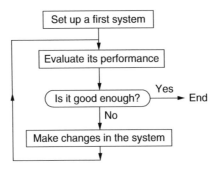

Figure 1.4 Lens design flow chart.

it, and so on. The process may be illustrated by a simple flow chart (Figure 1.4). These four steps will be considered in turn. Throughout this book, a plethora of guidance for design techniques is presented. In Chapter 17, the elements of automatic lens design are discussed along with a brief discussion of the historical evolution of methods of ray tracing and performing optimization.

1.2.1 Sources of a Likely Starting System

In some cases, such as a simple telescope doublet, a lens design can be generated from first principles by a series of logical operations followed in a prescribed order. This is, however, exceptional. Far more often we obtain a likely starting system by one of the following means:

1. A mental guess. This may work well for an experienced designer but it is hopeless for a beginner.
2. A previously designed lens in the company files. This is the most usual procedure in large companies, but most firms not strongly involved in lens development will not have such files.
3. Purchase of a competing lens and analysis of its structure. This is laborious and time-consuming, but it has often been done, especially in small firms with very little backlog of previous designs to choose from.
4. A search through the patent files or of a (commercial) lens design database.

There are literally thousands of lens patents on file, but often the examples given are incomplete or not very well-corrected; such a starting point may require a great deal of work before it is usable, not to mention the necessity of avoiding the claims in the patent itself! A classic book by Cox[22] includes an analysis of 300 lens patent examples, which many lens designers have found quite useful. Today, there are tens of thousands of patents on lens designs, which makes a conventional patent search a rather daunting endeavor. Fortunately there are few databases that can be of significant assistance to the lens designer in looking for a potential starting point.[23,24]

1.2.2 Lens Evaluation

This is generally performed by tracing a sufficient number of rays through the lens by accurate trigonometrical methods. At first only two or three rays are required, but as the design progresses more rays must be added to provide an adequate evaluation of the system. There are a variety of graphs that can be plotted to represent the various aberrations, and a glance at these will often suggest to the designer what is wrong with the system. In addition to ray error plots, the ray data can be used for a number of purposes including analysis of wavefront error, encircled energy, line scans, optical transfer function, point spread function, and so on (see Section 8.4).

At the time of the first edition of this book, it was unthinkable to be able to perform most of these complex analyses on anything less than a mainframe computer, and then at a nontrivial cost. Today, such analyses can be performed on a laptop costing under a thousand dollars, in a very timely manner, and the cost per run is essentially nil if the costs of the laptop, software license, and annual support are ignored.

1.2.3 Lens Appraisal

It is often very difficult to decide whether or not a given lens system is sufficiently well-corrected for a particular application.[25] The usual method is to trace a large number of rays from a point source in a uniformly distributed array over the vignetted entrance pupil of the lens, and then plot a "spot diagram" of the points at which these rays pierce the image plane. It may be necessary to trace several hundred rays before a realistic appearance of the point image is obtained (see Section 8.4). Chromatic errors can be included in the spot diagram by tracing sets of rays in several wavelengths, the spacing of the rays as they enter the lens being adjusted in accordance with the weight to be assigned to that wavelength in the final image.

To interpret the significance of a spot diagram, some designers calculate the diameters of circles containing 10, 20, 30, ..., 100% of the rays, and thus plot a graph of "encircled energy" at each obliquity. An alternative procedure is to regard the spot diagram as a point spread function, and by means of a Fourier transform convert it into a curve of MTF (modulation transfer function) plotted against spatial frequency. Such a graph contains very much information both as to the resolving power of the lens and the contrast in the image of coarse objects. Moreover, in calculating the MTF values, diffraction effects can be taken into account, the result being the most comprehensive representation of lens performance that can be obtained. If the lens is then constructed with dimensions agreeing exactly with the design data, it is possible to measure the MTF experimentally and verify that the lens performance has come up to the theoretical expectations.

1.2.4 System Changes

When working by hand or with a small computer, the designer will have to decide what changes he should make to remove the residual aberrations in his lens. This is often a very difficult problem, and in the following chapters many hints are given as to suitable modifications that should be tried even when using a lens design program. Often a designer will make small trial changes in some of the lens parameters and determine the rate of change, or "coefficient," of each aberration with respect to each change. The solution of a few simultaneous equations will then indicate some reasonable changes that might be tried, although the extreme nonlinearity of all optical systems makes this procedure not as simple as one would like.

Today there are many programs for use on a high-speed computer in which a large number of aberrations are changed simultaneously by varying several lens parameters, using a least-squares technique. In spite of the enormous amount of computation required in this process, it can be performed remarkably cheaply on today's personal computers (see Chapter 17). A skew ray trace through a spherical surface would take an experienced human computer using a Marchant mechanical calculator about 500 seconds per ray surface (pre-1955). Today, the time has been reduced using a multiprocessor personal computer to about less than 10 ns or about fifty billion times faster!

1.3 OPTICAL MATERIALS

The most common lens material is, of course, optical glass, but crystals and plastics are frequently used, while mirrors can be made of essentially anything that is capable of being polished. Liquid-filled lenses have often been proposed, but for many obvious reasons they were practically never used until recently.[26,27,28,29] Optical materials in general have been discussed by Kreidl and Rood[30] and others.[31,32]

1.3.1 Optical Glass

There are several well-known manufacturers of optical glass, and their catalogs give an enormous amount of information about the glasses that are available; in particular, the Schott catalog is virtually a textbook of optical glasses and their properties.

Optical glasses are classified roughly as crowns, flints, barium crowns, and so on, but the boundaries of the various classes are not tightly standardized (see Figure 5.5). Optically, glasses differ from one another in respect to refractive index, dispersive power, and partial dispersion ratio, while physically they differ in color, density, thermal properties, chemical stability, bubble content, striae, and ease of polishing.

Glasses vary enormously in cost, over a range of at least 300 to 1 from the densest lanthanum crowns to the most common ordinary plate glass, which is good enough for many simple applications. One of the lens designer's most difficult problems is how to make a wise choice of glass types, and in doing so he must weigh several factors. A high refractive index leads to weaker surfaces and therefore smaller aberration residuals, but high-index glasses are generally expensive, and they are also dense so that a pound of glass makes fewer lenses. If lens quality is paramount, then of course any glass can be used, but if cost is important the lower-cost glasses must be chosen.

The cost of material in a small lens is likely to be insignificant, but in a large lens it may be a very serious matter, particularly as only a few types are made in large pieces (the so-called "massive optics"), and the price per pound is likely to vary as much as the cube of the weight of the piece. It is perhaps surprising to note that in a lens of 12 in. diameter made of glass having a density of 3.5, each millimeter in thickness adds nearly 0.75 lb to the weight.

The color of glass is largely a matter of impurities, and some manufacturers offer glass with less yellow color at a higher price. This is particularly important if good transmission in the near ultraviolet is required. A trace of yellow color is often insignificant in a very small or a very thin lens and, of course, in aerial camera lenses yellow glass is quite acceptable because the lens will be used with a yellow filter anyway.

It will be found that the cost of glass varies greatly with the form of the pieces, whether in random slabs or thin rolled sheets, whether it is annealed, and whether it has been selected on the basis of low stria content. Some lens makers habitually mold their own blanks, and then it is essential to give these blanks a slow anneal to restore the refractive index to its stable maximum value; this is the value stated by the manufacturer on the melt sheet supplied with the glass.

A most useful feature of modern lens design programs is their inclusion of extensive catalogs of the optical properties of glasses available from the various suppliers as well as many plastics and materials useful in the infrared.

1.3.2 Infrared Materials

Infrared-transmitting materials are a study in themselves, and many articles have appeared in books and journals listing these substances and their properties.[33] With few exceptions, they are not generally usable in the visible, however, because of light scatter at the crystal boundaries. An example of an exception is CLEARTRAN™ which is a water-free zinc sulfide material with transmittance from about 0.4 to 12 μm.

1.3.3 Ultraviolet Materials

For the ultraviolet region of the spectrum we have only relatively few materials that include UV-grade fused silica, crystal quartz, calcium fluoride, magnesium fluoride, sapphire, and lithium fluoride, with a few of the lighter glasses when in thin sections. With the advent of integrated circuits, the demand for finer and finer optical resolution to make masks to produce the integrated circuits and to image onto the silicon wafer, significant design and fabrication effort has been expended over the past several decades. Often these optical systems are catadioptric (see Chapter 15), but sometimes they are purely refractive. It should also be realized that these lenses are very, very expensive due to the cost of materials, fabrication, and alignment.[34,35,36]

1.3.4 Optical Plastics

In spite of the paucity of available types of optical plastics suitable for lens manufacture, plastics have found extensive application in this field since World War I and particularly since the early 1950s.[37,38,39] Since that time hundreds of millions of plastic lenses have been fitted to inexpensive cameras, and they are now used regularly in eyeglasses and many other applications. Plastic triplets of $f/8$ aperture were first introduced by the Eastman Kodak Company in 1959, the "crown" material being methyl methacrylate and the "flint" a copolymer of styrene and acrylonitrile. The refractive indices of available optical plastics are typically very low, so that they fall into the region below the old crown–flint line, along with liquids and a few special titanium flints. The presently available optical plastics are shown in Table 1.3 and properties of frequently used plastic optical materials are provided in Table 1.4.

These refractive index and dispersion data are not highly precise since they depend on such factors as the degree of polymerization and the temperature. The spectral dispersion curves for acrylic, polystyrene, and polycarbonate modeled in the optical design programs CODE V, OSLO, and ZEMAX showed nontrivial differences (up to about 0.005).[40] This is an example where the lens designer should take care to be certain the optical material data are adequately valid for the intended purpose.

Table 1.3

Currently Available Plastic Optical Materials

Plastic	Trade Name	Nd	V-value
Allyl diglycol carbonate	CR-39	1.498	53.6
Polymethyl methacrylate	Lucite/PMMA	1.492	57.8
Polystyrene		1.591	30.8
Copolymer styrene-methacrylate	Zerlon	1.533	42.4
Copolymer methylstyrene-methyl methacrylate	Bavick	1.519	
Polycarbonate	Lexan	1.586	29.9
Polyester-styrene		1.55	43
Cellulose ester		1.48	47
Copolymer styrene acrylonitrile	Lustran	1.569	35.7
Amorphous polyethylene terephthalate	APET	1.571	
Proprietary	LENSTAR	1.557	
Pentaerythritol tetrakis thioglycolate	PETG	1.563	
Polyvinyl chloride	PVC	1.538	
Polymethyl a-chloroacrylate		1.517	57
Styrene acrylnitrile	SAN	1.436	
Poly cyclohexyl methacrylate		1.506	57
Poly dimethyl itaconate		1.497	62
Polymethylpentene	TPX	1.463	
Poly diallyl phthalate		1.566	33.5
Polyallyl methacrylate		1.519	49
Polyvinylcyclohexene dioxide		1.53	56
Polyethylene dimethacrylate		1.506	54
Poly vinyl naphthalene		1.68	20
Glass resin (Type 100)		1.495	40.5
Cyclic olefin copolymer	COC/COP	1.533	30.5
Acrylic	PMMA	1.491	57.5
Methyl methacrylate styrene copolymer	NAS	1.564	
Blend of KRO3 & SMMA	NAS-21 Novacor	1.563	33.5
Polyolefin	Zeonex	1.525	56.3

Table 1.4

Properties of Frequently Used Plastic Optical Materials

Properties	Acrylic (PMMA)	Polycarbonate (PC)	Polystyrene (PS)	Cyclic Olefin Copolymer	Cyclic Olefin Polymer	Ultem 1010 (PEI)
Refractive index						
N_F (486.1 nm)	1.497	1.599	1.604	1.540	1.537	1.689
N_d (587.6 nm)	1.491	1.585	1.590	1.530	1.530	1.682
N_C (656.3 nm)	1.489	1.579	1.584	1.526	1.527	1.653
Abbe value	57.2	34.0	30.8	58.0	55.8	18.94
Transmission (%) Visible spectrum 3.174 mm thickness	92	85–91	87–92	92	92	36–82
Deflection temp						
3.6°F/min @ 66 psi	214°F/101°C	295°F/146°C	230°F/110°C	266°F/130°C	266°F/130°C	410°F/210°C
3.6°F/min @ 264 psi	198°F/92°C	288°F/142°C	180°F/82°C	253°F/123°C	263°F/123°C	394°F/201°C
Max continuous service temperature	198°F 92°C	255°F 124°C	180°F 82°C	266°F 130°C	266°F 130°C	338°F 170°C
Water absorption % (in water, 73°F for 24 hrs)	0.3	0.15	0.2	<0.01	<0.01	0.25
Specific gravity	1.19	1.20	1.06	1.03	1.01	1.27
Hardness	M97	M70	M90	M89	M89	M109
Haze (%)	1 to 2	1 to 2	2 to 3	1 to 2	1 to 2	–
Coeff of linear exp cm X 10^{-5}/cm/°C	6.74	6.6–7.0	6.0–8.0	6.0–7.0	6.0–7.0	4.7–5.6
dN/dT X 10^{-5}/°C	–8.5	–11.8 to –14.3	–12.0	–10.1	–8.0	–
Impact strength (ft-lb/in) (Izod notch)	0.3–0.5	12–17	0.35	0.5	0.5	0.60
Key advantages	Scratch resistance Chemical resistance High Abbe Low dispersion	Impact strength Temperature resistance	Clarity Lowest cost	High moisture barrier High modulus Good eletrical properties	Low birefringence Chemical resistance Completely amorphous	Impact resistance Thermal and chemical resistance High index

Source: Reprinted by permission of G-S Plastic Optics.

The advantages of plastic lenses are:

1. Ease and economy of manufacture in large quantities.
2. Low cost of the raw material.
3. The ability to mold the mount around the lens in one operation.
4. Lens thicknesses and airspaces are easier to maintain.
5. Aspheric surfaces can be molded as easily as spheres.
6. A dye can be incorporated in the raw material if desired.

The disadvantages are:

1. The small variety and low refractive index of available plastics.
2. The softness of the completed lenses.
3. The high thermal expansion (eight times that of glass).
4. The high temperature coefficient of refractive index (120 times that of glass).
5. Plane surfaces do not mold well.
6. The difficulty of making a small number of lenses because of mold cost.
7. Plastics easily acquire high static charges, which pick up dust.
8. Plastic lenses cannot be cemented[41] and can be coated only with some
   difficulty.[42]

In spite of these issues, plastic lenses have proved to be remarkably satisfactory in many applications, including low-cost cameras, and as manufacturing and materials technologies advance, so will the variety of applications. In some cases, glass and plastic lenses have been used together effectively in optical systems.

1.4 INTERPOLATION OF REFRACTIVE INDICES

If we ever need to know the refractive index of an optical material for a wavelength other than those given in the catalog or used in measurement, some form of interpolation must be used, generally involving an equation connecting n with λ. A simple relation, which is remarkably accurate throughout the visible spectrum, is Cauchy's formula[43]:

$$n = A + B/\lambda^2 + C/\lambda^4$$

Indeed, the third term of this formula is often so small that when we plot n against $1/\lambda^2$ we obtain a perfectly straight line from the red end of the visible almost down to the blue-violet. For many glasses the curve is so straight that a very large graph may be plotted, and intermediate values picked off to about one in the fourth decimal place.

To use this formula, and the similar one due to Conrady,[44] namely,

$$n = A + B/\lambda + C/\lambda^{7/2} \tag{1-1}$$

It is necessary to set up three simultaneous equations for three known refractive indices and solve for the coefficients A, B, and C. In this way indices may be interpolated in the visible region to about one in the fifth decimal place.

Extrapolation is, however, not possible since the formulas break down beyond the red end of the spectrum.

Toward the end of the last century, several workers, including Sellmeier, Helmholtz, Ketteler, and Drude, tried to develop a precise relationship between refractive index and wavelength based on resonance concepts.[45] The one most generally employed is

$$n^2 = A + \frac{B}{\lambda^2 - C^2} + \frac{D}{\lambda^2 - E^2} + \frac{F}{\lambda^2 - G^2} + \cdots \tag{1-2}$$

In this formula the refractive index becomes infinite when λ is equal to C, E, G, and so on, so that these values of λ represent asymptotes marking the centers of absorption bands. Between asymptotes the refractive index follows the curve indicated schematically in Figure 1.5.

For most glasses and other transparent uncolored media, two asymptotes are sufficient for interpolation purposes, one representing an ultraviolet absorption and the other an infrared absorption. The visible spectrum is then covered by values of λ lying between the two absorption bands.

Expanding Eq. (1-2) by the binomial theorem, we obtain an approximate form of this equation, namely,

$$n^2 = a\lambda^2 + b + c/\lambda^2 + d/\lambda^4 + \ldots$$

in which the coefficient a controls the infrared indices (large λ) while coefficients c, d, and so on, control the ultraviolet indices (small λ). If the longer infrared is

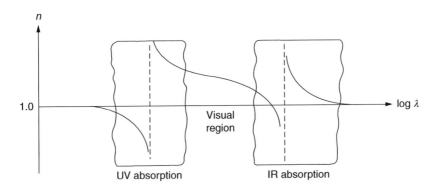

Figure 1.5 Schematic relationship between the refractive index of a glass and the log of the wavelength.

of importance in some particular application, then it is advisable to add one or more terms of the type $e\lambda^4 + f\lambda^6$, and so on.

Herzberger[46] proposed a somewhat[47] different formula, namely,

$$n = A + B\lambda^2 + \frac{C}{\lambda^2 - \lambda_0^2} + \frac{D}{\left(\lambda^2 - \lambda_0^2\right)^2}$$

in which A, B, C, D are coefficients for any given glass, and λ_0 has a fixed value for all glasses. He found that a suitable value is given by $\lambda_0^2 = 0.035$, or $\lambda_0 = 0.187$. This takes care of the ultraviolet absorption, and the near infrared is covered by the $B\lambda^2$ term. If the infrared is more important, another infrared term should be added.

In the first edition of this book, the then current Schott glass catalog contained a six-term expression used for smoothing the stated index data. It was

$$n^2 = A_0 + A_1\lambda^2 + A_2/\lambda^2 + A_3/\lambda^4 + A_4/\lambda^6 + A_5/\lambda^8$$

which provided a very high degree of control in the blue and ultraviolet regions, but it is not valid much beyond 1 μm in the infrared. Since then, Schott has adopted the Sellmeier dispersion formula[48] given by

$$n(\lambda) = \sqrt{1 + \frac{B_1\lambda^2}{\lambda^2 - C_1} + \frac{B_2\lambda^2}{\lambda^2 - C_2} + \frac{B_3\lambda^2}{\lambda^2 - C_3}}.$$

It should be noted that Schott now uses a nine-digit glass code where the first three digits represent the refractive index, the next three the Abbe value, and the final three the density of the glass. For example, the glass code for SF6 is 805254.518. Then $n_d = 1.805$ (note that 1.000 is added to the first three digits), $v_d = 25.4$ (second three digits are divided by 10), and the density is 5.18 (third three digits are divided by 100).

The Bausch and Lomb Company[49] has used the following seven-term formula for its interpolation:

$$n^2 = a + b\lambda^2 + c\lambda^4 + \frac{d}{\lambda^2} + \frac{e\lambda^2}{\left(\lambda^2 - f\right) + g\lambda^2/\left(\lambda^2 - f\right)}$$

This is an awkward nonlinear type of relationship involving a considerable computing problem to determine the seven coefficients for any given type of glass.

1.4.1 Interpolation of Dispersion Values

When using the $(D - d)$ method of achromatism (Section 5.9.1), it is necessary to know the Δn values of the various glasses for the particular spectral region that is being used. For achromatism in the visible, the Δn is usually taken to be $(n_F - n_C)$, but for any other spectral region a different value of Δn must be used. Indeed, a change in the relative values of Δn is really the only factor that determines the spectral region for the achromatism.

To calculate Δn we must differentiate the (n, λ) interpolation formula. This gives us the value of $dn/d\lambda$, which is the slope of the (n, λ) curve at any particular wavelength. The desired value of Δn is then found by multiplying $(dn/d\lambda)$ by a suitable value of $\Delta\lambda$. Actually, the particular choice of $\Delta\lambda$ is unimportant since we shall be working toward a zero value of $\Sigma (D - d) \Delta n$, but if we are expecting to compare a residual of $\Sigma (D - d) \Delta n$ with some established tolerance, it is necessary to adopt a value of $\Delta\lambda$ that will yield a Δn having approximately the same magnitude as the $(n_F - n_C)$ of the glass.

As an example, suppose we are using Conrady's interpolation formula, and we wish to achromatize a lens about some given spectral line. Then by differentiating Eq. (1-1), we get

$$\frac{dn}{d\lambda} = -\frac{b}{\lambda^2} - \frac{7}{2}\frac{c}{\lambda^{9/2}} \qquad (1\text{-}3)$$

This formula contains the b and c coefficients of the particular glass being used, and also the wavelength λ at which we wish to achromatize, say, the mercury g line.

Suppose we are planning to use Schott's SK-6 and SF-9 types. Solving Eq. (1-1) for two known wavelengths, we find

Glass	b	c	$dn/d\lambda$ at the g line
SK-6	0.0124527	0.000520237	−0.142035
SF-9	0.0173841	0.001254220	−0.275885

For wavelength 0.4358 μm, we find for these two glasses that $\Delta n = 0.010369$ and 0.020140, respectively, using the arbitrary value of $\Delta\lambda = -0.073$. These values should be compared with the ordinary $\Delta n = (n_F - n_C)$ values, which for these glasses are 0.01088 and 0.01945 respectively. It is seen that the flint dispersion has increased relative to the crown dispersion, which is characteristic of the blue end of the spectrum.

1.4.2 Temperature Coefficient of Refractive Index

If the ambient temperature in which the lens is to be used is liable to vary greatly, we must consider the resulting change in the refractive indices of the materials used. For glasses this usually presents no problem since the temperature coefficient of refractive index is very small, on the order of 0.000001 per °C.[50] However, for crystals it is likely to be much greater, and for plastics it is very large:

- fluorite: 0.00001 per °C
- plastics: 0.00014 per °C

Thus over a normal temperature range, say from 0 to 40°C, the refractive index of plastic lenses changes by 0.0056, quite enough to alter the focus significantly. In a reflex camera this would be overcome during the focusing operation before making the exposure, but in a fixed-focus or rangefinder camera, or one depending on the use of a focus scale, something must be done to avoid this temperature effect. One way that has been proposed is to place all or most of the lens power in a glass element, using the plastic elements only for aberration correction.

Another suggestion is to mount the lens in a compensated mount of two materials having very different coefficients of expansion, so that as the temperature changes, one airspace of the lens is altered by just the right amount to restore the image position on the film. The thermal expansion of plastics is also large, but this is immaterial if the camera body is also made of plastic, since a temperature change then merely expands or contracts the entire apparatus, leaving the image always in the same plane.

1.5 LENS TYPES TO BE CONSIDERED

Lenses fall into several well-defined and well-recognized types, many of which will be considered in this book. They are

1. Lenses giving excellent definition only on axis
   (a) Telescope doublets (low aperture)
   (b) Microscope objectives (high aperture)
2. Lenses giving good definition over a wide field
   (a) Photographic objectives
   (b) Projection lenses
   (c) Flat-field microscope objectives
3. Lenses covering a finite field with a remote stop
   (a) Eyepieces, magnifiers, and loupes
   (b) Viewfinders
   (c) Condensers
   (d) Afocal Galilean or anamorphic attachments
4. Catadioptric (mirror–lens) systems
5. Varifocal and zoom lenses

Each of these types, and indeed every form of lens, requires an individual and specific process for its design. Some lenses contain many refracting surfaces while some contain few. In some lenses there are so many available parameters that almost any glass can be used; in others the choice of glass is an important degree of freedom. Some lens systems favor a high relative aperture but cover only a small angular field, while other types are just the reverse.

Several classical lens types are considered in this book and the design of a specific example of each is shown in detail. The reader is strongly advised to follow through these designs carefully, since they employ a number of well-recognized techniques that can often be usefully applied to other design situations.

Some of the procedures that have been utilized in the examples in this book are as follows:

1. Lens bending
2. Shift of power from one element to another
3. Single and double graphs, to vary one or two lens parameters simultaneously
4. Symmetry, for the automatic removal of the transverse aberrations
5. Selection of stop position by the $(H' - L)$ plot
6. Achromatism by the $(D - d)$ method
7. Selection of glass dispersions at the end of a design
8. The matching principle for the design of a high-aperture aplanat
9. Use of a "buried surface" for achromatism
10. Reduction of the Petzval sum by a variety of methods
11. Use of a narrow airspace to reduce zonal spherical aberration
12. Introduction of vignetting to cut off bad rim rays
13. Solution of four aberrations by the use of four simultaneous equations
14. Application of aspheric surfaces for aberration control

Since this book is primarily directed toward the needs of the beginner, no reference has been made to the more complex modern photographic objectives. This omission includes particularly high aperture lenses of the Double-Gauss and Sonnar types, and wide-angle lenses such as the Biogon and reversed telephoto. Zoom lenses and afocal and anamorphic attachments have been omitted for the same reason. Today these complex systems are invariably designed with the aid of an optimization program on a computer. Throughout the following chapters, additional guidance is occasionally provided in highlighted sections denoted as Designer Note.

ENDNOTES

[1] Rudolf Kingslake, *Optical System Design*, Academic Press, Orlando (1983).

[2] Warren J. Smith, *Modern Optical Engineering, Fourth Edition*, McGraw-Hill, New York (2008).

[3] F. Twyman, *Prism and Lens Making*, Hilger and Watts, London (1952).

[4] Arthur S. De Vany, *Master Optical Techniques*, Wiley, New York (1981).

[5] D. F. Horne, *Optical Production Technology, Second Edition*, Adam Hilger, Bristol (1983).

[6] Hank H. Karow, *Fabrication Methods for Precision Optics*, Wiley, New York (1993).

[7] W. Zschommler, *Precision Optical Glassworking*, SPIE Press, Bellingham (1984).

[8] R. Barry Johnson and Michael Mandina, "Aspheric glass lens modeling and machining," *Proc. SPIE*, 5874:106–120 (2005).

[9] George Curatu, "Design and fabrication of low-cost thermal imaging optics using precision chalcogenide glass molding," *Proc. SPIE*, 7060:706008–706008-7 (2008).

[10] Gary Herrit, "IR optics advance—Today's single-point diamond-turning machines can produce toroidal, cylindrical, and spiral lenses," *OE Magazine* DOI:10.1117/2.5200510.0010 (2005).

[11] *Advanced Optics Using Aspherical Elements*, Rudiger Hentschel; Bernhard Braunecker; Hans J. Tiziani (Eds.), SPIE Press, Bellingham (2008).

[12] Qiming Xin, Hao Liu, Pei Lu, Feng Gao, and Bin Liu, "Molding technology of optical plastic refractive-diffractive lenses," *Proc. SPIE*, 6722:672202 (2007).

[13] The center thicknesses for many infrared lenses are as small as 2.5%. Cost, weight, and internal transmittance are often driving factors.

[14] Paul R. Yoder, Jr., *Mounting Lenses in Optical Instruments*, SPIE Press, Bellingham, TT21 (1995).

[15] Paul R. Yoder, Jr., *Design and Mounting of Prisms and Small Mirrors in Optical Instruments*, SPIE Press, Bellingham, TT32 (1995).

[16] Paul R. Yoder, Jr., *Opto-Mechanical Systems Design, Third Edition*, SPIE Press, Bellingham (2005).

[17] Keith B. Doyle, Victor L. Genberg, and Gregory J. Michels, *Integrated Optomechanical Analysis*, SPIE Press, Bellingham, TT58 (2005).

[18] *Handbook of Optical Properties, Volume I: Thin Films for Optical Coatings*, Rolf E. Hummel and Karl H. Guenther (Eds.), CRC Press, Boca Raton (1994).

[19] L. Martinu, "Optical coating on plastics," in *Optical Interference Coatings*, OSA Technical Digest Series, paper MF1 (2001).

[20] *Optical Glass: Description of Properties 2009*, Schott and Duryea (2009); available at *http://www.us.schott.com/advanced_optics/english/download/pocket_catalogue_1.8_us.pdf*.

[21] William S. Beich, "Injection molded polymer optics in the 21st century," *Proc. SPIE*, 5865:58650J (2005).

[22] A. Cox, *A System of Optical Design*, pp. 558–661, Focal Press, London and New York, (1964).

[23] ZEBASE, ZEMAX Development Corp., Bellevue, WA (2009). [Collection of more than 600 lens designs.]

[24] LensVIEW™, Optical Data Solutions, Inc. (distributed by Lambda Research Corp.) (2009). [Contains more than 30,000 individual designs from U.S. and Japanese patent literature and from the classic *Zeiss Index of Photographic Lenses*.]

[25] J. M. Palmer, *Lens Aberration Data*, Elsevier, New York (1971).

[26] Ian A. Neil, "Ultrahigh-performance long-focal-length lens system with macro focusing zoom optics and abnormal dispersion liquid elements for the visible waveband," *Proc. SPIE*, 2539:12–24, (1995).

27 James Brian Caldwell and Iain A. Neil, "Wide-Range, Wide-Angle Compound Zoom with Simplified Zooming Structure," U.S. Patent 7227682 (2007).

28 James H. Jannard and Iain A. Neil, "Liquid Optics Zoom Lens and Imaging Apparatus," U.S. Patent Application 20090091844 (2009).

29 James H. Jannard and Iain A. Neil, "Liquid Optics with Folds Lens and Imaging Apparatus," U.S. Patent Application 20090141365 (2009).

30 N. J. Kreidl and J. L. Rood, "Optical materials," in *Applied Optics and Optical Engineering*, Vol. I, pp. 153–200, R. Kingslake (Ed.), Academic Press, New York (1965).

31 Solomon Musikant, *Optical Materials: An Introduction to Selection and Application*, Marcel Dekker, New York (1985).

32 Heinz G. Pfaender, *Schott Guide to Glass*, Van Nostrand Reinhold, New York (1983).

33 *Handbook of Infrared Materials*, Paul Klocek (Ed.), Marcel Dekker, New York (1991).

34 John A. Gibson, "Deep Ultraviolet (UV) Lens for Use in a Photolighography System," U.S. Patent 5,031,977 (1991).

35 David R. Shafer, Yug-Ho Chang, and Bin-Ming B. Tsai, "Broad Spectrum Ultraviolet Inspection Systems Employing Catadioptric Imaging," U.S. Patent 6,956,694 (2001).

36 Romeo I. Mercado, "Apochromatic Unit-Magnification Projection Optical System," U.S. Patent 7,148,953 (2006).

37 H. C. Raine, "Plastic glasses," in *Proc. London Conf. Optical Instruments 1950*, W. D. Wright (Ed.), p. 243. Chapman and Hall, London (1951).

38 Donald Keys, "Optical plastics," Section 3 in *Handbook of Laser Science and Technology, Supplement 2: Optical Materials*, Marvin J. Weber (Ed.), CRC Press, Boca Raton (1995).

39 Michael P. Schaub, *The Design of Plastic Optical Systems*, SPIE Press, Bellingham (2009).

40 Nina G. Sultanova, "Measuring the refractometric characteristics of optical plastics," *Optical and Quantum Electronics*, 35:21–34 (2003).

41 R. Barry Johnson and Michael J. Curley have demonstrated the bonding of acrylic and polysulfone flat plates using Norland Optical Adhesive 68. The bond withstood a temperature range of −15 to 60 °C. Spectral transmittance was not degraded noticeably. Private communications (2008).

42 Plastics, such as polycarbonate, acrylic, styrenes, Ultem, and Zeonex, can be coated with single-layer or multilayer coatings. Refer to Evaporated Coatings, Inc. (2008).

43 A. L. Cauchy, *Mémoire sur la Dispersion de la Lumière*, J. G. Calve, Prague (1836).

44 A. E. Conrady, *Applied Optics and Optical Design, Part II*, p. 659, Dover, New York (1960).

45 P. Drude, *The Theory of Optics*, p. 391, Longmans Green, New York and London (1922).

46 M. Herzberger, "Colour correction in optical systems and a new dispersion formula," *Opt. Acta (London)*, 6:197 (1959).

47 M. Herzberger, *Modern Geometrical Optics*, p. 121, Wiley (Interscience), New York (1958).

48 W. Sellmeier, *Annalen der Physik und Chemie*, 143:271 (1871).

49 N. J. Kreidl and J. L. Rood, "Optical materials," in *Applied Optics and Optical Engineering*, R. Kingslake (Ed.), Vol. I, p. 161, Academic Press, New York (1965).

50 F. A. Molby, "Index of refraction and coefficients of expansion of optical glasses at low temperatures," *JOSA*, 39:600 (1949).

Chapter 2

Meridional Ray Tracing

2.1 INTRODUCTION

It is reasonable to assume that anyone planning to study lens design is already familiar with the basic facts of geometrical and physical optics. However, there are a few points that should be stressed to avoid confusion or misunderstanding on the part of the reader.

2.1.1 Object and Image

All lens design procedures are based on the principles of geometrical optics, which assumes that light travels along rays that are straight in a homogeneous medium. Light rays are refracted or reflected at a lens or mirror, where they proceed to form an image. Due to the inherent properties of refracting and reflecting surfaces and the dispersion of refracting media, the image of a point is seldom a perfect point but is generally afflicted with aberrations. Further, owing to the wave nature of light, the most perfect image on a point is always, in fact, a so-called Airy disk, a tiny patch of light of the order of a few wavelengths in diameter surrounded by decreasingly bright rings of light.

It should be remembered that both objects and images can be either "real" or "virtual." The object presented to the first surface of a system is, of course, always real. The second and following surfaces may receive converging or diverging light, indicating respectively a virtual or real object for that surface. It must never be forgotten that in either case the refractive index to be applied to the calculation is that of the space containing the entering rays at the surface under consideration. This is known as the object space for that surface.

Similarly, the space containing the rays emerging from a surface is called the image space, and real or virtual images are considered to lie in this space. Because of the existence of virtual objects and virtual images we must regard the object and image spaces as overlapping to infinity in both directions. It is also a commonly accepted convention that light from the source propagates initially from left to right.

2.1.2 The Law of Refraction

Over several millennia, attempts to uncover the secret of mathematically describing the refraction of light remained undiscovered. About 1621, Snell successfully provided the needed fundamental equation and insight that allowed optics to have a firm foundation. Descartes published Snell's discovery in 1637, appropriately crediting Snell. During the past four centuries, numerous methods have been developed to trace rays through specifically shaped and free-form surfaces. The well-known Snell's law is generally written

$$n' \sin I' = n \sin I$$

where I and I' are, respectively, the angles between the incident and refracted rays and the normal at the point of incidence, while n and n' are the refractive indices of the media containing the incident and refracted rays, respectively.

Although Snell's law is an elegantly simple equation, its actual application often requires clever use of geometrical constructs. The second part of the law of refraction is that the incident ray, the refracted ray, and the normal at the point of incidence all lie in one plane called the *plane of incidence*. This part of the law becomes important in the tracing of skew rays (see Chapter 8). Computations historically were made by using trigonometric and logarithmic tables, Newton's method for determining the square root, and very talented human computers. The time to trace a skew ray through a single refractive surface was significant and nontrivial for even a meridional ray.

A more generalized form of Snell's law useful for tracing rays in three dimensions is expressed in vector form. Letting r and r' be unit vectors along the incident and refracted rays respectively, and n being a vector along the interface normal, the vector form of Snell's law is given by $n'(r' \wedge n) = n(r \wedge n)$. A good human computer of yesteryear could hand-compute the path of a meridional ray, with six-place accuracy, at a speed of 40 to 60 seconds per ray-surface.[1] For the past several decades, ray tracing has been accomplished almost exclusively using digital computers that can today trace rays billions of times faster than the human computer.

Refractive index is the ratio of the velocity of light in air to its velocity in the medium, and the refractive indices of all transparent media vary with wavelength, being greater for blue light than for red. The refractive index of vacuum relative to air is about 0.9997, which must occasionally be taken into account if a lens is to be used in vacuum. In addition, the refractive indices of air and transparent media are a function of their temperature and the imposed pressure. For example, an infrared lens made of germanium has a remarkably different refractive index when used at room temperature or when cooled by liquid nitrogen.

For reflection we merely write $n' = -n$; this is because I' at a mirror surface is equal to I but with opposite sign. Thus, if a clockwise rotation takes us from the normal to the incident ray, it will require an equal counterclockwise rotation to go from the normal to the reflected ray.

2.1.3 The Meridional Plane

In this book we shall consider almost entirely centered systems, that is, lenses in which the centers of curvature of spherical surfaces, and the axes of symmetry of aspheric surfaces, all lie on a single optical axis. Such systems are also referred to as rotationally symmetric systems. An object point lying on this axis is called an axial object, while one lying off-axis is called an extraaxial or off-axis object point. The plane containing an extraaxial object point and the lens axis is known as the meridional plane; it constitutes a plane of symmetry for the whole system (see Chapter 4).[2]

2.1.4 Types of Rays

Geometrical optics is based on the concept of rays of light, which are assumed to be straight lines in any homogeneous medium and which are bent at a surface separating two media having differing refractive indices. We often need to trace the path of a ray through an optical system, which will generally contain a succession of refracting or reflecting surfaces separated by given distances along the axis. A rough graphical procedure is available for rapid ray tracing, but for more precision it is necessary to use a set of trigonometric formulas executed in succession.

Rays in general fall into three classes: *meridional, paraxial,* and *skew*. For a rotationally symmetric system, meridional rays lie in the plane containing the lens axis and an object point lying to one side of the axis. This plane is called the meridional plane. If the object point lies on the axis, all rays are necessarily meridional.

An important limiting class of rays that has many applications are the so-called paraxial rays, which lie throughout their length so close to the optical axis that their aberrations are negligible. The ray tracing formulas for paraxial rays contain no trigonometric functions and are therefore well-suited to algebraic manipulation. A paraxial ray is really only a mathematical abstraction, for if the diaphragm of a real lens were stopped down to a very small aperture in an effort to isolate only paraxial rays, the depth of focus would become so great that no definite image could be located, although the theoretical image position can be calculated as a mathematical limit. Nevertheless, in the next chapter, it is shown that a paraxial ray can actually be considered at finite heights and angles.

Skew rays, on the other hand, do not lie in the meridional plane, but they pass in front of or behind it and pierce the meridional plane at the diapoint. A skew ray never intersects the lens axis. Skew rays are much more difficult to trace than meridional rays, and we shall not refer to them again.

If the object point lies on the lens axis, we trace only axial rays. However, for an extraaxial object point there are two kinds of rays to be traced, namely meridional rays, which lie in the meridional plane, shown in the familiar ray diagram of a system, and skew rays, which lie in front of or behind the meridional plane and do not intersect the axis anywhere. Each skew ray pierces the meridional plane at the object point and also at another point in the image space known as the *diapoint* of the ray. The paths of two typical skew rays are shown diagrammatically in Figure 2.1.

Axial rays and meridional rays can be traced by relatively simple trigonometric formulas, or even graphically if very low precision is adequate. Skew rays, on the other hand, are much more difficult to trace, the procedure being discussed in Chapter 8.

For an oblique ray in the meridional plane it is useful to consider two limiting rays very close to the traced ray, one slightly above or below it in the meridional plane, and the other a sagittal (skew) ray lying just in front of or behind the traced ray. These are used in the calculation of astigmatism (see Chapter 11).

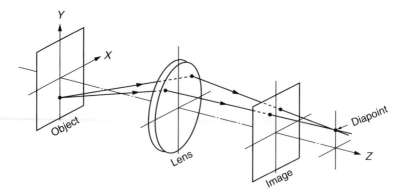

Figure 2.1 A typical pair of skew rays.

2.1.5 Notation and Sign Conventions

This is a very vexed subject, as every lens designer seemingly has his own preferred system, which never seems to agree with that used by others. In spite of the efforts of several committees that have been appointed since World War II, no standard system has been established. Indeed, at the time of the writing of this book, there is still no consistency between all lens design programs. In the first edition we adhered strictly to Conrady's notation except for the signs of the aberrations.

In Conrady's day it was customary to regard all the properties of a single positive lens as positive, whereas today it is universal to regard undercorrected aberrations as negative and overcorrected aberrations as positive. This change in the prevailing attitude leads to a reversal of the sign of all Conrady's aberration expressions, requiring care on the part of any reader who is familiar with the earlier writings on practical optics. In the first edition of this book, a left-handed Cartesian coordinate system was used while in this second edition the standard right-handed Cartesian coordinate system is utilized. Readers attempting to compare material from the first edition or Conrady's books with the second edition should exercise care.

So far as meridional rays are concerned, the origin of coordinates is placed at the vertex A of a refracting or reflecting surface, with distances measured along the axis (the Z axis) as positive to the right and negative to the left of this origin (Figure 2.2). Transverse distances Y in the meridional plane are considered positive if above the axis and negative below it. For skew rays, distances X in the third dimension perpendicular to the meridional plane are generally considered positive when behind that plane, because then the X and Y dimensions occupy their normal directions when viewed from the image space looking back into the lens. However, in a centered system all X dimensions are symmetrical about the meridional plane, so that any phenomenon having a $+X$ dimension is matched by a similar phenomenon having an identical $-X$ dimension, as if the whole of the X space were imaged by a plane mirror lying in the meridional plane itself.

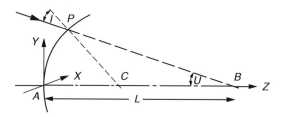

Figure 2.2 A typical meridional ray incident on a spherical surface.

For the angles in the first edition, we regarded the slope U of a meridional ray as positive if a clockwise rotation takes us from the axis to the ray, and the angle of incidence I as positive if a counterclockwise rotation takes us from the normal to the ray. These angle conventions are admittedly inconsistent, and there was a strong move at that time to reverse the sign of U. Unfortunately this change leads to the introduction of as many minus signs as it removes, and worse still, it becomes impossible to draw an all-positive diagram for use when deriving computing formulas. In Conrady's system the paraxial ray height y is equal to (lu), but in the proposed new system this becomes $(-lu)$. The presence of these negative signs is not an inconvenience, and we shall therefore not use Conrady's angle conventions. In the second edition, the angles are consistent with the right-handed Cartesian coordinate system; that is, a ray having a positive slope angle is considered positive.

Finally, all data relating to the portion of a ray lying in the space to the left of a surface, usually the object space, are represented by unprimed symbols, while data referring to the portion of a ray lying in the space to the right of a surface are denoted by primed symbols. In a mirror system where the object and image spaces overlap, data of the entering ray are unprimed while those of the reflected ray are primed, even though both rays lie physically on the same side of the mirror. Mirror systems are considered in Chapter 15.

2.2 GRAPHICAL RAY TRACING

For many purposes, such as in the design of condenser lenses, a graphical ray trace is entirely adequate. The procedure is based on Snell's construction; it has been described by Dowell[3] and van Albada.[4] It is illustrated in Figure 2.3. Having made a large-scale drawing of the lens, we add a series of concentric circles at any convenient place on the paper about a point O, of radii

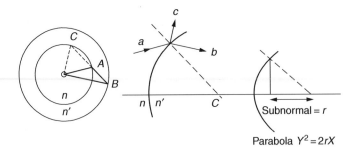

Figure 2.3 Graphical ray tracing.

proportional to the refractive indices of all the materials in the system.[5] A convenient scale for these circles is 10 cm radius for air and 16 cm radius for a glass of index 1.6.

Having drawn the incident ray on the lens diagram, a line is drawn through *O* parallel to the incident ray to cut at *A* the index circle corresponding to the index of the medium containing the incident ray. A line is next drawn through *A* parallel to the normal at the point of incidence, to cut the circle corresponding to the index of the next medium at *B*; then *OB* will be the direction of the refracted ray in the medium *B*.

This process is repeated for each refracting surface in the system. Mirrors can be handled by drawing the normal line right across the diagram to intersect the same index circle on the opposite side (point *C*). It is convenient to draw the index circles in ink, and to indicate rays by little pencil marks labeled with the same letters as the rays on the lens diagram. Some workers have made a practice of erasing each mark after the next mark has been made, to avoid confusion. System changes can be made conveniently by laying a sheet of tracing paper over the diagram and marking the changes on the new paper; this permits the previous system to be seen as well as the changes.

A ray can be traced graphically through an aspheric surface if the direction of the normal is known. A parabolic surface is particularly simple, since the subnormal of a parabola is equal to the vertex radius (Figure 2.3). Graphical ray tracing is rapid and easy, and at any time the ray can be traced accurately by trigonometry to confirm the graphical trace. It also enables the designer to keep track of the lens diameters and thicknesses as he moves along. A more complicated graphical ray trace ascribed to Thomas Young is given in Chapter 11. Paraxial rays can also be traced graphically as discussed in Chapter 3.

An alternative graphical ray tracing method is illustrated in Figure 2.4. In this case two circles having radii proportional to the ratio of the refractive indexes are drawn with the centers of the circles located at the intersection of

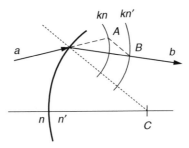

Figure 2.4 Alternative graphical ray tracing method.

32 Meridional Ray Tracing

the incident ray "a" with the surface. The actual radius of each circle is determined by the selection of an arbitrary constant "k" to make the circles of convenient size for drawing. Actually, only arcs of the circles are needed to be drawn as illustrated in the figure. The process is as follows.

First, a line is drawn from the center of curvature of the refractive surface to the point of intersection of the incident ray with the surface. This line is the normal to the surface. Next extend the incident ray until it intersects the arc having radius kn at point A. Now draw a line starting at point A that is parallel to the normal line and to the intersection of this line with the arc having radius of kn' at point B. The refracted ray is now drawn from the surface intersection point through B. In some cases this ray tracing method is found to be easier and often more accurate, in part, because graphical spatial transfers are minimized.

2.3 TRIGONOMETRICAL RAY TRACING AT A SPHERICAL SURFACE

The path of a meridional ray through a single spherical refracting surface can be traced with high accuracy by various well-established procedures that will now be described. The ray emerging from one surface is then transferred to the next surface, where the whole process is repeated until the ray emerges into the final image space.

We define a meridional ray by its slope angle U, which is reckoned positive if a counterclockwise rotation takes us from axis to ray, and by its perpendicular distance Q from the surface vertex. The distance Q is reckoned positive if the ray passes above the surface vertex.

A spherical refracting surface is defined by its radius of curvature r, which is considered positive if the center of curvature lies to the right of the surface, and by the refractive indices n and n' of the media lying to left and right of the surface, respectively. The distance measured along the axis from one surface to the next is given by d and is reckoned positive if the light is proceeding from left to right.

The first step in the ray tracing process is to calculate the angle of incidence I between ray and normal, and this is reckoned positive if a counterclockwise rotation takes us from the normal to the ray. All the data of the incident ray are expressed by plain symbols, and the corresponding data for the refracted ray are given in primed symbols. Figure 2.5a shows that for a spherical surface with radius $r = PC = AC$, the line CN being drawn parallel to the ray shows that the perpendicular distance Q from A is given by

$$Q = r \sin I - r \sin U,$$

from which

$$\sin I = (Q/r) + \sin U, \tag{2-1}$$

or $\sin I = Qc + \sin U$ if the surface curvature c is given instead of its radius r ($c = 1/r$). We next apply the law of refraction to determine the angle of refraction I':

$$\sin I' = \frac{n}{n'} \sin I.$$

The third ray-tracing equation is found from the obvious fact that the central angle PCA in Figure 2.5a is the same for both the entering and emerging rays, or

$$PCA = I - U = I' - U',$$

from which

$$U' = U + I' - I.$$

The final equation is found by adding primes to the first relationship, giving

$$Q' = r(\sin I' - \sin U').$$

With these four equations we can determine the U' and Q' of the refracted ray, given the U and Q of the incident ray and the data of the surface: r, n, and n'.

These equations are perfectly general provided that the radius of curvature of the surface is finite. They obviously cannot be applied to a plane surface because then, in the fourth equation, we find that $I' = U'$, and r is infinite, so we have the product of ∞ and 0, which is indeterminate. Consequently, for a plane we must develop a separate set of equations.

From Figure 2.5b we see that $Y = Q/\cos U = Q'/\cos U'$, so

$$\sin U' = \frac{n}{n'} \sin U \quad \text{and} \quad Q' = \frac{\cos U'}{\cos U} Q.$$

In writing a computer program to trace meridional rays, our first act must be to test the value of $c = 1/r$, and if it is zero, we use the plane surface equations, whereas if it is finite, we use the finite radius equations.

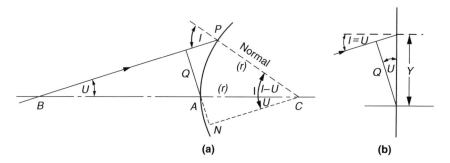

Figure 2.5 Refraction of a meridional ray: (a) at a sphere, and (b) at a plane.

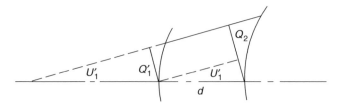

Figure 2.6   Transfer to the next surface: $Q_2 = Q'_1; + d \sin U'_1$.

In both cases the transfer to the next surface is the same. The transfer equation can be derived from Figure 2.6, where we see that

$$Q_2 = Q'_1 + d \sin U'_1.$$

Example

As an example in the use of the ray-tracing equations, we will trace a ray entering parallel to the axis at height 3.172 through the lens shown in Figure 2.7. This is a typical $f/1.6$ projection lens used for many years for projecting 16-mm and 8-mm movie films in a home projector. In Table 2.1, we start by listing the lens data across the page, followed by the Q and Q' values, and then the angles. The value of the incident ray height Y and the sag Z are given as shown. The height Y and the sag Z are found by

$$Y = r \sin(I - U) \text{ and } Z = r[1 - \cos(I - U)].$$

Throughout this book it is anticipated that calculations will be performed on a small pocket electronic calculator where sines and arcsines are given to eight or ten significant figures, electronic spreadsheet, or one of many software programs that trace rays. Only some of the computed quantities need be recorded, therefore, and angles will be stated to five decimals of a degree, or 1/28 sec of arc. Obviously this precision is much higher than that to which optical parts can be manufactured, but since we often calculate aberrations as the small difference

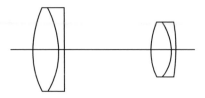

Figure 2.7   Example $f/1.6$ projection lens.

Table 2.1
Example of Ray Tracing

r		8.572		−7.258		∞		5.735		−3.807		−16.878	
c		0.1166589		−0.1377790		0		0.1743679		−0.2626740		−0.0592487	
d			2.4		0.4		7.738		1.8		0.4		
n	1.0		1.52240		1.61644		1.0		1.51625		1.61644		1.0
Marginal ray f/1.6													
Q		3.172		2.905252		2.902741		1.665901		1.299741		1.254617	
Q′		3.224772		2.941579		2.880377		1.663256		1.319817		1.200372	
I		21.71821		−32.23657				7.67363		−32.91271		−13.72930	
I′		14.06750		−30.15782				5.05236		−30.64266		−22.55920	
U	0.0		−7.65070		−5.57196		−9.02988		−11.65115		−9.38110		−18.21100
Y		3.172		3.020		2.917		1.648		1.381		1.280	
Z		0.608		−0.658		0		0.242		−0.259		−0.049	
Paraxial ray													
y		1.0		0.903927		0.891744		0.510797		0.397768		0.376798	
u	0.0		−0.040031		−0.030456		−0.049231		−0.062794		−0.052426		−0.098505

Marginal L' = 3.840978, paraxial l' = 3.825163, and focal length = 10.151767.

between two very nearly equal large numbers, this extra precision is quite useful. Currently, it is rather uncommon to manually trace rays since computer software is readily available to compute the propagation of rays through an optical system; however, understanding how to trace rays through an optical system can be of value when the other ray tracing tools are not available.

There are two special cases that should be recognized:

(a) If $\sin I$ is greater than 1.0, this indicates that the radius is so short that the ray misses the surface altogether.
(b) If $\sin I'$ is greater than 1.0, this indicates total internal reflection.

2.3.1 Program for a Computer

When programming this procedure for a computer, it is of course possible to use available sine and arc sine subroutines, but it is generally much quicker to work through the square root, remembering

$$\sin(a+b) = \sin a \cos b + \cos a \sin b$$

and

$$\cos(a+b) = \cos a \cos b - \sin a \sin b$$

Given Q, $\sin U$, and $\cos U$, the equations to be programmed are

$$\left.\begin{aligned}
\sin I &= Qc + \sin U \\
\cos I &= (1 - \sin^2 I)^{1/2} \\
\sin(I - U) &= -\sin U \cos I + \cos U \sin I \\
\cos(I - U) &= \cos U \cos I + \sin U \sin I
\end{aligned}\right\} \text{(A)}$$

$$\sin(-I') = -(n/n') \sin I$$

$$\left.\begin{aligned}
\cos I' &= [1 - \sin^2(-I')]^{1/2} \\
\sin U' &= -\sin(I - U) \cos(I') + \cos(I - U) \sin(I') \\
\cos U' &= \cos(I - U) \cos(-I') - \sin(I - U) \sin(-I')
\end{aligned}\right\} \text{(B)}$$

$$G = Q/(\cos U + \cos I)$$

$$Q' = G(\cos U' + \cos I')$$

Transfer:

$$Q_2 = Q'_1 - d \sin U'_1$$

Note that the three equations in (A) and (B) are identical with different numbers substituted. It is therefore convenient to write a "cosine cross-product subroutine" to handle the three equations, and substitute the appropriate numbers each time it is used. Remember, of course, that the cosine of a negative angle is positive. When using this routine, it is necessary to carry over both sin U' and cos U' to become sin U and cos U at the next surface.

2.4 SOME USEFUL RELATIONS

2.4.1 The Spherometer Formula

The relation between the height Y and the sag Z of a spherical surface of radius r is often required. It is evident from Figure 2.8 that $r^2 = Y^2 + (r - Z)^2$; hence

$$Z = \frac{Z^2 + Y^2}{2r} = r - \sqrt{r^2 - Y^2}$$

This can be expanded by the binomial theorem to give

$$Z = \frac{Y}{2}\left(\frac{Y}{r}\right) + \frac{Y}{8}\left(\frac{Y}{r}\right)^3 + \frac{Y}{16}\left(\frac{Y}{r}\right)^5 + \cdots \tag{2-2}$$

Because r can become infinite, it is generally better to express Z in terms of the surface curvature c rather than the radius r. Writing $c = 1/r$ gives

$$Z = \frac{cY^2}{1 + \sqrt{1 - c^2 Y^2}}. \tag{2-3}$$

This expression never becomes indeterminate. For a plane surface, $c = 0$ and of course $Z = 0$ also. Note that the first term in Eq. (2-2) is parabolic; that is, $Z = \frac{Y^2}{2r}$. In other words, a sphere and a parabola have essentially the same geometric shape when $Y/r \ll 1$. The parabolic approximation for the sag of a surface is useful to remember as it has many practical applications and can serve as a quick "sanity check." Figure 2.9 shows the percent sag error between a spherical surface and a parabolic surface as a function of the ratio Y/r. For a given Y value, the sag for the parabola is always *less* than that of the sphere. Notice that the error is about 1% for Y/r of 0.2.

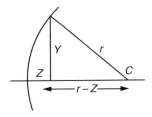

Figure 2.8 The spherometer formula.

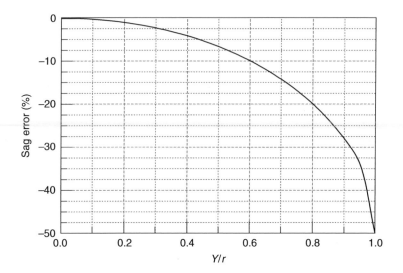

Figure 2.9   Sag error between spherical and parabolic surfaces.

2.4.2 Some Useful Formulas

There are a number of useful relations that can be readily derived between the quantities involved in ray tracing at a spherical surface. Some of them are

$$G = r \tan \tfrac{1}{2}(I - U) = PA^2/2Y$$

$$(\text{chord})PA = 2r \sin \tfrac{1}{2}(I - U) = 2G \cos \tfrac{1}{2}(I - U)$$

$$Y = PA \cos \tfrac{1}{2}(I - U) = PA^2(\cos U + \cos I)/2Q$$

$$Z = PA \sin \tfrac{1}{2}(I - U) = PA^2(\sin I - \sin U)/2Q$$

$$Z = Y \tan \tfrac{1}{2}(I - U) = Y(\sin I - \sin U)/(\cos U + \cos I)$$

The following relations also involve the refraction of a ray at a surface:

$$n \sin U - n' \sin U' = Y \left[\frac{n' \cos U' - n \cos U}{r - Z} \right]$$

$$= Y \left[\frac{n' \cos I' - n \cos I}{r} \right]$$

$$nL \sin U - n'L' \sin U' = r(n \sin U - n' \sin U') = n'Q' - nQ$$

$$n' \cos U' - n \cos U = \cos(U + I)(n' \cos I' - n \cos I)$$

$$\tan \tfrac{1}{2}(I + I') = \tan \tfrac{1}{2}(I - I')(n' + n)/(n' - n)$$

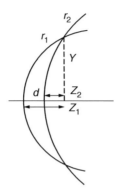

Figure 2.10 Axial separation d such that the two adjacent surfaces intersect at a diameter $2Y$.

2.4.3 The Intersection Height of Two Spheres

If we decide to make two lenses touch at the edge as an aid to mounting, we must choose an axial separation such that the two adjacent surfaces intersect at a diameter lying between the clear aperture and trim diameter of the lenses, as illustrated in Figure 2.10. Or again, if we wish to reduce the thickness of a large lens to its absolute minimum, we must be able to calculate the thickness so that the lens surfaces intersect at the desired diameter, plus a small addition to provide sufficient edge thickness.

Given r_1, r_2, and the axial thickness d, we see by inspection of Figure 2.10 that

$$Z_1 = Z_2 + d.$$

We first calculate

$$A = (2r_2 + d)/(2r_1 - d).$$

Then $Z_2 = d/(A - 1)$ and $Z_1 = AZ_2 = (Z_2 + d)$, and the intersection height Y is given by

$$Y = (2r_1Z_1 - Z_1^2)^{1/2} = (2r_2Z_2 - Z_2^2)^{1/2}$$

Example

If $r_1 = 50$, $r_2 = 250$, and $d = 3$, we find that $A = 503/97 = 5.18556$. Then $Z_2 = 0.71675$ and $Z_1 = 3.71675$, giving $Y = 18.917$.

2.4.4 The Volume of a Lens

To calculate the volume of a lens, and hence its weight, we divide the lens into three parts, the two outer spherical "caps" and a central cylinder. The volume of each of the caps is found by the standard formula

$$\text{volume} = \frac{1}{3}\pi\, Z^2(3r - Z)$$

or, by eliminating r, we have

$$\text{volume} = \frac{1}{2}\pi Y^2 Z + \frac{1}{6}\pi Z^3. \tag{2-4}$$

For many purposes, only the first term of Eq. (2-4) need be used, showing that the "average" thickness of the cap is approximately $\frac{1}{2}Z$. Hence, the lens has approximately the volume of a cylinder of thickness $\frac{1}{2}Z_1 + d - \frac{1}{2}Z_2$, remembering that *each Z must have the same sign as its corresponding r*.

Example

As an example, consider the lens sketched in Figure 2.11 having $r_1 = 20$, $r_2 = 10$, diam. $= 16$, and edge thickness $= 6$. The surface sags are found to be $Z_1 = 1.6697$ and $Z_2 = 4.00$. The three volumes to be added up are shown in Table 2.2. The error in the approximate calculation is only 3%, even for such a very deeply curved lens.

2.4.5 Solution for Last Radius to Give a Stated U'

In some cases we need to determine the last radius of a lens to yield a specified value of the emerging ray slope U', given the Q and U of the incident ray at the surface and the refractive indices n and n'.

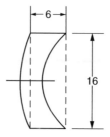

Figure 2.11   The volume of a lens.

Table 2.2

Computation of the Volume of a Lens

	Accurate by Eq. (2.4)	Approximate
Convex cap	54.2π	53.4π
Cylinder	384.0π	384.0π
Concave cap	-138.7π	-128.0π
Volume	299.5π	309.4π

Since $I' = I + (U' - U)$,

$$\sin I' = \sin I \cos(U' - U) + \cos I \sin(U' - U)$$

and dividing by $\sin I$ gives

$$\sin I'/\sin I = n/n' = \cos(U' - U) + \operatorname{ctn} I \sin(U' - U).$$

Hence,

$$\tan I = \frac{\sin(U' - U)}{(n/n') - \cos(U' - U)}. \tag{2-5a}$$

Then knowing I we calculate r by

$$r = Q/(\sin I - \sin U). \tag{2-5b}$$

2.5 CEMENTED DOUBLET OBJECTIVE

In many portions of this book, we will use the cemented doublet objective shown in Figure 2.12 as the basis for discussing a variety of topics such as spherical aberration, chromatic aberration, coma, and so on. The prescription for this lens is as follows:

$r_1 = 7.3895$ $c_1 = 0.135327$

$\quad\quad\quad\quad\quad\quad\quad\quad\quad\quad\quad\quad d_1 = 1.05 \quad\quad n_1 = 1.517$

$r_2 = -5.1784$ $c_2 = -0.19311$

$\quad\quad\quad\quad\quad\quad\quad\quad\quad\quad\quad\quad d_2 = 0.40 \quad\quad n_2 = 1.649$

$r_3 = -16.2225$ $c_3 = -0.06164$

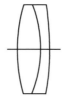

Figure 2.12 A cemented doublet objective.

This lens has a focal length of 12 and will often be used with a marginal ray entering parallel to the optical axis at a height of 2.0. The *f*-number for this lens is *f*/3 (in this simple case, focal length divided by the beam diameter of 4.0).

2.6 RAY TRACING AT A TILTED SURFACE

So far we have considered only a lens system in which the centers of curvature all lie on a single axis. However, it is sometimes required to consider the effect of a slight tilt of a single surface or element in order to compute a "tilt tolerance" for use in the factory. Special formulas are necessary to trace a meridional ray through such a tilted surface.

2.6.1 The Ray Tracing Equations

Suppose the center of curvature of a tilted surface lies at a distance δ to one side of the lens axis. The angular tilt α of the surface is then given by $\sin \alpha = -\delta/r$, the angle α being reckoned positive for a clockwise tilt. The vertex remains on the original optical axis and not spatially displaced. The distance δ is positive if above the optical axis and negative if below.

In Figure 2.13a, P is the point of incidence of the ray at the tilted surface, C is the center of curvature of the surface distance δ below the axis, and angle PCA is clearly equal to $I - \alpha - U$. We draw a line through C parallel to the ray, which intersects the perpendicular AL at H. Thus, Q is equal to $LH + HA$. Angle PCH is equal to I, where LH is $r \sin I$, and the length $HA = r \sin(\angle HCA)$, where $\angle HCA = \angle PCA - I = -(U + \alpha)$. Consequently,

$$Q = r \sin I + r \sin(-U - \alpha) \text{ or } \sin I = Qc - \sin(-U - \alpha).$$

To complete the derivation, we turn to Figure 2.13b. Here angle PCA is bisected to intersect the vertical line PN at O. By the congruence of the two triangles POC and AOC, we see that $PO = OA = G$. Angle $APO = \angle OJA = \angle PAO = \theta$.

However, $\theta = \angle ACJ + \angle JAC = \frac{1}{2}(I - \alpha - U) + \alpha = \frac{1}{2}(I + \alpha - U)$. Therefore, angle $AON = 2 \angle APO = (I + \alpha - U)$, where

$$Y = PN = G[1 + \cos(I + \alpha - U)]$$
$$Z = AN = G \sin(I + \alpha - U)$$

To relate Q and G, we draw the usual perpendicular from A onto the ray at L and draw a line through O parallel to the ray intersecting Q at the point K. Then

$$Q = LK + KA = G \cos U + G \cos(\angle KAO)$$

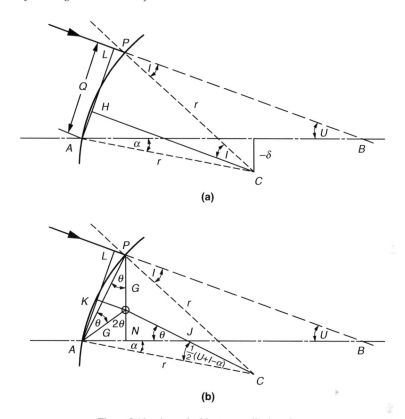

Figure 2.13 A ray incident on a tilted surface.

However,

$$\angle KAO = \angle KAN - \angle NAO = (90° + U) - (90° - 2\theta) = 2\theta + U = I + \alpha.$$

Therefore,

$$Q = G[\cos U + \cos(I + \alpha)]$$

or

$$G = Q/[\cos U + \cos(I + \alpha)]$$

The ray tracing equations therefore become

$$\sin I = Qc - \sin(-\alpha - U)$$
$$\sin I' = (n/n') \sin I$$
$$U' = U + I' - I$$

Short radius only: $Q' = [\sin I' + \sin(-\alpha - U')]/c$

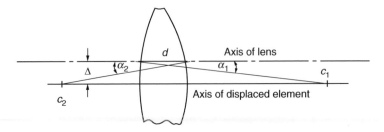

Figure 2.14   A decentered lens.

Universal: $G = Q/[\cos U + \cos(I + \alpha)]$
$$Q' = G[\cos U' + \cos(I' + \alpha)]$$

The transfer to the next surface is normal. In using these equations, it is advisable to list the unusual angles as they arise. They are $-\alpha - U$, $I + \alpha$, $I' + \alpha$, and $I + \alpha - U$ for calculating Y and Z. It should be noted that a ray running along the axis is refracted at a tilted surface, regardless of the surface power, and sets off in an inclined direction, so that *paraxial rays have no meaning*. Calculation of astigmatism through a tilted surface is covered in Chapter 11.

A lens element that has been displaced laterally by an amount Δ without otherwise being tilted possesses two tilted surfaces, as indicated in Figure 2.14, with respect to the optical axis of the system. The tilt of the first surface is $\alpha_1 = \arcsin(\Delta/r_1)$ and the tilt of the second surface is $\alpha_2 = \arcsin(\Delta/r_2)$, the Δ being reckoned negative if the lens is displaced below the axis, as shown in this diagram. Care must be taken to compute the axial separations d along the main axis of the system and not along the displaced axis of the decentered lens element. For small displacements such as might occur by accident this is no problem, but if a lens has been deliberately displaced for some reason, this point must be carefully watched.

2.6.2 Example of Ray Tracing through a Tilted Surface

Consider the cemented doublet lens, as described in Section 2.5, having the following prescription, focal length of 12, and a marginal ray height of 2.0.

$r_1 = 7.3895$

$\qquad\qquad d_1 = 1.05 \qquad n_1 = 1.517$

$r_2 = -5.1784$

$\qquad\qquad d_2 = 0.40 \qquad n_2 = 1.649$

$r_3 = -16.2225$

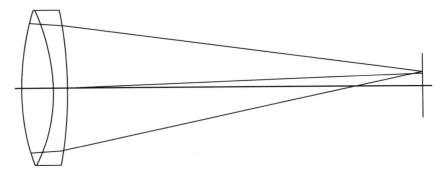

Figure 2.15   A cemented doublet objective lens with final surface tilted.

Now imagine that the rear surface has been tilted clockwise by $\alpha = 3°$ as shown in Figure 2.15. We shall have to trace the axial ray, the upper marginal ray, and the lower marginal ray because all three of these rays are treated differently by a tilted surface.

To understand what has happened as a result of tilting the rear surface by $3°$, we calculate the height at which each emerging ray crosses the paraxial focal plane:

upper marginal ray: 0.429515
axial ray: 0.334850
lower marginal ray: 0.461098

In Figure 2.16 we have plotted on a large scale this situation as compared with the case before the surface was tilted. It is clear that the entire image has been raised, and there is a large amount of coma introduced by the tilting. Even small tilts and decenters can ruin the image quality of an otherwise good lens system. A lens designer should pay particular attention to tilt and decenter sensitivities during the design process. Most modern lens design computer programs provide some means to aid the designer in achieving sensitivity objectives.

2.7 RAY TRACING AT AN ASPHERIC SURFACE

An aspheric surface can be defined in several ways, the simplest being to express the sag of the surface from a plane as follows:

$$Z = a_2 Y^2 + a_4 Y^4 + a_6 Y^6 + \dots$$

Only even powers of Y appear because of the axial symmetry. The first term is all that is required for a parabolic surface. To express a sphere in this way we

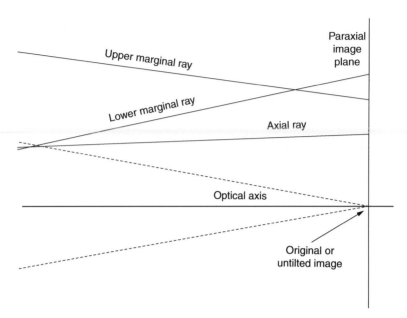

Figure 2.16   Result of tilting a lens surface.

use the power series given in Eq. (2-2), but a great many terms will be required if the sphere is at all deep, that is, has significant sag.

For many purposes it is better to express the asphere as a departure from a sphere:

$$Z = \frac{cY^2}{1 + (1 - c^2 Y^2)^{1/2}} + a_4 Y^4 + a_6 Y^6 + \cdots \tag{2-6}$$

Here c represents the curvature of the osculating sphere and a_4, a_6, ... are the aspheric coefficients.

If the surface is known to be a conic section, we may express it by

$$Z = \frac{cY^2}{1 + [1 - c^2 Y^2 (1 - e^2)]^{1/2}} \tag{2-7}$$

where c is the vertex curvature of the conic and e its eccentricity. The term $1 - e^2$ in this expression is called *a* conic constant, often designated as *p*, since it defines the shape of the surface.[6] In optics, the term *conic constant*, κ, is generally used to imply that $\kappa = -e^2$. Their values are shown in Table 2.3.

To trace a ray through an aspheric surface, we must first determine the *Y* and *Z* coordinates of the point of incidence. The asphere is defined by a relation

Table 2.3

Relationship of Conic Surface Type to Eccentricity and Conic Constants

Surface	Eccentricity	Conic Constant ρ	Conic Constant κ
Hyperbola	>1	<0	<−1
Parabola	1	0	−1
Prolate spheroid (small end of ellipse)	$0 < e^2 < 1$	<1	$-1 < \kappa < 0$
Sphere	0	1	0
Oblate spheroid (side of ellipse)	<0	>1	$0 <$

between Y and Z, while the incident ray is defined by its Q and U. Now it is clear from Figure 2.17 that

$$Q = Y \cos U - [Z] \sin U$$

where $[Z]$ is to be replaced by the expression for the aspheric surface, giving an equation for Y having the same order as the asphere itself.

To solve this equation, we first guess a possible value of Y, say $Y = Q$. We then evaluate the residual R as follows:

$$R = Y \cos U - [Z] \sin U - Q$$

Obviously the correct value of Y is that which makes $R = 0$. Now Newton's rule says that

$$\text{(a better } Y) = \text{(the original } Y) - (R/R')$$

where R' is the derivative of R with respect to Y, namely,

$$R' = \cos U - \sin U (dZ/dY)$$

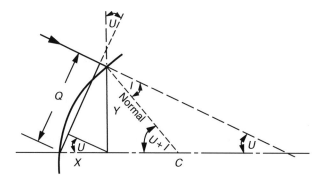

Figure 2.17 Ray trace through an aspheric surface.

A very few iterations of this formula will give us the value of Y that will make R less than any defined limit, such as 0.00000001. Knowing Y we immediately find Z from the equation of the asphere. We then proceed as follows:

The slope of the normal is dZ/dY. Hence,

$$\tan(I - U) = dZ/dY$$

$$\sin I' = (n/n') \sin I$$

$$U' = U + I' - I$$

$$Q' = Y \cos U' - Z \sin U'$$

The transfer to the next surface is accomplished as shown before.

Example

Suppose our asphere is given by

$$[Z] = 0.1\,Y^2 + 0.01\,Y^4 - 0.001\,Y^6$$

Then

$$dZ/dY = 0.2\,Y + 0.04\,Y^3 - 0.006\,Y^5$$

with $n = 1.0$ and $n' = 1.523$. If our entering ray has $U = -10°$ and $Q = 3.0$, then successive iterations of Newton's rule give the values in Table 2.4.

Hence

$$\tan(I - U) = (dZ/dY) = 0.343244, \text{ and } I - U = 18.94448°$$

But $U = -10°$. Therefore

$$I = 8.94448°, \quad I' = 5.85932°$$

$$U' = -13.08516°$$

$$Q' = Y \cos U' - Z \sin U' = 3.018913°$$

Table 2.4

Iterative Solution of Surface Intercept Coordinates

	Y	Z	dZ/dY	R	R'	R/R'
1	3.0	0.981	0.222	0.124772	1.023358	0.121924
2	2.878076	0.946119	0.344369	-0.001357	1.044607	-0.001299
3	2.879375	0.946566	0.343244	0		

ENDNOTES

[1] MIL-HDBK-141, *Optical Design*, Section 5.6.5.8, Defense Supply Agency, Washington, DC (1962).

[2] In this book's first edition, the term "meridian plane" was used rather than "meridional plane" which is used in the second edition. As will be further discussed in Chapter 4, the selection of meridional plane is arbitrary because of the rotational symmetry of the optical system. Once a nonaxial object point is placed, the meridional plane is defined.

[3] J. H. Dowell, "Graphical methods applied to the design of optical systems," *Proc. Opt. Convention*, p. 965 (1926).

[4] L.E.W. van Albada, *Graphical Design of Optical Systems*, Pitman, London (1955).

[5] Although one can use computer-based CAD programs to graphically trace rays, it can also be done using conventional drafting means as described; using such CAD programs affords improved speed and accuracy over the conventional methods.

[6] Warren J. Smith, *Modern Optical Engineering, Fourth Edition*, p. 514, McGraw-Hill, New York (2008).

Chapter 3

Paraxial Rays and First-Order Optics

Suppose we trace a number of meridional rays through a lens from a given object point, the incidence heights varying from the marginal ray height Y_m down to a ray lying very close to the lens axis. We then plot a graph (Figure 3.1) connecting the incidence height Y with the image distance L'. This graph will have two branches, the half below the axis being identical with that above the axis but inverted. The precision of the various point locations is good at the margin but drops badly when the ray is very close to the lens axis, and actually at the axis there is no precision at all. Thus by ordinary ray tracing we can plot all of this graph with the exception of the portion lying near the axis, and we cannot in any way find the exact point at which the graph actually crosses the axis. This failure is, of course, historically due to the limited precision of our mathematical tables and our computing procedures.

However, the exact point at which the graph crosses the axis can be found as a limit. A ray lying everywhere very close to the optical axis is called a "paraxial" ray, and we can regard the paraxial image distance l' as the limit toward which the true L' tends as the aperture Y is made progressively smaller, or

$$l' = \lim_{y \to 0} L'$$

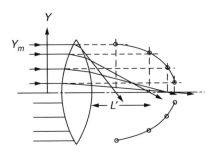

Figure 3.1 Plot of Y against L'.

3.1 TRACING A PARAXIAL RAY

Since all paraxial heights and angles are infinitesimal, we can determine their relative magnitudes by use of a new set of ray-tracing equations formed by writing sines equal to the angles in radians, and cosines equal to 1.0. Since infinitesimals have finite relative magnitudes, we may use any finite numbers to represent paraxial quantities, but we must remember to assume that each number is to be multiplied by a very small factor such as 10^{-50}, so that a para-xial angle written 2.156878 does not mean 2.156878 rad but 2.156878×10^{-50} rad. It is quite unnecessary to write the 10^{-50} every time, but its existence must be assumed if paraxial quantities are to have any meaning. Of course, the longitudinal paraxial data such as l and l' are not infinitesimals.

It should be understood that imagery formed using paraxial ray tracing is stigmatic (free from aberrations) since the paraxial heights and angles are infinitesimal. Consequently, the imagery formed by a physically realizable lens, when well corrected, will be in the same location along the optical axis as the paraxial image.

3.1.1 The Standard Paraxial Ray Trace

Once this is understood, we can derive a set of equations for tracing paraxial rays by modifying the equations in the early part of Section 2.3. Writing sines as angles and cosines as unity, and remembering that in the paraxial region both Q and Q' degenerate to the paraxial ray height y, we get

$$i = yc + u, \quad y = -lu = -l'u'$$
$$i' = (n/n')i \quad \text{(paraxial law of refraction)} \tag{3-1}$$
$$u' = u + i' - i = i - yc = i' - yc$$

with the transfer $y_2 = y_1 + du_1'$.

> It should be observed that the convention is used where paraxial quantities are written with lowercase letters to distinguish them from true heights and angles, which are written in uppercase letters, such as are used in computing the path of a real ray.

As an example, using the lens data given in Section 2.5 for a cemented doublet, Table 3.1 contains the data for a paraxial ray traced through it with the starting data $y = 2.0$ and $u = 0$. As before, the paraxial image distance is found by dividing the last y by the emerging u', giving $l' = 11.285849$. This is slightly different from the marginal L', which was found to be 11.293900. The difference is caused by spherical aberration.

Table 3.1

Tracing a Paraxial Ray through a Cemented Doublet

c	0.1353271		−0.1931098		−0.0616427	
d		1.05		0.4		
n		1.517		1.649		
y	2		1.9031479		1.8809730	
i	0.2706542		−0.4597566		−0.1713856	
i'	0.1784141		−0.4229538		−0.2826148	
u	0	−0.0922401		−0.0554373		−0.1666665

3.1.2 The ($y − nu$) Method

Because of the linear nature of paraxial relationships, we can readily submit the paraxial ray-tracing equations to algebraic manipulation to eliminate some or all of the paraxial angles, which are actually only auxiliary quantities. For example, to eliminate the angles of incidence i and i', we multiply the first part of Eq. (3-1) by n and the corresponding expression for the refracted ray by n', giving

$$ni = nu + nyc, \quad n'i' = n'u' + n'yc$$

Now the law of refraction for paraxial rays is merely $ni = n'i'$; hence equating these two expressions gives

$$n'u' = nu + y(n − n')c \qquad (3\text{-}2)$$

This formula can be used to trace paraxial rays, in conjunction with the transfer

$$y_2 = y_1 + (d/n)(n_1' u_1') \qquad (3\text{-}3)$$

It will be noticed that, written in this way, Eqs. (3-2) and (3-3) are of the same form. That is, in each equation the new value is found by taking the former value and adding to it the product of the other variable multiplied by a constant. This leads to a remarkably convenient and simple ray-tracing procedure known as the ($y − nu$) method. In Table 3.2 we have traced the paraxial ray of Table 3.1 by this new set of equations.

The operating procedure is as follows. To calculate each number, be it a y or a nu, we take the previous y or nu and add to it the product of the next number to the right multiplied by the constant located immediately above it. Thus, starting with y_1 and $(nu)_1$, we first find $(nu)_1' = (nu)_1 + y_1(n − n')_1 c_1$. Then for y_2 we take y_1 and add to it the product of $(nu)_1'$ and d/n, and so on, in a zigzag manner right through to the last surface. The closing equation is, of course,

$$l' = (\text{last } y)/[\text{last } (nu)']$$

Table 3.2

The $(y - nu)$ Method for Tracing Paraxial Rays

c	0.1353271		−0.1931098		−0.0616427	
d		1.05		0.4		
n		1.517		1.649		
$-\phi = (n - n')c$	−0.0699641		0.0254905		−0.0400061	
d/n		0.6921556		0.2425713		
y	2		1.9031479		1.8809730	$l' = 11.285856$
nu	0		−0.1399282		−0.0914160	−0.1666664
l	∞		20.632549		33.929774	
l'	21.682549		34.329774		11.285856	

The numbers in the $(y - nu)$ ray-tracing table obviously resemble perfectly the corresponding numbers in Table 3.1, where the paraxial ray was traced by conventional means. The amount of work involved in the $(y - nu)$ method is about the same as by the direct method, but there are many advantages in tracing rays this way, as we shall see.

Since the image distance l' is the same for all paraxial rays starting out from the same object point, we may select any value we please for either the starting y or the starting nu, but not both, since they are related by $y = -lu$. Many designers always use $y_1 = 1.0$ and calculate the appropriate value of $(nu)_1$. Thus if an object is located at 50 units to the left of the first surface, we could take $y_1 = 1.0$ and $(nu)_1 = 0.02$, remembering that l is negative if the object lies to the left of the surface. A positive l implies a virtual object lying to the right of the first lens surface when the entering rays come in from the left.

When tracing a paraxial ray backwards from right to left, we must subtract each product from the previous value instead of adding it. Thus for right-to-left work we have

$$nu = n'u' - y(n - n')c, \quad y_1 = y_2 - (d/n)(nu)_2$$

3.1.3 Inverse Procedure

One advantage of the $(y - nu)$ method over the straightforward procedure using i and i' is that we can, if we wish, invert the process and work upward from the ray data to the lens data. Thus if we know from some other considerations the succession of y and nu values, we can calculate the lens data by inverting Eqs. (3-2) and (3-3) giving

$$\phi = \frac{n' - n}{r} = \frac{(nu)' - nu}{y} \quad \frac{d}{n} = \frac{y_2 - y_1}{nu}$$

This is often an extremely useful procedure, which cannot be performed when using the straightforward ray trace.

3.1.4 Angle Solve and Height Solve Methods

When making changes in a lens, it is sometimes desired to maintain either the height of incidence of a paraxial ray at a particular surface by a change in the preceding thickness, or to maintain the paraxial ray slope after refraction by a change in the curvature of the surface. Both of these can be achieved by an inversion of Eqs. (3-2) and (3-3). Thus for a height solve we determine the prior surface separation by

$$d = (y_2 - y_1)/u_1'$$

and for an angle solve we use

$$c = [(nu)' - nu]/y(n - n')$$

The last formula is particularly useful if we wish to maintain the focal length of a lens by a suitable choice of the last radius. It should be noted that this formula is the paraxial equivalent of Eq. (2.5), obtained by writing i for tan I, $(u - u')$ for $\sin(U - U')$, 1.0 for $\cos(U - U')$, u and i for sin U and sin I, respectively, and y for Q.

Although it is possible to maintain the focal length of a lens having many elements by changing any of the curvatures preceding the final curvature, it will not generally be found prudent to do so. As will be explained later, using an internal curvature for focal length control during the design process will generally upset the optimization because of extrinsic aberration contributions being transferred and other factors.

3.1.5 The (l, l') Method

In the derivation of Eq. (3-1) we eliminated the angles of incidence as being unnecessary auxiliaries. Actually we can go further and also eliminate the ray slope angles u and u'. To do this we divide Eq. (3-1) by y and note that $l = y/u$, while $l' = y/u'$. These substitutions give the well-known expression

$$\frac{n'}{l'} + \frac{n}{l} = \frac{n' - n}{r} = \phi \tag{3-4}$$

In computations, this is used in the form

$$l' = \frac{n'}{\phi - (n/l)}$$

where

$$\phi = (n' - n)/r = (n' - n)c$$

The transfer now is merely

$$l_2 = l'_1 - d$$

Remember that l and l' refer to the portions of a ray lying to the left and right of a surface, respectively. Of course, in the spaces between surfaces the ray almost never reaches the optical axis, so that neither the l nor the l' is actually realized.

3.1.6 Paraxial Ray with All Angles

There are, of course, other ways to trace a paraxial ray. For instance, we can trace a paraxial ray with all angles by using these equations in order. Given the l and y of the incident ray and that $c = 1/r$ and t is the distance between surfaces, we have

$$u = -y/l$$
$$i = yc + u$$
$$i' = \frac{n}{n'}i$$
$$u' = i' - yc$$
$$l' = -\frac{y}{u'}$$

with the transfer

$$l_2 = l'_1 - t_1.$$

These equations can be collected together to give

$$u' = (yc + u)\frac{n}{n'} - yc.$$

The transfer is now

$$y_2 = y_1 + t_1 u'_1.$$

3.1.7 A Paraxial Ray at an Aspheric Surface

In tracing a paraxial ray, the aspheric terms have no effect and we need to consider only the vertex curvature of the surface. This is given by the coefficient of the second-order term in the power series expansion. In the paraxial region, the surface equation for both spherical and parabolic surfaces is the same; however, for typical finite dimensions they are not the same.

3.1.8 Graphical Tracing of Paraxial Rays at Finite Heights and Angles

As mentioned in the introduction to this section, Eqs. (3-1), (3-2), and (3-3) were derived on the assumption that the y and u are too small to form stigmatic imagery. Nevertheless, we will now show that it is possible to trace paraxial rays at finite heights and angles, which is both a remarkable and a very useful reality.

Figure 3.2 depicts a single refracting surface that images object O at a distance d from the surface to the image O' located at d' along the optical axis. The refracting surface in the paraxial region is represented as a plane. Consider now a ray A from O having an angle u and height at the refracting surface of y. The refracted ray intercepts the optical axis at O' at angle u'. Since u and u' in Eqs. (3-1), (3-2), and (3-3) were assumed to be very small, it follows from the geometry shown in Figure 3.2 that they can be replaced by $\tan u = y/d$ and $\tan u' = -y/d'$, respectively. It is noted that the expansion series for both $\tan u$ and $\sin u$ have the same first-order term, namely u.

Recalling that $n'u' = nu + yc(n - n')$, we can now substitute the preceding values for u and u', which yields that

$$n'\left(\frac{-y}{d'}\right) = n\left(\frac{y}{d}\right) + \frac{y}{r}(n - n').$$

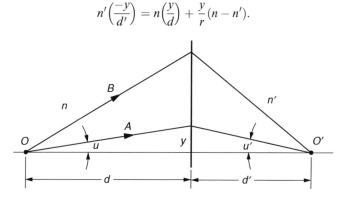

Figure 3.2 Tracing a paraxial ray through a refracting surface.

Dividing both sides by y gives

$$\frac{n}{d} + \frac{n'}{d'} = \frac{n' - n}{r}$$

which is recognized as the imaging equation for a paraxial surface. The significance of this equation is that it is independent of u, u', and y, which means that *any* ray, such as B, from the object O will be imaged at O'; that is, paraxial imaging is stigmatic or free of aberrations. For objects located off of the optical axis, the same stigmatic imaging property can be readily shown.

In Section 2.2, graphical ray tracing of real rays was presented. Tracing paraxial rays graphically is accomplished by replacing circles representing refractive index with tangent planes located proportionally to the refractive indexes. Also, spherical surfaces are replaced by tangent planes. Figure 3.3a shows the geometry for constructing the ray path. The incident ray intersects the refracting surface at D, as illustrated in Figure 3.3b. Next, draw a line parallel to the optical axis and a pair of planes orthogonal to this line at distances proportional to the n and n'. Now project the incident ray from D to intersect the n-plane at A and then draw a line from A that is parallel to the surface normal CD. Finally, the refracted ray is the line drawn from D through B on the n'-plane.

We also see that $y_1 = n \tan u$ and $y_2 = n' \tan u'$. Since line AB in Figure 3.3a is parallel to line DC in Figure 3.3b, by similar triangles,

$$\frac{y_1 - y_2}{y} = \frac{n' - n}{r}.$$

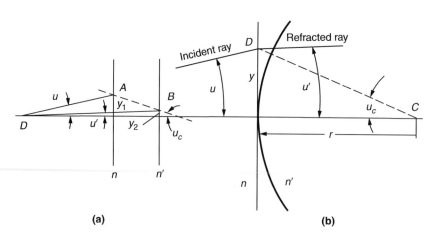

Figure 3.3   Graphical paraxial ray tracing.

Now substituting y_1 and y_2 into this equation, we obtain

$$n' \tan u' = n \tan u + yc(n - n').$$

The importance of this equation is that it shows that the standard paraxial ray trace equations

$$y' = y + \frac{d}{n}(nu) \text{ and } n'u' = nu + yc(n - n')$$

can be used to trace paraxial rays incident at any finite height and finite angle with respect to the optical axis as long as the angles u and u' are interpreted to mean $\tan u$ and $\tan u'$. This understanding is important when using optical design programs. For example, if we desire to control the focal length of a lens by solving for the final surface curvature (see Section 3.1.4) to achieve a specific final marginal ray slope angle, then we use $\tan u_{final}$ in the paraxial curvature solve.

The linear nature of paraxial optics is convenient for rapidly making layouts of optical systems. When using paraxial equations, one should remember that the angles in the equations should be interpreted as $\tan u$.

3.1.9 Matrix Approach to Paraxial Rays

It has been pointed out by Gauss and others that the similarity between the paraxial equations for nu and y suggests a simple matrix formulation for these relations.[1,2,3] The rules of matrix algebra are simple. Suppose we have two simultaneous equations in x and y such as

$$A = ax + by$$
$$B = cx + dy$$

Then in matrix notation we can write

$$\begin{bmatrix} A \\ B \end{bmatrix} = \begin{bmatrix} a & b \\ c & d \end{bmatrix} \begin{bmatrix} x \\ y \end{bmatrix}.$$

Furthermore, the product of two matrices is another matrix, of which the elements are

$$\begin{bmatrix} a & b \\ c & d \end{bmatrix} \begin{bmatrix} e & f \\ g & h \end{bmatrix} = \begin{bmatrix} ae + cg & af + ch \\ be + dg & bf + dh \end{bmatrix}$$

To apply matrix notation to the case of a paraxial ray through a lens, we note that, for the first lens surface,

$$(nu)'_1 = (nu)_1 - y_1\phi_1$$
$$y_1 = y_1$$

In matrix notation these formulas become

$$\begin{bmatrix} (nu)'_1 \\ y_1 \end{bmatrix} = \begin{bmatrix} 1 & -\phi_1 \\ 0 & 1 \end{bmatrix} \begin{bmatrix} (nu)_1 \\ y_1 \end{bmatrix}$$

This square matrix is known as the *refraction matrix* for the first surface. The transfer to the next surface is performed by

$$(nu)_2 = (nu)'_1 \quad \text{and} \quad y_2 = y_1 + (nu)'_1 (t/n)'_1,$$

which in matrix notation becomes

$$\begin{bmatrix} (nu)_2 \\ y_2 \end{bmatrix} = \begin{bmatrix} 1 & 0 \\ (t/n)'_1 & 1 \end{bmatrix} \begin{bmatrix} (nu)'_1 \\ y_1 \end{bmatrix}.$$

This square matrix is known as the *transfer matrix* from surface 1 to surface 2. But the last matrix here is the left side of the above refraction matrix. Substituting this into the last relation gives

$$\begin{bmatrix} (nu)_2 \\ y_2 \end{bmatrix} = \begin{bmatrix} 1 & 0 \\ (t/n)'_1 & 1 \end{bmatrix} \begin{bmatrix} 1 & -\phi_1 \\ 0 & 1 \end{bmatrix} \begin{bmatrix} (nu)'_1 \\ y_1 \end{bmatrix}.$$

We can verify this by multiplying the two square matrices together. This gives

$$\begin{bmatrix} (nu)_2 \\ y_2 \end{bmatrix} = \begin{bmatrix} 1 & -\phi_1 \\ (t/n)'_1 & 1-\phi_1(t/n)'_1 \end{bmatrix} \begin{bmatrix} (nu)'_1 \\ y_1 \end{bmatrix}$$

which correctly represents the two equations

$$(nu)_2 = (nu)_1 - y_1\phi_1$$
$$y_2 = y_1 + \left[(nu)_1 - y_1\phi_1 \right](t/n)'_1$$

We can extend this argument to an optical system containing any number k of surfaces, giving

$$\begin{bmatrix} (nu)'_k \\ y_k \end{bmatrix} = \underbrace{\begin{bmatrix} 1 & -\phi_k \\ 0 & 1 \end{bmatrix}}_{\substack{\text{refraction at} \\ \text{surface } k}} \underbrace{\begin{bmatrix} 1 & 0 \\ (t/n)'_{k-1} & 1 \end{bmatrix}}_{\substack{\text{transfer from} \\ \text{surface } (k-1) \text{ to } k}} \underbrace{\begin{bmatrix} 1 & -\phi_{k-1} \\ 0 & 1 \end{bmatrix}}_{\substack{\text{refraction at} \\ \text{surface } k-1}} \cdots$$

$$\underbrace{\begin{bmatrix} 1 & 0 \\ (t/n)'_1 & 1 \end{bmatrix}}_{\substack{\text{transfer from} \\ \text{surface 1 to 2}}} \underbrace{\begin{bmatrix} 1 & -\phi_1 \\ 0 & 1 \end{bmatrix}}_{\substack{\text{refraction at} \\ \text{surface 1}}} \begin{bmatrix} (nu)_1 \\ y_1 \end{bmatrix}$$

The product of all the square matrices, taken in order, is another square matrix which is a pure property of the lens. It can be written as

$$\begin{bmatrix} B & -A \\ -D & C \end{bmatrix}$$

with the property that the determinant of this matrix, $BC - AD$, equals 1.0. The four quantities A, B, C, and D are called the Gauss constants of the lens. Here A is the lens power, B the ratio of the front focal distance to the front focal length, and C the ratio of the back focal distance to the rear focal length. We find D by $(BC - 1)/A$. Knowing the four elements of this matrix, we can immediately find the values of $(nu)'_k$ and y_k for any ray defined by its entering values of $(nu)_1$ and y_1.

As an example we will take the doublet in Section 2.5. We find that for this lens the Gauss constants are

$$\begin{aligned} A &= 0.0833332 = 1/f \\ B &= 0.9800774 = -FF/f \\ C &= 0.9404865 = BF/J' \\ D &= -0.9390067 = (BC-1)/A \end{aligned}$$

Using this matrix, if $(nu)_1 = 0.02$ and $y_1 = 1.0$, for example, we find that $(nu)'_3 = -0.063732$ and $y_3 = 0.959267$, both agreeing perfectly with the results of a direct paraxial ray trace.

In practice it is generally easiest to find the lens power and the positions of the focal points by tracing right-to-left and left-to-right paraxial rays through the lens and then to determine the Gauss constants by their meanings given above. Then for any ray defined by its (nu) and y, we have for the emerging ray

$$(nu)'_k = B(nu)_1 - Ay_1$$
$$y_k = -D(nu)_1 + Cy_1$$

A Single Thick Lens

Since $\phi_1 = (n - 1)c_1$ and $\phi_2 = (1 - n)c_2$, we see that the Gauss constants of a single thick lens are

$$\begin{aligned} A &= \phi_1 + \phi_2 - (t/n)\phi_1\phi_2 \\ B &= 1 - (t/n)\phi_2 \\ C &= 1 - (t/n)\phi_1 \\ D &= -(t/n). \end{aligned}$$

Example

Suppose $r_1 = 5.0$ and $r_2 = -10.0$ for a biconvex lens, with $t = 1.5$ and $n = 1.52$. Then $(n-1)/n = 0.343401$; hence $1/f' = 0.151512$. This gives

$$f' = 6.600137$$
$$FF = -6.260163$$
$$BF = 5.920189$$

hence,

$$l_{pp} = 0.339974$$
$$l'_{pp} = -0.679948$$

and the Gauss constants are

$$A = 0.151512$$
$$B = 0.948490$$
$$C = 0.896980$$
$$D = -0.984898$$

with $BC - AD = 1$ (as it should).

A Succession of Separated Thin Lenses

If we apply the matrix notation to a succession of thin lenses separated by air, the refraction matrix becomes

$$\begin{bmatrix} 1 & -\phi \\ 0 & 1 \end{bmatrix}$$

for each thin lens, and the transfer matrix becomes

$$\begin{bmatrix} 1 & 0 \\ d & 1 \end{bmatrix}$$

for each space between lenses. Thus, for a system of two thin lenses, A and B, separated by a distance d, we have

$$\begin{bmatrix} u'_B \\ y_B \end{bmatrix} = \begin{bmatrix} 1 & -\phi_B \\ 0 & 1 \end{bmatrix} \begin{bmatrix} 1 & 0 \\ d & 1 \end{bmatrix} \begin{bmatrix} 1 & -\phi_A \\ 0 & 1 \end{bmatrix} \begin{bmatrix} u_1 \\ y_1 \end{bmatrix}$$

Since the product of these three matrices must be equal to $\begin{bmatrix} B & -A \\ -D & C \end{bmatrix}$ for two thin lenses we have

$$A = \phi_1 + \phi_2 - d\phi_1\phi_2$$
$$B = 1 - d\phi_2$$
$$C = 1 - d\phi_1$$
$$D = -d.$$

3.2 MAGNIFICATION AND THE LAGRANGE THEOREM

3.2.1 Transverse Magnification

Consider first a single refracting surface as in Figure 3.4. Let B, B' be a pair of axial conjugate points, their distances from the surface being l and l', respectively. We now place a small object at B and draw a paraxial ray from the top of the object to the vertex of the surface. The ray will be refracted there, the slope angles θ and θ' being the angles of incidence and refraction. Hence $n\theta = n'\theta'$, and therefore $nh/l = n'h'/l'$. Multiplying both sides by y gives

$$hnu = h'n'u' \tag{3-5}$$

This important relationship is called the theorem of Lagrange, or sometimes the Smith–Helmholtz theorem. Because the h', n', and u' on the right-hand side of one surface are, respectively, equal to the same quantities on the left-hand side of the next surface, it is clear that the product hnu is invariant for all the spaces between surfaces, including the object space and the image space. This product is called the Lagrange invariant or, more often nowadays, the optical invariant.

Since this theorem applies to the original object and also the final image, it is clear that the image magnification is given by

$$m = h'/h = nu/n'u'$$

For a lens in air, the magnification is merely u_1/u'_k (assuming that there are k surfaces in the system). The fact that the ratio of the *nu* values at the object

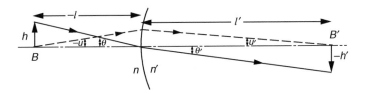

Figure 3.4 The Lagrange relationship.

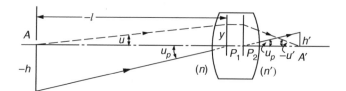

Figure 3.5   The Lagrange equation relationship for a distant object.

and at the image determines the magnification is one of the reasons it is usually preferred to trace a paraxial ray by the $(y - nu)$ method.

In the case where the object is located at infinity, Eq. (3-5) is indeterminate ($u = 0$ and $h = \infty$). Figure 3.5 shows a very distant object where $h'n'u' = (h/l)\, n(lu)$. As $l \rightarrow \infty$, the ratio (h/l) tends toward $\tan u_p$ as (lu) becomes $-y$, where y is the height on the first principal plane (see Section 3.3). Since $y/u' = -f'$ (posterior or rear focal length), then

$$h' = -\frac{n}{n'}\frac{y}{u'}\tan u_p = \frac{n}{n'}f'\tan u_p$$

where u_p is the slope of an entering parallel beam of light. Therefore the image height for an infinitely distant object is equal to the focal length times the angular subtense of the object.

DESIGNER NOTE

Recall that the anterior (front) focal length is $f = -(n/n')f'$, so using the preceding equation, we see that $h' = -f\tan u_p$. This can be interpreted by considering a ray passing through the anterior focal point with an angle u_p. It is evident that the ray will exit the lens parallel to the optical axis at a height h'. Since for small angles the tangent equals the angle in radians, we have two relationships to express focal length, namely,

$$f' = -\frac{y}{u'} \text{ and } f' = \frac{h'}{u_p}.$$

Both of these equations refer to paraxial rays. Although the following points will be covered in Chapter 4 and others, it should be observed that if the focal length varies with the lens aperture, the lens suffers coma. Should the focal length vary with obliquity, then distortion is present.

3.2.2 Longitudinal Magnification

If an object has a small longitudinal dimension δl along the lens axis, or if it is moved along the axis through a small distance δl, then the corresponding longitudinal image dimension is $\delta l'$, and the longitudinal magnification $\overline{m}$ is given by $\overline{m} = \delta l'/\delta l$. By differentiating Eq. (3-4) we find

$$-n'\delta l'/l'^2 = -n\,\delta l/l^2$$

and multiplying both sides by y^2 gives

$$n'\delta l'u'^2 = n\,\delta l\,u^2$$

This is the longitudinal equivalent of the Lagrange equation, and the product $n\,\delta l\,u^2$ is also an invariant. The longitudinal magnification is found to be

$$\overline{m} = \delta l'/\delta l = nu^2/n'u'^2 = (n'/n)m^2 \tag{3-6}$$

so that for a lens in air, $\overline{m} = m^2$. Hence longitudinal magnification is always positive, meaning that if the object is moved a short distance from left to right the image will move from left to right also. On the other hand, for a mirror, the signs of n and n' are equal and opposite, so that when the object moves from left to right the image must move from right to left.

In situations where the ordinary magnification m is large, such as in a microscope objective, the longitudinal magnification will be very large, which explains the small depth of field noticed in a microscope. On the other hand, in a camera the ordinary magnification is small, so the longitudinal magnification is very small, accounting for the great depth of field noticed in most cameras.

Even when the object and image longitudinal dimensions are large, a useful expression for the longitudinal magnification can be derived. Using the Newtonian imaging equation, the magnifications of objects A and B, as shown in Figure 3.6, are related to their image distances z and z' from the rear focal point of the lens by using this equation

$$z'_A = -f'm_A \text{ and } z'_B = -f'm_B + 81$$

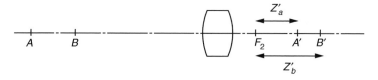

Figure 3.6 Longitudinal magnification.

When an object moves from A to B, the change in image distance $A'B'$ is then given by

$$A'B' = (z'_A - z'_B) = f'(m_A - m_B).$$

The corresponding change in object distance AB is likewise given by

$$AB = f'\left(\frac{1}{m_B} - \frac{1}{m_A}\right).$$

Therefore, the large-scale longitudinal magnification $A'B'/AB$ is given by

$$\overline{m} = \frac{f'(m_A - m_B)}{f'\left(\dfrac{1}{m_B} - \dfrac{1}{m_A}\right)} = m_A m_B.$$

When the longitudinal displacements AB and $A'B'$ are so small that magnification m hardly changes, then the above equation for longitudinal magnification reduces to $\overline{m} = m^2$. The following example illustrates one application of m and $\overline{m}$.

Example

Consider that a spherical object of radius r_o is to be imaged as shown in Figure 3.7. The equation of the object is $r_o^2 = y_o^2 + z^2$, where z is measured along the optical axis and is zero at the object's center of curvature. Letting the surface sag as measured from the vertex plane of the object be denoted as ζ_o, the equation of the object becomes $r_o^2 = (r_o - \zeta_o)^2 + y_o^2$ since $z = r_o - \zeta_o$. In the region near the optical axis, $\zeta_o^2 \ll r_o^2$, which implies that $r_o \approx y_o^2/2\zeta_o$. The image of the object is expressed in the transverse or lateral direction by $y_i = m y_o$ and in the longitudinal or axial direction by $\zeta_i = \overline{m}\zeta_o = \zeta_o m^2(n_i/n_o)$.

In a like manner, the image of the spherical object is expressed as $r_i \approx (y_i)^2/2\zeta_i$. By substitution, the sag of the image is expressed by

$$r_i \equiv \frac{n_o y_o^2}{2 n_i \xi_o} = r_o\left[\frac{n_o}{n_i}\right].$$

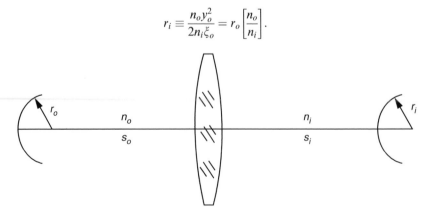

Figure 3.7 Imaging of a spherical object.

Hence, in the paraxial region about the optical axis, *the radius of the image of a spherical object is independent of the magnification* and depends only on the ratio of the refractive indices of the object and image spaces.

3.3 THE GAUSSIAN OPTICS OF A LENS SYSTEM

In 1841, Professor Carl Friedrich Gauss (1777–1855) published his famous treatise on optics (*Dioptrische Untersuchungen*) in which he demonstrated that, so far as paraxial rays are concerned, *a lens of any degree of complexity can be replaced by its cardinal points, namely, two principal points and two focal points*, where the distances from the principal points to their respective focal points are the focal lengths of the lens. Gauss realized that imagery of a rotationally symmetric lens system could be expressed by a series expansion where the first order provided the ideal or stigmatic image behavior and the third and higher orders were the aberrations. He left the computation of the aberrations to others.

To understand the nature of these cardinal-point terms, we imagine a family of parallel rays entering the lens from the left in a direction parallel to the axis (Figure 3.8). A marginal ray such as *A* will, after passing through the lens, cross the axis in the image space at *J*, and so on down to the paraxial ray *C*, which crosses the axis finally at F_2.

If the entering and emerging portions of all of these rays are extended until they intersect, we can construct an "equivalent refracting locus" as a surface of revolution about the lens axis, to contain all the equivalent refracting points for the entire parallel beam. The paraxial portion of this locus is a plane perpendicular to the axis and known as the *principal plane*, and the axial point itself is called the *principal point, P_2*. The paraxial image point F_2, which is conjugate to an axial object point located at infinity, is called the *focal point*, and the longitudinal distance from P_2 to F_2 is the posterior *focal length* of the lens, marked f'.

A beam of parallel light entering parallel to the axis from the right will similarly yield another equivalent refracting locus with its own principal point P_1

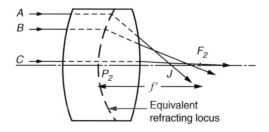

Figure 3.8 The equivalent refracting locus.

and its own focal point F_1, the separation from P_1 to F_1 being known as the anterior focal length f. The distance from the rear lens vertex to the F_2 point is the *back focal distance/length* or more commonly the *back focus* of the lens, and of course the distance from the front lens vertex to the F_1 point is the *front focus* of the lens. For historical reasons the focal length of a compound lens has often been called the *equivalent focal length*, or EFL, but the term *equivalent* is redundant and will not be used here.[4]

3.3.1 The Relation between the Principal Planes

Proceeding further, we see in Figure 3.9 that a paraxial ray A traveling from left to right is effectively bent at the second principal plane Q and emerges through F_2, while a similar paraxial ray B traveling from right to left along the same straight line will be effectively bent at R and cross the axis at F_1. Reversing the direction of the arrows along ray BRF_1 yields two paraxial rays entering from the left toward R, which become two paraxial rays leaving from the point Q to the right; thus Q is obviously an image of R, and the two principal planes are therefore conjugates. Because R and Q are at the same height above the axis, the magnification is $+1$, and for this reason the principal planes are sometimes referred to as *unit planes*.

When any arbitrary paraxial ray enters a lens from the left it is continued until it strikes the P_1 plane, and then it jumps across the *hiatus* between the principal planes, leaving the lens from a point on the second principal plane at the same height at which it encountered the first principal plane (see Figure 3.10).

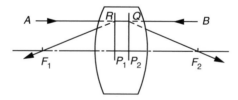

Figure 3.9   The principal planes as unit planes.

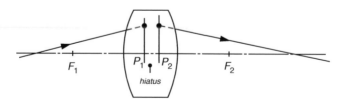

Figure 3.10   A general paraxial ray traversing a lens.

3.3.2 The Relation between the Two Focal Lengths

Suppose a small object of height h is located at the front focal plane F_1 of a lens (Figure 3.11). We draw a paraxial ray parallel to the axis from the top of this object into the lens; it will be effectively bent at Q and emerge through F_2 at a slope ω'. A second ray from R directed toward the first principal point P_1 will emerge from P_2 because P_1 and P_2 are images of each other, and it will emerge at the slope ω' because R is in the focal plane and therefore all rays starting from R must emerge parallel to each other on the right-hand side of the lens. From the geometry of the figure, $\omega = -h/f$ and $\omega' = h/f'$; hence,

$$\omega'/\omega = -f/f' \tag{3-7}$$

We now move the object h along the axis to the first principal plane P_1. Its image will have the same height and will be located at P_2. We can now apply the Lagrange theorem to this object and image, knowing that a paraxial ray is entering P_1 at slope ω and leaving P_2 at slope ω'. Therefore, by the Lagrange equation,

$$hn\omega = hn'\omega' \text{ or } \omega'/\omega = n/n' \tag{3-8}$$

Equating Eqs. (3-7) and (3-8) tells us that

$$f/f' = -n/n'$$

The two focal lengths of any lens, therefore, are in proportion to the outside refractive indices of the object and image spaces. For a lens in air, $n = n' = 1$, and the two focal lengths are equal but of opposite sign. This negative sign simply means that if F_1 is to the right of P_1 then F_2 must lie to the left of P_2. It does *not* mean that the lens is a positive lens when used one way round and a negative lens when used the other way round. The sign of the lens is the same as the sign of its posterior focal length f'. For a lens used in an underwater housing, $n = 1.33$ and $n' = 1.0$; hence, the anterior focal length is 1.33 times as long as the posterior focal length.

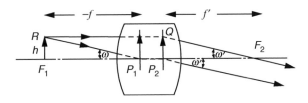

Figure 3.11 Ratio of the two focal lengths.

3.3.3 Lens Power

Lens power is defined as

$$P = \frac{n'}{f'} = -\frac{n}{f}$$

Thus for a lens in air the power is the reciprocal of the posterior focal length. Focal length and power can be expressed in any units, of course, but if focal length is given in meters, then power is in diopters. Note also that the power of a lens is the same on both sides no matter what the outside refractive indices may be.

Applying Eq. (3-2) to all the surfaces in the system and summing, we get

$$\text{power} = P = \frac{(nu')_k}{y_1} = \sum \frac{y}{y_1}\left(\frac{n'-n}{r}\right) \tag{3-9}$$

The quantity under the summation is the contribution of each surface to the lens power. The expression in parentheses, namely, *(n' − n)/r*, is the *power* of a surface which is also called *surface power*.

3.3.4 Calculation of Focal Length

1. By an Axial Ray

If a paraxial ray enters a lens parallel to the axis from the left at an incidence height y_1 and emerges to the right at a slope u' (see Figure 3.12a), then the posterior focal length is $f' = y_1/u'$. The anterior focal length f is found similarly by tracing a parallel paraxial ray right to left, and of course we find that $f = -f'$ if the lens is in air. The distance from the rear lens vertex to the second principal plane is given by

$$l'_{pp} = l' - f'$$

and similarly

$$l_{pp} = l - f$$

2. By an Oblique Ray

The Lagrange equation can be modified for use with a very distant object in the following way. In Figure 3.12b, let A represent a very distant object and A' its image. As the object distance l becomes infinite, the image A' approaches the rear focal point. Then by the Lagrange equation, the following equation applies:

$$h'n'u' = hnu = (h/l)n(lu) = -ny_1 \tan \phi$$

or

$$h' = -\left(\frac{n}{n'}\right)f' \tan \phi = f \tan \phi \qquad (3\text{-}10)$$

where f is the anterior focal length of the lens, no matter what the outside refractive indices may be. This equation forms the basis of the current ANSI definition of focal length. Actually this relation is obvious from a consideration of Figure 3.12c where a paraxial ray is shown entering a lens through the anterior focal point at a slope angle ϕ.

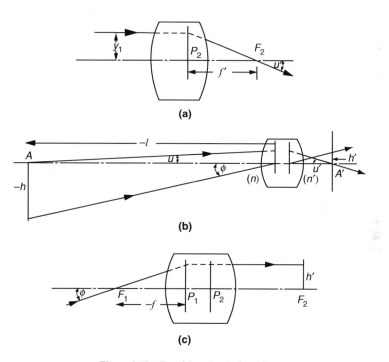

Figure 3.12 Focal-length relationships.

3.3.5 Conjugate Distance Relationships

It is easy to show by similar triangles that if the distances of object and image from the corresponding focal points of a lens are x and x', then

$$m = -f/x = -x'/f'$$

hence,

$$xx' = ff' \tag{3-11}$$

This relationship is called *Newton's equation* or the *Newtonian imaging equation.* Similarly, if the distances of object and image from their respective principal points are p and p', then

$$n'/p' - n/p = n'/f' = -n/f = \text{lens power and } m = np'/n'p \tag{3-12}$$

For a lens in air this becomes simply

$$\frac{1}{p'} - \frac{1}{p} = \frac{1}{f'} \quad \text{and} \quad m = \frac{p'}{p} \tag{3-13}$$

It is often convenient to combine the last two equations for the usual case of a positive lens forming a real image of a real object. Furthermore, if we then ignore all signs and regard all dimensions as positive, with a positive magnification, we get

$$f' = \frac{pp'}{p + p'}, \quad p = f'\left(1 + \frac{1}{m}\right), \quad p' = f'(1 + m) \tag{3-14}$$

These relations are often expressed verbally as "Object distance is $[1 + (1/m)]$ focal lengths, and image distance is $(1 + m)$ focal lengths."

Combining these we get an expression for the object-to-image distance D as

$$D = f'\left(2 + m + \frac{1}{m}\right) \tag{3-15}$$

Inverting this we can calculate the magnification when we are given f' and D:

$$m = \tfrac{1}{2} k = 1 \pm \left(\tfrac{1}{4} k^2 - k\right)^{1/2} \quad \text{where} \quad k = D/f' \tag{3-16}$$

It is important to understand that p and x refer to that section of the ray that lies to the left of the lens, no matter whether that ray actually crosses the axis to the left of the lens, and no matter whether that ray defines the "object" or the "image" in any particular situation. Similarly, p' and x' refer to the section of a ray lying to the right of the lens. The p' and x' are positive if they lie to the right of their origins, namely, the second principal point and the second focal point, respectively.

3.3.6 Nodal Points

Professor Johann Benedict Listing (1808–1882) was one of eight of Gauss' doctoral students and received his degree in 1834. Listing was appointed professor of physics at Göttingen in 1839 and began to study the optics of the human eye.

He published *Beiträge zur physiologischen Optik* in 1845, which became a classic. In this work, Listing introduced the concept of nodal points in a lens system because he needed a means to describe a simple model of the eye. He determined that conjugate points having unit angular magnification exist and named them *Knotenpunkte* or knot points. In the 1880s, they became known as nodal points. Listing also derived the imaging equations using nodal points (N_1 and N_2) and proved that the distances P_1P_2 and N_1N_2 are equal and that nodal points are also in the set of cardinal points.

Since the nodal points of a lens are a pair of conjugate points on the lens axis having unit angular magnification, any paraxial ray directed toward the first nodal point emerges from the second nodal point at the same slope at which it entered. In Figure 3.13, ray A enters the first nodal point N_1 at a slope angle ω and exits N_2 at the same angle. Consider now an object of height h located at N_1. Application of the Lagrange invariant leads to

$$h'n'\omega = hn\omega,$$

hence the magnification for these conjugate points is given by

$$m = \frac{h'}{h} = \frac{n}{n'}.$$

In a manner equivalent to the principal planes, the above can be interpreted that nodal planes have a magnification of $\dfrac{n}{n'}$ as illustrated in Figure 3.13. It is easy to show (Figure 3.14) that the equation at the top of the next page applies:

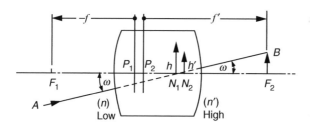

Figure 3.13　Nodal points of a lens.

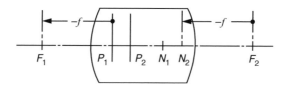

Figure 3.14　The principal and nodal points.

$$F_1P_1 = F_2N_2 = f \quad \text{and} \quad F_1N_1 = P_2F_2 = f'.$$

Listing consequently showed that there are actually six cardinal points that are a property of the lens *and* the mediums in which the lens is immersed. If the lens is in air (or in general, if the mediums are the same, for example water), the two focal lengths are equal, and the nodal points coincide respectively with the two principal points.

As mentioned in Section 3.3.1, the principal planes are conjugate planes with unit lateral magnification. When $n \neq n'$, a ray incident at the first principal point with angle ω will exit the second principal point with angle $\dfrac{n}{n'}\omega$. Although imaging can be done solely using $\dfrac{n}{n'}$ and either the principal point and planes, or the nodal point and planes, it is customary and easier to use a mixed set: principal planes and nodal points. In this manner, the lateral and angular magnifications are both unity. This is particularly useful when performing graphical ray tracing. It should be understood that even though the cardinal points are valid only in the paraxial region, they very often are useful for practical lenses and moderate angles and heights.

An important application of the nodal points is as an experimental method to determine the focal length of a lens. Figure 3.15 shows a lens mounted on a rotatable stage with its axis orthogonal to the optical axis of the lens being evaluated. This stage allows the rotational axis to be situated anywhere along the lens' optical axis. A microscope or a TV camera is used to view the image of a distant point source (parallel rays of light). The light from the light source is aligned with the optical axis as is the microscope. The lens is then rotated (see black dot within the small circle on the axis in Figure 3.15) and the position of the image is observed. When in the position that is shown in the figure, the image will move to the left and right.

Once the lens is shifted such that the rotation axis is coincident with the second nodal point N_2, the image will be stationary. The distance from the rotation axis to the image is therefore the focal length (N_2P_2). The reason that this works is that when the lens is rotated about N_2 by an angle θ, the nodal ray leaving N_2 follows along the original optical axis (and that of the microscope). The

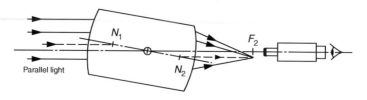

Figure 3.15   Nodal slide to determine focal length.

corresponding entering nodal ray will lay away from the original optical axis by an amount $N_1 N_2 \sin \theta$. The first nodal point can be located in a like manner by reversing the lens in the rotating stage. Should the measurement be made when $n \neq n'$, rotation about N_2 will yield the anterior focal length and rotation about N_1 will yield the posterior focal length.

In addition to the principal planes and nodal points, there are also anti-principal planes and anti-nodal points (also called negative-principal planes and negative-nodal points). Anti-principal planes are conjugate planes having negative unit lateral magnification, while anti-nodal points are conjugate points having negative unit angular magnification. When a lens is immersed in the same medium, the anti-nodal points are located a distance $\pm f$ from F_1 and F_2. An example is a thin lens being used at negative unity magnification ($m = -1$) with the object located at $-2f$ and the image at $2f$.

3.3.7 Optical Center of Lens

Consider the nodal ray passing through the thick lens shown in Figure 3.16. As previously explained, a ray aimed at the first nodal point will pass through the lens undeviated (although translated) and appear to emerge from the lens from the second nodal point. The *optical center* of the lens is where the nodal ray intersects the optical axis.[5,6,7] The optical center OC location can be determined by realizing that the ratio of the height at each surface is equal to the ratio of the respective radii, that is,

$$\frac{y_1}{y_2} = \frac{r_1}{r_2}$$

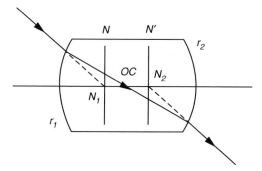

Figure 3.16 Optical center.

Letting the distance from the first surface vertex to the OC be t_1 and the distance from the second surface vertex to the OC be t_2, then the lens thickness t is given by $t = t_1 - t_2$. It is also evident that

$$\frac{y_1}{y_2} = \frac{t_1}{t_2}.$$

Solving for t_1 in terms of radii, the optical center is located at

$$t_1 = \frac{tr_1}{r_1 - r_2} = \frac{t}{1 - \frac{r_2}{r_1}} = \frac{t}{1 - \frac{c_1}{c_2}}.$$

A remarkable property of the optical center is its wavelength independence (n does not appear in the preceding equation), which means that the OC spatial position is fixed. In contrast, the spatial positions of the six cardinal points are a function of wavelength because of their dependence on n.

The location of the optical center can occur before, between, or after the nodal points. For example, for a symmetrical bi-convex lens ($r_1 = -r_2$), the optical center lies exactly at the center of the lens and between the nodal points. For a lens having $r_1 = 20, r_2 = 5, t = 8$, and $N = 1.5$, both nodal points and the optical center are located behind the lens in the order N_1, N_2, and OC. In the first example, a nodal ray transversing the lens physically crosses the optical axis at the optical center while in the second example it does not. In the second case, back-projecting the transversing nodal ray locates the intersection with the optical axis. Also, when the radii have equal value and sign, the optical center is located at infinity.

The optical center point (plane) is conjugate with the nodal points (planes); however, while the nodal points are related by unit angular magnification, the nodal-point to optical-center magnification (m_{OC}) is not necessarily unity. In general, m_{OC} is the ratio of the nodal ray slope angles at the first nodal point and the optical center. For a single thick lens, the magnification m_{OC} can be readily shown to be given by

$$m_{OC} = \frac{r_1 - r_2}{N(r_1 - r_2) - t(N - 1)}.$$

It is noted that as $t \to 0$, $m_{OC} \to \frac{1}{N}$ and as $t \to r_1 - r_2$, $m_{OC} \to 1$ for all N.

All rotationally symmetric lenses have an optical center just as they possess the six cardinal points. Since the optical center is conjugate with N_1 and N_2, *the optical center can also justifiably be considered a cardinal point.* Should the aperture stop be located at the optical center, then the entrance pupil will be located at the first nodal point and the exit pupil will be located at the second nodal point with a unity pupil magnification. This statement is true whether the lens is of symmetrical or unsymmetrical design. When $n \neq n'$, the exit pupil magnification will be $\dfrac{n}{n'}$ rather than unity. The meaning of the aperture stop and entrance/exit pupils will be discussed in detail in Chapter 8.

3.3.8 The Scheimpflug Condition

When an optical system as shown in Figure 3.17 images a tilted object, the image will also be tilted. By employing the concept of lateral and longitudinal magnifications, it can be easily shown that the intersection height of the object plane with the first principal plane P_1 of the lens must be the same as the intersection height of the image plane with the second principal plane P_2 of the lens. This principle was first described by Captain Theodor Scheimpflug of the Austrian army in the early twentieth century and is known as the *Scheimpflug condition*. This can be proved for the paraxial region in the following manner.

Referring to Figure 3.17, axial point object A at the center of the tilted planar object is imaged on the optical axis at A' and point B at the bottom end of the tilted planar object is imaged at B'. A plane passing through A and B will intersect the first principal plane at C. In a like manner, a plane passing through A' and B' will intersect the second principal plane at D. The intersection heights P_1C and P_2D are given by

$$P_1C = \frac{y}{z}s \text{ and } P_2D = \frac{y'}{z'}s'$$

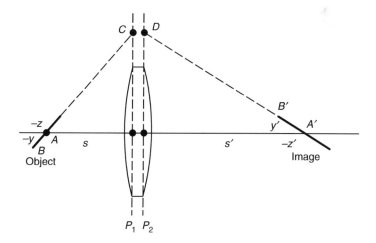

Figure 3.17 Imaging of a tilted object illustrating the Scheimpflug condition.

where y and z are the coordinates of B with respect to A and s is the distance from P_1 to A. Similarly, the image space coordinates are denoted with primes. Now, $y' = my$ and $s' = ms$, and $z' = m^2z$ using the longitudinal magnification relationship; hence $P_1C = P_2D$ proving the Scheimpflug condition.

Consider the following examples of the utility of understanding the Scheimpflug condition. Tilting a camera when taking a photograph of a building causes a defect known as keystone distortion, which is observed to have parallel lines in the scene appearing as converging lines in the film negative or digital image. This defect can be corrected when making a print by tilting the easel and enlarger lens appropriately so that the film plane, easel plane, and principal planes of the enlarger lens intersect in accordance with the Scheimpflug condition. The sharpness of the imagery will then also be as good as possible.

A projector whose film or LCD/DLP is not parallel with the screen will show the keystone defect. The simple way to correct this problem is to tilt the screen to be parallel with the projector's projection plane. The intersections are at infinity in accordance with the Scheimpflug condition. Some projectors provide a means to tilt the projection plane to compensate for keystone introduced by the physical relationship of the screen and projector. It is noted that some modern digital projectors compensate by distorting the shape of the imagery being projected, but this does not allow sharp focusing over the screen and also degrades the displayed resolution.

3.4 FIRST-ORDER LAYOUT OF AN OPTICAL SYSTEM

Most optical systems, as opposed to a specific objective lens, are assembled first from a series of "thin" lens elements at finite separations, and it is therefore of interest to collect here a few useful relations governing the properties of a single thick lens and a set of thin lenses.

3.4.1 A Single Thick Lens

By setting up the familiar $(y - nu)$ table for the two surfaces of a single thick lens, it is easy to show that

$$\text{power} = \frac{1}{f'} = (N-1)\left(\frac{1}{r_1} - \frac{1}{r_2} + \frac{t}{N}\frac{N-1}{r_1 r_2}\right)$$

where N is the refractive index of the glass. The back focus is given by

$$l' = f'\left(1 - \frac{t}{N}\frac{N-1}{r_1}\right)$$

and the rear principal plane is located at

$$l'_{pp} = l' - f' = -f'\left(\frac{t}{N}\frac{N-1}{r_1}\right)$$

Similar relations exist for the front focal length and front focal distance. The hiatus or separation between the two principal planes is

$$Z_{pp} = t + l'_{pp} - l_{pp} = t(N-1)/N$$

For common crown glass with a refractive index of approximately 1.5, the value of Z_{pp} is about $t/3$.

3.4.2 A Single Thin Lens

If a lens is so thin that, within the precision in which we are interested, we can ignore the thickness for any calculations, then we can regard it as a *thin lens*. For accurate work, of course, no lens is thin. Nevertheless, the concept of a thin lens is so convenient in the preliminary layout of optical systems that we often use thin-lens formulas in the early stages of a design and insert thickness for the final studies.

The power of a thin lens is the sum of the powers of its component surfaces, or component elements if it is a multielement thin system. This is because an entering ray remains at the same height y throughout the thin system. Hence for a single lens,

$$\text{power} = \frac{1}{f'} = (N-1)\left(\frac{1}{r_1} - \frac{1}{r_2}\right)$$

and for a thin system,

$$\text{power} = \sum 1/f$$

3.4.3 A Monocentric Lens

A lens in which all the surfaces are concentric about a single point is called *monocentric*. The nodal points of such a lens are, of course, at the common center because any ray directed toward this center is undeviated. Hence the principal and nodal points, as well as the optical center, also coincide at the common center. The image of a distant object is also a sphere centered about the same common center, of radius equal to the focal length. Monocentric systems can be entirely refracting or may include reflecting surfaces.

3.4.4 Image Shift Caused by a Parallel Plate

It is easy to show (Section 6.4) that if a parallel plate of transparent material is inserted between a lens and its image, the image will be displaced further from the lens by an amount

$$s = t\left(1 - \frac{1}{N}\right)$$

Thus, if $N = 1.5$, s will be one-third the thickness of the plate. The image magnification is unity, and this is a well-known method for displacing an image longitudinally without altering its size.

A prism lying between a lens and its image also displaces the image by this distance measured along the ray path in the prism; however, the actual physical image displacement will depend on the folding of the ray path inside the prism, and it is possible to devise such a prism that it may be inserted or removed without any physical shift of the final image.

3.4.5 Lens Bending

One of the most powerful tools available to the lens designer is *bending*; that is, changing the shape of an element without changing its power. If the lens is thin, we know that its focal length is given by

$$\frac{1}{f'} = (N - 1)\left(\frac{1}{r_1} - \frac{1}{r_2}\right)$$

We may write $c_1 = 1/r_1$ and $c_2 = 1/r_2$. Then $c = c_1 - c_2$ and we have

$$1/f' = (N - 1)(c_1 - c_2) = (N - 1)c$$

So long as we retain the value of c, we can obviously select any value of c_1 and solve for c_2. If our thin system contains several thin elements, we can state c_1 and then find the other radii in the following manner:

With c_1 as a given, then $c_2 = c_1 - c_a$, $c_3 = c_2 - c_b$, and so on

Alternatively, we can take the data of a given lens and change each surface curvature by the same amount Δc. Then

$$\text{new } c_1 = \text{old } c_1 + \Delta c$$
$$\text{new } c_2 = \text{old } c_2 + \Delta c$$

$$\vdots$$

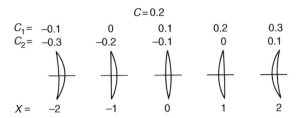

Figure 3.18 Bending a single thin lens.

In Figure 3.18 is shown a series of bendings of a lens in which $c = 0.2$, and we start with a bending having $c_1 = -0.1$. We then add $\Delta c = 0.1$ each time, giving the set of lens shapes shown here. Note that a positive bending bends the top and bottom of the lens to the right, whereas a negative bending turns them to the left.

A convenient dimensionless *shape parameter* X has been used to express the shape of a single lens. It is defined by

$$X = \frac{r_2 + r_1}{r_2 - r_1} = \frac{c_1 + c_2}{c_1 - c_2}$$

Then if we are given f' and X, we can solve for the surface curvatures of a thin lens by

$$c_1 = \tfrac{1}{2} c(X + 1) \quad \text{and} \quad c_2 = \tfrac{1}{2} c(X - 1)$$

or

$$c_1 = \frac{x + 1}{2f'(N - 1)} \quad \text{and} \quad c_2 = \frac{X - 1}{2f'(N - 1)}$$

Note that for an equiconvex or equiconcave lens, $X = 0$. A plano lens has an X value of $+1.0$ or -1.0, while X values greater than 1.0 indicate a meniscus element. X is always positive when the lens is bent to the right and negative to the left.

If the lens to be bent is thick, and especially if it is compound, we can bend it by applying the same Δc to all the surfaces except the last, and then solve the last radius to give the desired lens power by holding the final u'. This is an angle solve problem, discussed in Section 3.1.4. However, if the lens is a single thick element, we can still use the X notation for the lens shape if we wish. For a thick lens of focal length f', we find that

$$r_1 = (N - 1) \frac{f' \pm [f'^2 + (f't/N)(X + 1)(X - 1)]^{1/2}}{X + 1}$$

or

$$c_1 = \frac{-N \pm [N^2 + (Nt/f')(X+1)(X-1)]^{1/2}}{t(N-1)(X-1)}$$

We can then find r_2 or c_2 by the relation

$$r_2 = r_1 \left(\frac{X+1}{X-1}\right) \quad \text{or} \quad c_2 = c_1 \left(\frac{X-1}{X+1}\right)$$

Example

If $f' = 8$, $t = 0.8$, $N = 1.523$, and $X = 1.2$, then the thin-lens formulas give $c_1 = 0.26291$ and $c_2 = 0.02390$. If the thickness is taken into account, the thick-lens formulas give $c_1 = 0.26103$ and $c_2 = 0.02373$. The effect of the finite thickness is remarkably small, even for a meniscus lens such as this.

3.4.6 A Series of Separated Thin Elements

In the case of a series of separated thin elements we cannot merely add the lens powers to get the power of the system because the y at each element varies with the separations. Instead we must use the result of Eq. (3-9), namely,

$$\text{power} = \sum (y/y_1)\phi$$

where ϕ is the power of each element.

The familiar $(y - nu)$ ray-tracing procedure can be conveniently applied to a series of separated thin lenses of power ϕ and separation d, noting that the refractive indices appearing in the $(y - nu)$ method are now all unity. The equations to be used are

$$u' = u - y\phi \quad \text{and} \quad y_2 = y_1 + d'_1 u'_1 \tag{3-17}$$

As an example, we will determine the power and image distance of the following system:

$$\phi_a = 0.125, \quad \phi_b = -0.20, \quad \phi_c = 0.14286$$

$$d'_a = 2.0, \, d'_b = 3.0$$

The $(y - u)$ table for this system is shown in Table 3.3.

Hence the focal length is $1/0.09286 = 10.769$, and the back focus is $0.825/0.09286 = 8.885$. Of course, as always, the (y, u) process is reversible, and if

Table 3.3

The $(y - u)$ Table for Example System

$-\phi$		−0.125		+0.2		−0.14286	
d			2.0		3.0		
y		1		0.75		0.825	
u	0	−0.125		+0.025		−0.09286	

we know what values of y and u the ray should have, we can readily work upwards in the table and find what lens system will give the desired ray path.

As another illustration, suppose we have two lenses at 2-in. intervals between a fixed object and image 6 in. apart, and we wish to obtain a magnification of −3 times. What must be the powers of the two lenses? We proceed to fill out what we know, in a regular (y, u) table, as shown in Table 3.4. Since the magnification is to be −3, the entering part of the ray must have minus three times the slope of the emerging part, and the two lenses must join up the two ray sections shown in Figure 3.19.

Obviously, the intermediate ray slope $u_b = (2 - 6)/(-2) = 2.0$. Then $\phi_a = (u_b + 3)/6 = 5/6 = 0.8333$, and $\phi_b = (1 - u_b)/2 = -0.5$. The required focal lengths are therefore 1.2 and −2 in., respectively.

A glance at Figure 3.19 will reveal that any lens system that joins the two sections of the ray will solve the problem; indeed, it could be done with a single lens located at the intersection of AB and CD, shown dashed. For this lens $f_a = 1.5$, $f_b = 4.5$, and $f' = 1.125$ inches.

Table 3.4

The $(y - u)$ Table for Two-Lens System at Finite Magnification

	ϕ		ϕ_a		ϕ_b		
	$-d$			−2			
$l = -2.0$	y		6		2		
	u	−3		(u_b)		1	$l' = 2.0$

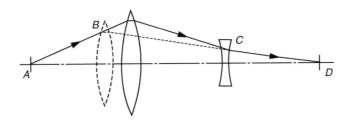

Figure 3.19   A two-lens system at finite magnification.

3.4.7 Insertion of Thicknesses

Having laid out a system of thin lenses to perform some job, we next have to insert suitable thicknesses. A scale drawing of the lenses (assumed equiconvex or equiconcave) will indicate suitable thicknesses, but we must then scale the lenses to their original focal lengths. We next calculate the positions of the principal points of each element, and adjust the air spaces so that the principal-point separations are equal to the original thin-lens separations. If this operation is correctly performed, tracing a paraxial ray from infinity will yield exactly the same focal length and magnification as in the original thin system.

3.4.8 Two-Lens Systems

Figure 3.20 illustrates the general imaging problem where an image is formed of an object by two lenses at a specified magnification and object-to-image distance. Most imaging problems can be solved by using two *equivalent lens* elements. An equivalent lens can comprise one lens or multiple lenses and may be represented by the principal planes and power of a single thick lens. All distances are measured from the principal points of each equivalent lens element. For simplicity, the lenses shown in Figure 3.20 are thin lenses. If the magnification m, object-image distance s, and lens powers ϕ_a and ϕ_b are known, then the equations for s_1, s_2, and s_3 are given by

$$s_1 = \frac{\phi_b(s - s_2) - 1 + m}{m\phi_a + \phi_b}$$

$$s_2 = \frac{s}{2}\left[1 \pm \sqrt{1 - \frac{4\left[sm(\phi_2 + \phi_b) + (m - 1)^2\right]}{s^2 m\phi_a\phi_b}}\right]$$

$$s_3 = s - s_1 - s_2$$

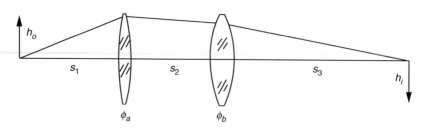

Figure 3.20   General imaging problem.

The equation for s_2 indicates that zero, one, or two solutions may exist.

If the magnification and the distances are known, then the lens powers can be determined by

$$\phi_a = \frac{s + (s_1 + s_2)(m - 1)}{m s_1 s_2}$$

and

$$\phi_b = \frac{s + s_1(m - 1)}{s_2(s - s_1 - s_2)}$$

It can be shown that only certain pairs of lens powers can satisfy the magnification and separation requirements. Commonly, only the magnification and object-image distance are specified with the selection of the lens powers and locations to be determined. By using the preceding equations, a plot of regions of all possible lens power pairs can be generated. Such a plot is shown as the shaded region in Figure 3.21 where s = 1 and $m = -0.2$.

Examination of this plot can assist in the selection of lenses that may likely produce better performance by, for example, selecting the minimum power lenses. The potential solution space may be limited by placing various physical constraints on the lens system. For example, the allowable lens diameters can dictate the maximum powers that are reasonable. Lines of maximum power can then be plotted to show the solution space.[8]

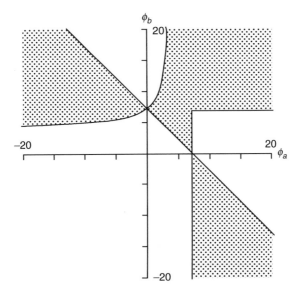

Figure 3.21 Potential power pairs shown in shaded regions.

When s_1 becomes very large compared to the effective focal length *efl* of the lens combination, the optical power of the combination of these lenses is expressed by

$$\phi_{ab} = \phi_a + \phi_b - s_2\phi_a\phi_b$$

The effective focal length is ϕ_{ab}^{-1} or

$$f_{ab} = \frac{f_af_b}{f_a + f_b - s_2}$$

and the back focal length is given by

$$bfl = f_{ab}\left(\frac{f_a - s_2}{f_a}\right)$$

The separation between lenses is expressed by

$$s_2 = f_a + f_b - \frac{f_af_b}{f_{ab}}$$

Figure 3.22 illustrates the two-lens configuration when thick lenses are used. The principal points for the lens combination are denoted by P_1 and P_2, P_{a_1} and

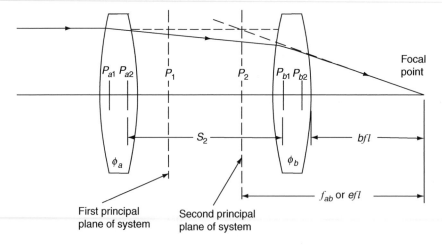

Figure 3.22 Combination of two thick lenses illustrating the principal points of each lens and the system, the f_{ab} or *efl*, and the *bfl*. Distances are measured from the principal points with the exception of the *bfl*. (*Source:* Adapted from Vol. 2, Chapter 1, Figure 18, *Handbook of Optics, Second Edition.* Used with permission.)

P_{a_2} for lens *a,* and P_{b_1} and P_{b_2} for lens *b.* With the exception of the back focal length, all distances are measured from the principal points of each lens element or the combined lens system as shown in the figure. For example, s_2 is the distance from P_{a_2} to P_{b_1}. The *bfl* is measured from the final surface vertex of the lens system to the focal point.

3.5 THIN-LENS LAYOUT OF ZOOM SYSTEMS

A zoom lens is one in which the focal length can be varied continuously by moving one or more of the lens components along the axis, the image position being maintained in a fixed plane by some means, either optical or mechanical. If the focal length is varied but the image is not maintained in a fixed plane, the system is said to be varifocal. The latter type is convenient for projection lenses and the lenses on a reflex camera, in which the image focus is observed by the operator before the exposure is made. A true zoom lens must be used in a movie camera or in any situation in which it is necessary to be sure that the focus is maintained during a zoom.

3.5.1 Mechanically Compensated Zoom Lenses

A zoom camera lens is usually composed of a Donders-type afocal system mounted in front of an ordinary camera lens (Figure 3.23). To vary the focal length, the middle negative component is moved along the axis, the focal position being maintained by simultaneously moving either the front or the rear component by an in-and-out cam.

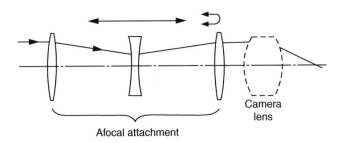

Figure 3.23 A mechanically compensated zoom system.

Example

Suppose we wish to design a symmetrical Donders telescope in which the magnifying power can be varied over a range of 3:1. The magnification of the negative component must therefore vary from $\sqrt{3}$ to $1/\sqrt{3}$, or from 1.732 to 0.577. The focal length of the negative component is found by

$$\text{focal length} = \frac{\text{shift of lens}}{\text{change in magnification}}$$

Suppose $f_a = f_c = 4$ in., and $f_b = -1.0$ in. A series of lens locations are shown in Table 3.5. The last column of the table, image shift, indicates the required movement of either the front or the rear component of the afocal Donders telescope to maintain the image at infinity, so that the telescope can then be mounted in front of a camera set to receive parallel light. Focusing on a near object must be performed by moving the front component axially; otherwise, the zoom law will not hold for a close object. Of course, if this is to be a projection lens, there is no need to maintain the afocal condition or to provide any focusing adjustment for near objects.

The focal length of the camera lens attached to the rear of the Donders telescope can have any value, and it is generally best to use as large an afocal attachment as possible to reduce the aberrations. The early zoom lenses of this type were equipped with simple achromatic doublets for the zoom components.

Table 3.5

Image and Component Movement of the Afocal Donders Telescope

Data of middle component			Thin-lens separations		
Magnification	Object dist.	Image dist.	Front	Rear	Image shift
1.732	1.577	−2.732	2.423	1.268	−0.309
1.4	1.714	−2.400	2.286	1.600	−0.114
1.0	2.000	−2.000	2.000	2.000	0
0.7	2.429	−1.700	1.571	2.300	−0.129
0.577	2.732	−1.577	1.268	2.423	−0.309

3.5.2 A Three-Lens Zoom

In this *mechanically compensated* system once more we have three components, plus–minus–plus, with no fixed lens in the rear (Figure 3.24). The first lens is fixed, and the second and third lenses move in opposite directions. The focal length of the system is equal to the focal length of lens *a* multiplied by the magnifications of lenses *b* and *c*. It is therefore highly desirable that *b* and *c* should both magnify or both demagnify together; otherwise the action of *c* will tend to undo the action of *b*.

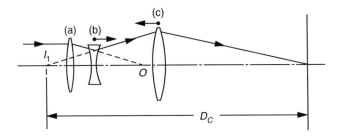

Figure 3.24　Layout of a three-lens zoom with mechanical compensation.

When the negative-power lens b is at unit magnification, the image I_1 will lie as far to the right (towards lens b) as possible. When b moves to the right its magnification will increase, and while this is occurring lens c should be moved to the left so that its magnification will also increase. The computation procedure is simple. Lens a being fixed, its image is also fixed at O. For each position of lens b, its image can be located, and so the object-to-image distance D_c for lens c can be found. Equation (3-16) is then employed to calculate m_c and hence the conjugate distances of the third lens.

Example

Let $f_a = 3.0$ with a very distant object, $f_b = -1.0$, and $f_c = 2.7$. The distance from lens a to the image plane is to be 10.0. Four typical positions of the lenses are indicated in Table 3.6. The focal length range is thus just over 3:1, although the range of the negative-lens magnification is only 2.3:1. The motions of the two lenses are indicated in Figure 3.25 on the next page. The focal length of lens a can be anything, and the original object distance can be anything, but the image produced by lens a must lie at 7 units in front of the final image plane for these data to be applicable. This type of zoom system is used in a zoom microscope, the objective lens alone producing a virtual object at the final image position.

Table 3.6

Positions of Lenses for Example Zoom Lens

		Separation					Separation			Focal
m_b	$1/m_b$	l_b	ab	l'_b	D_c	m_c	l_c	bc	l'_c	length
1.0	1.00	2.00	1.00	−2.0	11.00	1.3117	4.7584	2.7584	6.2416	3.935
1.5	0.67	1.67	1.33	−2.5	11.17	1.4426	4.5716	2.0716	6.5951	6.492
2.0	0.50	1.50	1.50	−3.0	11.50	1.6550	4.3314	1.3314	7.1686	9.930
2.3	0.435	1.435	1.565	−3.3	11.735	1.7864	4.2114	0.9114	7.5234	12.326

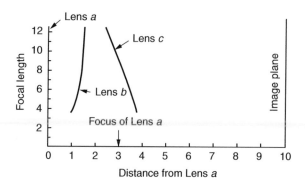

Figure 3.25 Lens motions in a three-lens zoom system.

3.5.3 A Three-Lens Optically Compensated Zoom System

This system was introduced in 1949 by Cuvillier[9] under the name Pan-Cinor. Two moving lenses are coupled together with a fixed lens between them. Generally the coupled lenses are both positive and the fixed lens is negative, but other arrangements are possible. If the powers and separations of the lenses are properly chosen, then the image will remain virtually fixed while the outer lenses are moved, without any need for a cam, hence the name *optical compensation*. To focus on a close object, it is necessary to move the inside negative lens or to vary the separation of the two moving lenses.

The thin-lens predesign of such a system is straightforward, although the algebra involved is complicated. In Figure 3.26 we see the system in its initial

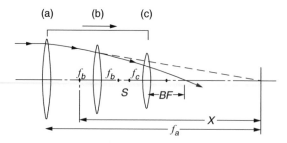

Figure 3.26 Layout of a three-lens optically compensated zoom system.

Table 3.7

Thin-Lens Predesign of a Three-Lens Optically Compensated Zoom System

ϕ	$1/f_a$		$1/f_b$		$1/f_c$	
$-d$		$-(f_a + f_b - X)$		$-(f_b + f_c + S)$		
y	1		$(X - f_b)/f_a$		$-\dfrac{f_b^2 + XS + Xf_c}{f_a f_b}$	
u	0	$1/f_a$		$X/f_a f_b$		$\dfrac{-f_b^2 + XS}{f_a f_b f_c}$

configuration. The separation of adjacent focal points of lenses a and b is X, as shown, and the separation of adjacent foci of lenses b and c is S. Then we can construct a table of the three lenses, and trace a paraxial ray by the $(y - u)$ method, as shown in Table 3.7.

The initial focal length is therefore

$$-f_a f_b f_c / (f_b^2 + XS) \qquad (3\text{-}18)$$

and the initial back focus is $f_c + f_c^2 X/(f_b^2 + XS)$. Note that the initial back focus is independent of f_a.

Suppose we now move the zoom section (lens a plus lens c) to the right by a distance D. Then X and S will both be increased by D, but to hold the image in a fixed position we require the back focus to be reduced by D. Thus

$$D = \text{(initial back focus)} \; - \; \text{(new back focus)}$$

$$= \left[f_c + \frac{f_c^2 X}{f_b^2 + XS} \right] - \left[f_c + \frac{f_c^2(X + D)}{f_b^2 + (X + D)(S + D)} \right] \qquad (3\text{-}19)$$

from which we get

$$f_b^4 + (f_b^2 + SX)(S + D)(X + D) + f_b^2(f_c^2 + SX) - f_c^2 X(X + D) = 0 \qquad (3\text{-}20)$$

Now, for this system to be an effective zoom lens, we require the image plane to lie in a fixed position for a shift D and also for a shift $2D$. Substituting $2D$ for D in Eq. (3-20) gives

$$f_b^4 + (f_b^2 + SX)(S + 2D)(X + 2D) + f_b^2(f_c^2 + SX) - f_c^2 X(X + 2D) = 0 \qquad (3\text{-}21)$$

Subtracting Eq. (3-20) from Eq. (3-21) gives

$$f_c^2 = \frac{(f_b^2 + SX)(X + S + 3D)}{X}$$

and substituting this into Eq. (3-19) gives

$$f_b^2 = \frac{X(X+D)(X+2D)}{S+2X+3D} \tag{3-22}$$

Thus for any set of values for X, S, and D we can solve for the powers of the two lenses b and c. However, we can simplify these expressions by introducing the "zoom range" R, which is the ratio of the initial to the final focal length. Using Eq. (3-18) we see that

$$R = \frac{f_b^2 + (X+2D)(S+2D)}{f_b^2 + XS}$$

which gives us

$$f_b^2 = \frac{(X+2D)(S+2D) - RXS}{R-1} \tag{3-23}$$

Combining Eqs. (3-22) and (3-23) we eliminate f_b and solve for S as a function of R, X, and D:

$$S^2[X(1-R)+2D] + S[2X^2(1-R)+3DX(3-R)+10D^2]$$
$$- (X+2D)[X(R-1)(X+D) - 2D(2X+3D)] = 0$$

For simplicity we can now normalize the system by writing $D = 1$, and then solving for S,

$$S = \frac{2X^2(R-1) + 3X(R-3) - 10 \pm [X(R+1)+2]}{2X(1-R)+4}$$

It will be found that the negative sign of the root gives useful systems, for which

$$S = \frac{X^2(R-1) + X(R-5) - 6}{2 - X(R-1)} \tag{3-24}$$

Then

$$f_b^2 = \frac{X+1}{R-1}(XR - X - 2), \quad f_b^2 = \frac{4R}{R-1} \cdot \frac{2+X+XR}{(2+X-XR)^2} \tag{3-25}$$

If R is greater than 1, the moving lenses will be positive, and if R is less than 1, the moving lenses will be negative. In order that the rear air space will be positive, where $d_b' = (f_b + f_c + S)$, we must select reasonable starting values for X. Approximate suitable values are

R:	5	4	3	2	0.5	0.4	0.3	0.2
X:	1.3	1.7	2.4	4.5	−7.0	−5.5	−4.5	−3.8

Example

Suppose we wish to lay out an optically compensated zoom having $R = 3$, with $X = 2.2$. Then Eqs. (3-24) and (3-25) give

$$S = 0.3, \quad f_b^2 = 3.84, \quad f_c^2 = 11.25$$

Since R is greater than 1, the two moving lenses will be positive and the fixed lens will be negative. Taking square roots gives

$$f_b = -1.95959 \text{ and } f_c = 3.35410$$

Assuming that the initial separation between lenses a and b is to be 3.0, we find that the focal length of the front lens must be 7.15959, and the rear air space d_b' will be initially 1.69451. Using the $(y - u)$ method, we calculate the data shown in Table 3.8.

Table 3.8

Performance of Example Zoom Lens

Shift of zoom components		Back focus	Image shift	Focal length
(Initial position)	−0.5	8.81839	−0.5357	
	0	8.85410	0	10.457
	0.5	8.41660	0.0625	
$D =$	1.0	7.85410	0	5.882
	1.5	7.31839	−0.0357	
	2.0	6.85410	0	3.486
	2.5	6.46440	0.1103	

It is clear that the image plane passes through the three designated positions corresponding to $D = 0$, 1, and 2, but it departs from that plane for all other values of D. These departures, commonly called loops, will be very noticeable if the system is made in a large size, but they can be rendered negligible if the zoom system is made fairly large and is used in front of a small fixed lens of considerable power, as on an 8-mm movie camera. It will be noticed, too, that the law connecting image distance with zoom movement is a cubic (Figure 3.27).

3.5.4 A Four-Lens Optically Compensated Zoom System

We can drastically reduce the sizes of the loops between the in-focus image positions by the use of a four-lens arrangement, as shown in Figure 3.28. Here we have a fixed front lens, followed by a pair of moving lenses coupled together

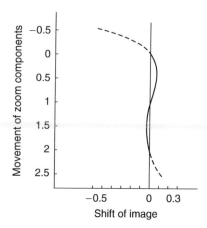

Figure 3.27 Image motion with a three-lens optically compensated zoom system.

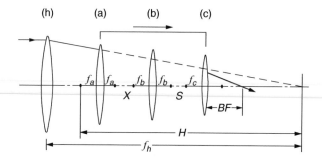

Figure 3.28 Layout of a four-lens optically compensated zoom system.

with a fixed lens between them. The algebraic solution for the powers and spaces of the four lenses is similar to that already discussed, but it is vastly more complicated.

We now designate the separations between adjacent focal points in the three airspaces by H, X, and S, the X and S serving the same functions as before. The initial lens separations are

$$d'_h = f_h + f_a - H \quad d'_a = f_a + f_b + X \quad d'_b = f_b + f_c + S$$

The initial focal length and the initial back focus are

$$\frac{f_h f_a f_b f_c}{f_a^2 S + HXS - f_b^2 H}, \quad f_c + f_c^2 \left(\frac{f_a^2 + HX}{f_a^2 S + HXS - f_b^2 H} \right)$$

respectively, the denominators being the same in each case. We now shift the moving elements by a distance D, so that S is increased by D, but H, X, and the back focus will be reduced by D. We then substitute $2D$ and $3D$ for D, and after three subtractions we obtain the relationship

$$f_c^2(f_a^2 + HX) + (f_b^2 H - f_a^2 S - HSX)(S - X - H + 6D) = 0 \qquad (3\text{-}26)$$

We can considerably simplify the problem by assuming that the moving lenses a and c have equal power. Then Eq. (3-26) becomes

$$f_a^4 - f_a^2[-HX + S(S - X - H + 6D)] + H(f_b^2 - SX)(S - X - H + 6D) = 0$$

We solve this for $f_b{}^2$ in terms of $f_a{}^2$, giving

$$f_b^2 = \frac{f_a^2 + HX}{H}\left[S - \frac{f_a^2}{S - X - H + 6D}\right]$$

Substituting $f_b{}^2$ in the original equation relating the back focus before and after the zoom shift, and noting that now $S = (X - 3D)$, we get

$$f_a^4 + f_a^2(2H - 3D)(H + X - 3D)$$
$$- H(H - D)(H - 2D)(H - 3D) = 0 \qquad (3\text{-}27)$$

and $$f_b^2 = \frac{f_a^2 + HX}{H}\left[\frac{f_a^2}{H - 3D} + (X - 3D)\right]$$

The focal-length range R, the ratio of the initial to the final focal lengths, is now

$$R = \frac{-f_a^2 X + X(H - 3D)(3D - X) + f_b^2(H - 3D)}{-(HX + f_a^2)(3D - X) - Hf_b^2}$$

This ratio R will be less than 1.0 if the moving lenses are negative.

Example

As an example we shall set up a system having the same range of focal lengths as in the last example, so that we can compare the sizes of the loops. We find that for this case we put $X = 3.5$, $D = 1$, and $H = 10.052343$. The equations just given yield

$$f_a^2 = 25.130858 \quad \text{or} \quad f_a = f_c = -5.0130687$$
$$f_b^2 = 24.380858 \quad \text{or} \quad f_b = 4.937698$$
$$R = 0.333333 \quad (S = 0.5)$$

The initial air spaces are

$$d'_h = f_h + f_a - H = 0.5 \,(\text{say}); \quad \text{hence} \quad f_h = 15.565412$$
$$d'_a = 3.424629$$
$$d'_b = 0.424629$$

We find that using these four lens elements, the overall focal length is negative, and so we must add a fifth lens at the rear to give us the desired positive focal lengths. To compare with the three-lens system in Section 3.5.3, we set the initial focal length at 3.486, which requires a rear lens having a focal length of 4.490131 located initially 4 units behind the fourth element. Tracing paraxial rays through this system, at a series of zoom positions, gives the data shown in Table 3.9 and is plotted in Figure 3.29 is on facing page.

It will be noticed that the loops are only about one-fiftieth of their former size, and that the error curve is now a quadratic. Obviously, with these very small errors, it would be quite reasonable to design a four-lens zoom of this type covering a much wider range of focal lengths, say 6:1 or even more, and this indeed has been done.

Table 3.9

Performance of Four-Element Optically-Compensated Zoom Lens

Shift of zoom components		Back focus	Image shift	Focal length
(Initial position)	−0.5	6.22805	−0.00383	
D	0	6.23188	0	3.486
	0.5	6.23267	0.00079	
D =	1.0	6.23188	0	5.026
	1.5	6.23120	−0.00068	
	2.0	6.23188	0	7.253
	2.5	6.23352	0.00164	
	3.0	6.23188	0	10.458
	3.5	6.21538	−0.01650	

3.5.5 An Optically Compensated Zoom Enlarger or Printer

Since the four-lens zoom discussed in Section 3.5.4 can be constructed with two equal positive lenses moving together between three negative lenses, it is obviously possible to remove the two outer negative lenses, leaving a three-lens zoom printer or enlarger system that has a quartic error curve. Equations (3-27) are now

$$f_a^4 + f_a^2(2H - 3)(H + X - 3) - H(H - 1)(H - 2)(H - 3) = 0$$

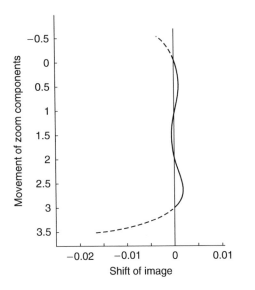

Figure 3.29 Image motion with a four-lens optically compensated zoom system.

and
$$f_b^2 = \left(X + \frac{f_a^2}{H}\right)\left[\frac{f_a^2}{H - 3} + (X - 3)\right] \qquad (3\text{-}28)$$

These can be used to set up a system, taking the positive root for $f_a = f_c$ and the negative root f_b. The H is the initial distance from the fixed object to the anterior focus of the front lens, the initial object distance being therefore $(H - f_a)$. The initial lens separations are respectively $(f_a + f_b + X)$ and $(f_a + f_b + X - 3)$.

Example

As an example, we will design such a zoom system with $H = -8$ and $X = 2$. The preceding formulas give $f_a = f_c = 6.157183$ and $f_b = -2.667455$. The separations are, respectively, 4.667455 and 1.667455 at the start; they will, of course, be increased or decreased as the zoom elements are moved to change the magnification. The overall distance from object to image is equal to $2(14.157183 + 4.667455) = 37.6493$. Tracing rays by the $(y - u)$ method gives the data shown in Table 3.10 on the next page.

The image shift is shown graphically in Figure 3.30. It will be noticed that as we are now moving a pair of positive components, the quadratic curve is in the opposite direction to that for the previous example, in which we moved a pair of negative lenses.

Table 3.10

Performance of Example Optically-Compensated Enlarger or Printer Zoom Lens

Shift of zoom components		Image distance	Desired image distance	Image shift	Magnification
(Initial position)	−0.5	17.762025	17.657184	+0.104841	
	0	17.157184	17.157184	0	−1.7520
	0.5	16.646875	16.657184	−0.010309	
$D =$	1.0	16.157184	16.157184	0	−1.2071
	1.5	15.661436	15.657184	+0.004252	
	2.0	15.157184	15.157184	0	−0.8285
	2.5	14.652316	14.657184	−0.004868	
	3.0	14.157184	14.157184	0	−0.5708
	3.5	13.680794	13.657184	+0.023610	

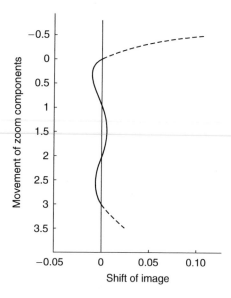

Figure 3.30   Image motion for a zoom enlarger system.

If it is desired to cover a wider range of magnifications, the value of H should be reduced, and if the lenses come too close together, then X can be somewhat increased. Obviously there is no magic about the size, and if a different object-to-image distance is required, the entire system can be scaled up or down as needed. The fixed negative component is very strong and in practice it is often divided into a close pair of negative achromats, but we leave this up to the designer.

ENDNOTES

[1] After Rudolf Kingslake, *Optical System Design*, Chapter 5.IV, Academic Press, New York (1983).

[2] W. Brower, *Matrix Methods in Optical Instrument Design*, Benjamin, New York (1964).

[3] H. Kogelnik, "Paraxial ray propagation," in *Applied Optics and Optical Engineering,* Vol. 7, p. 156, R. R. Shannon and J. C. Wyant (Eds.), Academic Press, New York (1979).

[4] Prior to the 1839 Petzval Portrait lens, the term *focal length* had meaning only for thin lenses, which was taken as the distance from the lens to the image formed when viewing a very distant object. For a long time, the term *equivalent focal length* was used for a complex lens to mean the focal length of the equivalent thin lens.

[5] H. Erfle, "Die optische Abbildung durch Kugelflaechen," Chapter III in *S. Czapski und O. Eppenstein Grundzuege der Theorie der Optischen Instrumente nach Abbe, Third Edition,* pp. 72–134, H. Erfle and H. Boegehold (Eds.), Barth, Leipzig (1924).

[6] H. Schroeder, "Notiz betreffend die Gaussischen Hauptpunkte," *Astron. Nachrichten,* 111:187–188 (1885).

[7] R. Barry Johnson, "Correctly making panoramic imagery and the meaning of optical center," *Current Developments in Lens Design and Optical Engineering IX,* Pantazis Z. Mouroulis, Warren J. Smith, and R. Barry Johnson (Eds.), *Proc. SPIE,* 7060:70600F (2008).

[8] R. Barry Johnson, James B. Hadaway, Tom Burleson, Bob Watts, and Ernest D. Park, "All-reflective four-element zoom telescope: Design and analysis," International Lens Design Conference, *Proc. SPIE,* 1354:669–675 (1990).

[9] R.H.R. Cuvillier, "Le Pan-Cinor et ses applications," *La Tech. Cinemat.,* 21:73 (1950); also U.S. Patent 2,566,485, filed January 1950.

Chapter 4

Aberration Theory

4.1 INTRODUCTION

In the preceding chapter, imaging was considered to be ideal or stigmatic. This means that rays from a point source P that pass through an optical system will converge to a point located at its Gaussian image P'. In a like manner, the portion of wavefronts from P passing through the optical system will converge as portions of spherical wavefronts toward P'. In other words, the point sources are mapped onto the image surface as point images according to the laws of Gaussian image formation presented in the prior chapter. Deviations from ideal image formation are the result of defects or aberrations inherent in the optical system.

As will be discussed in this chapter, it is possible that the actual image $\tilde{P}'$ is formed at a location other than at P' which can be caused by field curvature and distortion while still forming a stigmatic image. When an optical system fails to form a point image of a point source in the Gaussian image plane, the rays do not pass through the same location and the converging wavefront is no longer spherical as a consequence of the optical system suffering aberrations. In this chapter, a mathematical description of the aberrations for symmetrical optical systems will be presented primarily from the viewpoint of ray deviation errors rather than wavefront errors. In the following chapters, each of the aberrations will be treated in significant detail in addition to their control during the optical design process.

4.2 SYMMETRICAL OPTICAL SYSTEMS

Figure 4.1 illustrates the basic elements of a symmetric optical system. This system is invariant under an arbitrary rotation about its optical axis (OA) and under reflection in any plane containing OA. Both of these symmetry characteristics are necessary properties of a symmetrical optical system.[1] A right-hand

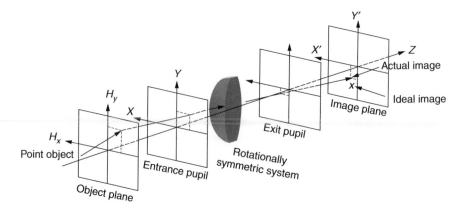

Figure 4.1   Basic elements of a symmetrical optical system.

Cartesian coordinate system is used where the optical axis is always taken to lie along the z-axis.[2] The ideal state of correction for a symmetrical optical system is when a system forms in the image plane (*IP*) normal to the optical axis a sharp and undistorted image of an object in the object plane (*OP*) orthogonal to the optical axis. These planes are designated as the image and object planes, respectively, and are conjugate since the optical system forms an image of one in the other. Unless otherwise specified, these planes should be considered to be orthogonal to the optical axis.

Consider for the moment an arbitrary point P in the object space of a symmetric system. In general the family of rays from P traversing the optical system will fail to pass through a unique point in the image space and the image of P formed by the system is said to be *astigmatic*, that is, to suffer from aberrations. If, on the other hand, all rays from P do pass through a unique point P' in the image space, the image of point P is said to be *stigmatic*.[3] From the definition of a symmetric system, it should be evident that if P' is the stigmatic image of some point P then the two points P and P' lie in a plane containing the optical axis. Now imagine that object points are constrained to lie in the object plane *OP* and that the images of all such points are stigmatic and that the object plane is stigmatically imaged by the system onto an image surface (in contrast to an image plane).

Again relying on the definition of a symmetric system, it is obvious that the stigmatic image of a plane object surface *OP*, which is normal to the optical axis of a symmetric system, is a surface of revolution about the optical axis. When this image surface of revolution is not planar, the imagery is considered to suffer an aberration or image defect known as *curvature of field* although there is no blurring of the image. Since the optical system is considered to be rotationally symmetric, we can arbitrarily select a reference plane that contains the optical axis. Referring to Figure 4.1, this plane is the *Y-Z* plane and is generally called the *tangential* or *meridional* plane.

Assume now that a stigmatic image of the object plane is formed in the image plane where the object has some geometrical shape. If the optical system forms an image having the same geometrical shape as the object to some scaling factor, the image is considered to be undistorted or be an accurate geometric representation of the object. Should the optical system form an image which is not geometrically similar to the object's shape, then the image is said to suffer distortion. When the system is free of distortion (undistorted), the ratio of image size to the corresponding object size is the *magnification m*, with the image for a positive lens being inverted and reverted with respect to the object. Let the object be a line extending from the origin of the object plane to the location denoted as point object in Figure 4.1 which has coordinates expressed as (H_x, H_y). The image size can be computed by

$$H'_x = mH_x \text{ and } H'_y = mH_y$$

since the line can be projected onto each axis and propagated independently without loss of generality since a paraxial skew ray is linearly separable into its orthogonal components.

It is evident from the preceding discussion that an ideal image of the object plane requires three conditions to be satisfied, namely, stigmatic image formation, no curvature of field, and no distortion. In contrast, an optical system having stigmatic image formation can still suffer the image defects of distortion and curvature of field.

As explained, an ideal optical system forms a perfect or stigmatic image which essentially means that rays emanating from a point source will be converged by the optical system to a point image, although curvature of field and distortion may be present. At this juncture, image quality will be discussed in strictly geometric terms. In later chapters, the impact of diffraction on image quality will be discussed.

The majority of this book addresses rotationally symmetric optical systems, their aberrations, and configurations. Figure 4.1 shows the generic geometry for such systems, which comprise five principal elements: the object plane, *entrance pupil*, lenses (including stop), *exit pupil*, and image plane.[4] A ray propagating through this system is specified by its object coordinates (H_x, H_y) and entrance pupil coordinates $(\rho_x, \rho_y) = \vec{\rho}$, or in polar coordinates (ρ, θ), as illustrated in Figure 4.2. This means that point P in the entrance pupil can be expressed by $X = \rho \cos(\theta)$ and $Y = \rho \sin(\theta)$ where θ is zero when $\vec{\rho}$ lies along the Y-axis.

This ray is incident on the image plane at (H'_x, H'_y) and displaced or aberrant from the ideal image location by $(\varepsilon_x, \varepsilon_y)$. Since the optical system is rotationally symmetric, the (point) object is assumed to always be located on the y-axis in the object plane, that is, $H \equiv (0, H_y)$. This means the ideal image is located along the y-axis in the image plane, that is, $h' = mH$ where m is the magnification. The actual

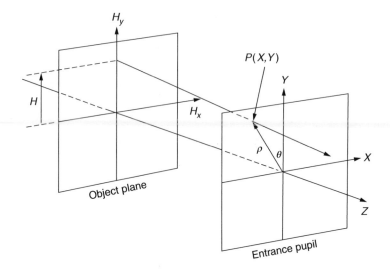

Figure 4.2   Entrance pupil coordinates of a ray.

image plane may be displaced a distance ξ from the ideal image plane. The ideal image plane is also called the Gaussian or paraxial image plane. The term image plane, as used in this book, means the planar surface where the image is formed which may be displaced from the ideal image plane by the defocus distance ξ.

A ray exiting the exit pupil, as shown in Figure 4.3, intersects the image plane at (X', Y') which in general does not pass through the ideal image

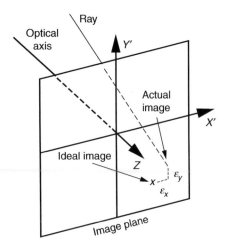

Figure 4.3   Image plane coordinates of ray suffering aberrations.

point shown in the figure as a consequence of aberrations. The object point is
located at $\vec{H} = H_x\hat{i} + H_y\hat{j}$ (see Figure 4.1) with the ideal image point being
located at (h'_x, h'_y) and the actual image being located at (X', Y').

$$X'(\rho, \theta, \vec{H}, \xi) \equiv \varepsilon_x(\rho, \theta, \vec{H}, \xi) + h'_x$$
$$Y'(\rho, \theta, \vec{H}, \xi) \equiv \varepsilon_y(\rho, \theta, \vec{H}, \xi) + h'_y$$

Using vector notation and ignoring the defocus parameter for the moment, the
ray aberration can be written as

$$\vec{\varepsilon}(\vec{\rho}, \vec{H}) = \vec{\varepsilon}_s(\vec{\rho}, \vec{H}) + \vec{\varepsilon}_c(\vec{\rho}, \vec{H})$$

where $\vec{\varepsilon}_s$ and $\vec{\varepsilon}_c$ are defined by

$$\vec{\varepsilon}_s(\vec{\rho}, \vec{H}) = \tfrac{1}{2}\left(\vec{\varepsilon}(\vec{\rho}, \vec{H}) - \vec{\varepsilon}(-\vec{\rho}, \vec{H})\right)$$
$$\vec{\varepsilon}_c(\vec{\rho}, \vec{H}) = \tfrac{1}{2}\left(\vec{\varepsilon}(\vec{\rho}, \vec{H}) + \vec{\varepsilon}(-\vec{\rho}, \vec{H})\right)$$

and $\vec{\varepsilon}_s$ and $\vec{\varepsilon}_c$ are called the symmetric and asymmetric aberrations as well as the
astigmatic and the comatic aberrations, respectively, of the ray $(\vec{\rho}, \vec{H})$.[5] The
importance of decomposing the ray aberration in this manner for our study of
lens design will become evident. Consider first the symmetric term $\vec{\varepsilon}_s$ which
means that the ray error will be symmetric about the ideal image location
assuming no distortion. Specifically this can be interpreted as $(\varepsilon_x, \varepsilon_y)$ for
$(\vec{\rho}, \vec{H})$ and $(-\varepsilon_x, -\varepsilon_y)$ for $(-\vec{\rho}, \vec{H})$. If a spot diagram of a point source is made
for an optical system suffering only astigmatic aberration, the pattern formed
will be symmetric.

In contrast, the comatic or asymmetric aberration $\vec{\varepsilon}_c$ is invariant when the
sign of $\vec{\rho}$ is changed. This means that rays $(\vec{\rho}, \vec{H})$ and $(-\vec{\rho}, \vec{H})$ will suffer the
identical image error $(\varepsilon_x, \varepsilon_y)$, that is, they each intercept the image plane at
the same location. Consequently, the comatic aberration creates an asymmetry
in the spot diagram. Further, it should be recognized that the astigmatic and
comatic aberration components are decoupled and can not be used to balance
one another. The importance of this knowledge in lens design will be explained
in more detail in the following chapters.

Since the optical system is rotationally symmetric, the object can be placed in
the meridional plane, or y-axis of the object plane, without the loss of generality
and the advantage of simplifying the computation and interpretation of the
resulting aberrations. Consequently, since the x-component is zero the object
is denoted by H and the ideal image by h'. The actual image coordinates now
become

$$X'(\rho, \theta, H, \xi) \equiv \varepsilon_x(\rho, \theta, H, \xi)$$
$$Y'(\rho, \theta, H, \xi) \equiv \varepsilon_y(\rho, \theta, H, \xi) + h'$$

for the specific ray coordinates ρ, θ, and H, and the image plane defocus ξ. It has been found useful to decompose the aberration into two elements with respect to how the aberrations transform under a change in the sign of ρ. These elements are called symmetric and asymmetric components and are orthogonal to one another. Since the object and image are located in the meridional plane, all rays emanating from the object point having entrance pupil coordinates $\vec{\rho} = (\rho, 0^o)$ necessarily lie in the meridional plane.[6] Consequently, $\varepsilon_x = 0$. It is common to plot the ray aberration for the meridional fan of rays with ρ being normalized (-1 to $+1$). The ordinate of the plot is the ray error measured from intercept of the principal ray.

Figure 4.4 provides an example of such a plot. In this case, $H \neq 0$ to allow illustration of the symmetric and asymmetric components of the ray aberration. As explained above, $\vec{\varepsilon}_s$ and $\vec{\varepsilon}_c$ represent these components. In this figure, the comatic and the stigmatic contributions for the total aberration are shown. Notice that the comatic aberration is symmetric about the $\rho = 0$ axis. In other words, any ray pair having entrance pupil coordinates of $(\rho, 0^o)$ and $(-\rho, 0^o)$ will have the same ray error, that is, $\varepsilon_y(\rho, 0, H) = \varepsilon_y(-\rho, 0, H)$. In contrast, the astigmatic aberration is asymmetric about the same axis. This means that any ray pair having entrance pupil coordinates of $(\rho, 0^o)$ and $(-\rho, 0^o)$ will suffer ray errors of equal and opposite sign, that is, $\varepsilon_y(\rho, 0, H) = -\varepsilon_y(-\rho, 0, H)$. Examination of the total aberration curve illustrates that it can be neither symmetric nor asymmetric. In this particular case, both the comatic and astigmatic aberrations comprise third- and fifth-order terms of opposite signs. The total aberration curve is simply the sum of the comatic and astigmatic values.

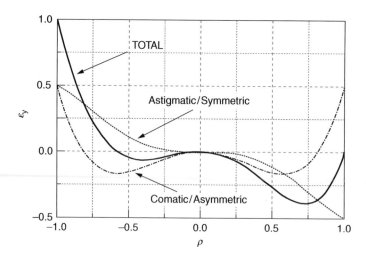

Figure 4.4 Ray aberration for a meridional fan of rays.

The interpretation of such plots for use in lens design will become evident in the following material.

Plots, such as that shown in Figure 4.4, are very useful during a lens design process; however, the meridional plots provide only a portion of the insight into the complete aberrations suffered by a particular lens design. Additional plots can be generated using non-meridional rays, which are generally called skew rays. The most common skew rays utilized have entrance pupil coordinates of $(\pm\rho, 90^o)$ and are commonly called sagittal rays. The name *sagittal* is generally given to the 90° and 270° skew rays that lie in a plane perpendicular to the meridional plane, containing the principal ray.

The *sagittal plane is not one single plane throughout a lens, but it changes its tilt after each surface refraction/reflection.* The point of intersection of a sagittal ray with the paraxial image plane may have both a vertical error and a horizontal error relative to the point of intersection of the principal ray, and both these errors can be plotted separately against some suitable ray parameter. This parameter is often the horizontal distance from the meridional plane to the point where the entering ray pierces the entrance pupil. The meridional plot, of course, has no symmetry, but the two sagittal ray plots do have symmetry. As a consequence, sagittal ray plots are often shown for only positive values of ρ since it is realized that

$$\varepsilon_x(\rho, 90^\circ, H, \xi) = -\varepsilon_x(-\rho, 90^\circ, H, \xi) \quad \text{and} \quad \varepsilon_y(\rho, 90^\circ, H, \xi) = \varepsilon_y(-\rho, 90^\circ, H, \xi).$$

It has been shown that the ray aberration can be decomposed into astigmatic and comatic components, which are orthogonal. These two components can be further decomposed. For the astigmatic component, it comprises spherical aberration, astigmatism, and defocus. In a like manner, the comatic component comprises coma and distortion. The following two equations for the ray errors ε_x and ε_y show this decomposition. The abbreviations for the various components will be utilized extensively in the following material.

$$\varepsilon_x(\rho, 0, H, \xi) = \underbrace{SPH_x(\rho, 0, 0) + AST_x(\rho, 0, H) + DF_x(\rho, 0, \xi)}_{\text{ASTIGMATIC COMPONENTS}}$$

$$+ \underbrace{CMA_x(\rho, 0, H)}_{\text{COMATIC COMPONENTS}}$$

$$\varepsilon_y(\rho, 0, H, \xi) = \underbrace{SPH_y(\rho, 0, 0) + AST_y(\rho, 0, H) + DF_y(\rho, 0, \xi)}_{\text{ASTIGMATIC COMPONENTS}}$$

$$+ \underbrace{CMA_y(\rho, 0, H) + DIST(H)}_{\text{COMATIC COMPONENTS}}$$

(4-1)

where SPH $\equiv$ spherical aberration, AST $\equiv$ astigmatism, CMA $\equiv$ coma, DIST $\equiv$ distortion, and DF $\equiv$ defocus. It should be recognized that the comatic component of ε_x does not contain a distortion term since it is assumed that the object lies in the meridional plane.

Being that the ray intercept error can be described as the linear combination of the astigmatic and comatic contributions, these contributions can be written as a power series in terms of H and ρ. Several conventions exist for expansion nomenclature; however, most follow that given by Buchdahl. Specifically, an aberration depending on ρ and H in the combination $\rho^{n-s}H^s$ is said to be of the type

- n^{th} order, s^{th} degree coma if $(n\text{-}s)$ is even, or
- n^{th} order, $(n\text{-}s)^{\text{th}}$ degree astigmatism if $(n\text{-}s)$ is odd.

For simplicity, the arguments of ε_x and ε_y are not explicitly shown unless needed for clarity, defocus is assumed zero, and recalling that the expansions are a function of θ for the general skew ray, the expansion of the ray errors are given by

$$\varepsilon_x = \underbrace{(\sigma_1\rho^3 + \mu_1\rho^5 + \tau_1\rho^7 + \ldots)\sin(\theta)}_{\text{SPHERICAL}}$$

$$+ \underbrace{(\sigma_2\rho^2 + \mu_3\rho^4 + \tau_3\rho^6 + \ldots)\sin(2\theta)H}_{\text{LINEAR or CIRCULAR COMA}}$$

$$+ \underbrace{(\mu_9\sin(2\theta)\rho^2 + (\tau_9\sin(2\theta) + \tau_{10}\sin(4\theta))\rho^4 + \ldots)H^3}_{\text{CUBIC COMA}}$$

$$+ \underbrace{(\tau_{17}\sin(2\theta)\rho^2 + \ldots)H^5}_{\text{QUINTIC COMA}} \qquad (4\text{-}2)$$

$$+ \underbrace{((\sigma_3 + \sigma_4)H^2 + \mu_{11}H^4 + \tau_{19}H^6 + \ldots)\sin(\theta)\rho}_{\text{LINEAR ASTIGMATISM}}$$

$$+ \underbrace{((\mu_5 + \mu_6\cos^2(\theta))H^2 + (\tau_{13} + \tau_{14}\cos^2(\theta))H^4 + \ldots)\sin(\theta)\rho^3}_{\text{CUBIC ASTIGMATISM}}$$

$$+ \underbrace{((\tau_5 + \tau_6\cos^2(\theta))H^2 + \ldots)\sin(\theta)\rho^5}_{\text{QUINTIC ASTIGMATISM}}$$

$$+ \cdots \text{HIGHER ORDER ABERRATIONS IN TERMS OF } \rho \text{ AND H.}$$

and

$$\varepsilon_y = \underbrace{(\sigma_1\rho^3 + \mu_1\rho^5 + \tau_1\rho^7 + \ldots)\cos(\theta)}_{\text{SPHERICAL}}$$

$$+ \underbrace{(\sigma_2(2+\cos(2\theta))\rho^2 + (\mu_2 + \mu_3\cos(2\theta))\rho^4 + (\tau_2 + \tau_3\cos(2\theta))\rho^6 + \ldots)H}_{\text{LINEAR or CIRCULAR COMA}}$$

$$+ \underbrace{((\mu_7 + \mu_8\cos(2\theta))\rho^2 + (\tau_7 + \tau_8\cos(2\theta) + \tau_{10}\cos(4\theta))\rho^4 + \ldots)H^3}_{\text{CUBIC COMA}}$$

$$+ \underbrace{((\tau_{15} + \tau_{16})\cos(2\theta)\rho^2 + \ldots)H^5}_{\text{QUINTIC COMA}}$$

$$+ \underbrace{((3\sigma_3 + \sigma_4)H^2 + \mu_{10}H^4 + \tau_{18}H^6 + \ldots)\cos(\theta)\rho}_{\text{LINEAR ASTIGMATISM}}$$

$$+ \underbrace{((\mu_4 + \mu_6\cos^2(\theta))H^2 + (\tau_{11} + \tau_{12}\cos^2(\theta))H^4 + \ldots)\cos(\theta)\rho^3}_{\text{CUBIC ASTIGMATISM}}$$

$$+ \underbrace{((\tau_4 + \tau_6\cos^2(\theta))H^2 + \ldots)\cos(\theta)\rho^5}_{\text{QUINTIC ASTIGMATISM}}$$

$$+ \underbrace{\sigma_5H^3 + \mu_{12}H^5 + \tau_{20}H^7 + \ldots}_{\text{DISTORTION}}$$

$$+ \cdots \text{HIGHER ORDER ABERRATIONS IN TERMS OF } \rho \text{ AND H.}$$

$$\tag{4-3}$$

The five σ, twelve μ, and twenty τ coefficients represent the third-, fifth-, and seventh-order terms, respectively. Even-order terms do not appear as a consequence of the rotational symmetry of the optical system. Further, there are actually five, nine, and 14 independent coefficients for the third-, fifth-, and seventh-order terms, respectively.[7]

There exist three identities between the μ coefficients, and six identities between the τ coefficients. These identities take the form of a linear combination of the nth-order coefficients being equal to combinations of products of the lower-order coefficients. If, for example, all of the third-order coefficients have been corrected to zero, then the following identities for the fifth-order coefficients exist: $\mu_2 - \frac{2}{3}\mu_3 = 0$; $\mu_4 - \mu_5 - \mu_6 = 0$; and $\mu_7 - \mu_8 - \mu_9 = 0$. Calculation of these coefficients is straightforward, although tedious, using the iterative process developed by Hans Buchdahl.[1] The third-order terms were first popularized by the publication of Seidel and are often referred to as the Seidel aberrations.[8] The fifth-order terms were first computed in the early twentieth century.[9]

In the late 1940s, Buchdahl published his work on how to calculate the coefficients to any arbitrary order. However, recent investigation into the historical work of Joseph Petzval, a Hungarian professor of mathematics at Vienna, has lead to the belief that he had developed in the late 1830s a computational scheme through fifth-order and perhaps to seventh-order for spherical aberration.[10] Conrady was well aware of the Petzval sum in addition to Petzval's greater contributions to optics as evidenced when he wrote:

> *[Petzval] who investigated the aberrations of oblique pencils about 1840, and apparently arrived at a complete theory not only of the primary, but also of the secondary oblique aberrations; but he never published his methods in any complete form, he lost the priority which undoubtedly would have been his. It is, however, perfectly clear from his occasional brief publications that he had a more accurate knowledge of the profound significance of the Petzval theorem than any of his successors in the investigation of the oblique aberrations for some eighty years after his original discovery.[11]*

Regrettably, the preponderance of his work was lost to posterity. The design and development for today's optical systems were made possible by theoretical understanding of optical aberrations through the contributions of numerous individuals. Although the subject is still evolving, serious research spans over four centuries.[12]

As an example, consider a meridional ray intersecting the paraxial image plane, and having entrance pupil coordinates of $(\rho, 90^0, H, 0)$. The ε_x and ε_y are given by

$$\varepsilon_x = \underbrace{(\sigma_1\rho^3 + \mu_1\rho^5 + \tau_1\rho^7 + \ldots)}_{\text{SPHERICAL}}$$

$$+ \ ((\sigma_3 + \sigma_4)H^2 + \mu_{11}H^4 + \tau_{19}H^6 + \ldots)\rho$$

$$+ \ (\mu_5 H^2 + \tau_{13}H^4 + \ldots)\rho^3$$

$$+ \ \underbrace{(\tau_5 H^2 + \ldots)\rho^5}_{\text{ASTIGMATISM}}$$

$$+ \cdots \text{HIGHER ORDER ABERRATIONS IN TERMS OF } \rho \text{ AND H.}$$

and

$$\varepsilon_y = (\sigma_2\rho^2 + (\mu_2 - \mu_3)\rho^4 + (\tau_2 - \tau_3)\rho^6 + \ldots)H$$

$$+ \ ((\mu_7 - \mu_8)\rho^2 + (\tau_7 - \tau_8 + \tau_{10})\rho^4 + \ldots)H^3$$

$$+ \ \underbrace{((\tau_{15} - \tau_{16})\rho^2 + \ldots)H^5}_{\text{COMA}}$$

$$+ \ \underbrace{\sigma_5 H^3 + \mu_{12}H^5 + \tau_{20}H^7 + \ldots}_{\text{DISTORTION}}$$

$$+ \cdots \text{HIGHER ORDER ABERRATIONS IN TERMS OF } \rho \text{ AND H.}$$

Observe that the sagittal term ε_x comprises only astigmatic contributions, while the meridional term ε_y contains only comatic contributions. The ability to isolate specific contributions of the ray error by proper selection of one or more rays will be exploited in the remainder of this chapter.

As previously explained, the actual ray height in the paraxial image plane can be considered to comprise two principal elements: the Gaussian ray height and the ray aberration, as illustrated in the aberration map shown in Figure 4.5 on the next page. The total aberration for a rotationally symmetric optical system comprises two orthogonal components, astigmatic and comatic. The astigmatic aberration is segmented into field independent and dependent components while the comatic aberration is divided into aperture independent and dependent components.

The field-independent astigmatic aberration has two contributions, which are defocus and spherical aberration. The defocus ξ is linearly dependent on the entrance pupil radius ρ while the spherical aberration is dependent on the odd orders of third and above of the entrance pupil radius, namely, $\rho^3, \rho^5, \ldots$. *The field-independent astigmatic aberration introduces a uniform aberration or blur over the optical system's field-of-view.*

Field-dependent astigmatic aberrations comprise two contributions which are linear astigmatism and oblique spherical aberration. Both of these aberrations are dependent on even orders of H, namely, $H^2, H^4, \ldots$. Linear astigmatism is linearly dependent on the entrance pupil radius ρ while the oblique spherical aberration is dependent on the odd orders of third and above of the entrance pupil radius. It should be noted that the defocus and linear astigmatism comprise the linearly-dependent entrance-pupil-radius components of the astigmatic aberration $(\rho; H^0, H^2, H^4, \ldots)$. In a like manner, spherical and oblique spherical aberrations comprise the higher-order terms in entrance-pupil-radius $(\rho^3, \rho^5, \ldots; H^0, H^2, H^4, \ldots)$.

Aperture-independent comatic aberration has two contributions, which are the Gaussian image height and distortion. Although the Gaussian image height is not considered an actual aberration, it is shown in the aberration map in a dashed box since the Gaussian image height is linearly proportional to H and aperture independent. Distortion is also aperture independent, but is dependent on the odd orders of third and above of H, namely, $H^3, H^5, \ldots$.

Aperture-dependent comatic aberration also has two contributions, which are linear coma and nonlinear coma. Linear coma is linearly dependent on the field angle and on even orders of the entrance pupil radius $(\rho^2, \rho^4, \ldots; H)$. Nonlinear coma has the same entrance pupil radius dependence as does linear coma, but is dependent on the odd orders of third and above of H in the same manner as distortion. Perhaps the most common element of nonlinear coma is referred to as elliptical coma; however, there are many other contributions to the nonlinear comatic aberration.

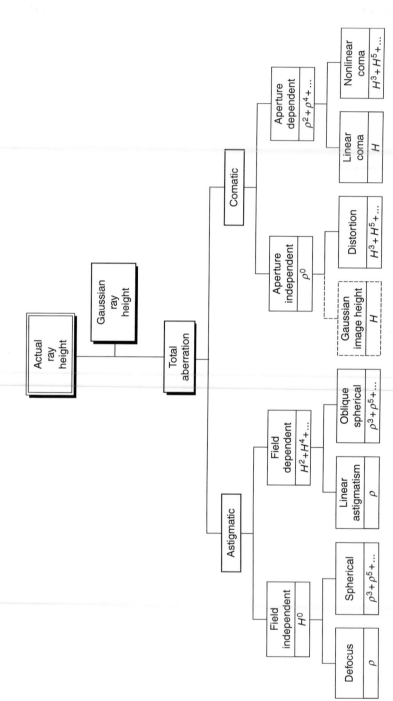

Figure 4.5 Ray aberration map showing the astigmatic and comatic elements comprising the total ray aberration.

The *aperture-dependent comatic aberration can be viewed as a variation of magnification from one zone to another zone of the entrance pupil.* It is also noted that because the astigmatic and comatic contributions are orthogonal, changing the location of the image plane from the paraxial location can impact the resultant astigmatic aberration while having no effect on the comatic contribution of the total aberration. In other words, the *defocus can change the astigmatic contribution to the total aberration while having no effect on the comatic contribution.* This will be discussed in more detail later in this chapter.

An interesting aspect of the Buchdahl aberration expansion is that the contribution for each coefficient is computed surface by surface and then summed to determine the value of the coefficient at the image plane. For example, σ_1 is the third-order spherical aberration coefficient. Its value for an optical system comprising n surfaces is computed as

$$\sigma_1 = \sum_{i=1}^{n} {}_i\sigma_1$$

Although there will be no attempt to compute the general set of Buchdahl aberration coefficients in this study, it is important to understand certain aspects of their relationship to the design process. It can be shown that these aberration coefficients have intrinsic and extrinsic contributions. The third-order aberration coefficients have only intrinsic contributions, which mean that the value of the aberration coefficients computed for any arbitrary surface are not dependent on the aberration coefficient values for any other surface. For the higher-order aberration coefficients, extrinsic contributions exist in addition to the intrinsic contributions. This means that aberration coefficients for the kth surface are to some extent dependent on the preceding surfaces while not at all dependent on the subsequent surfaces.

Two other characteristics of aberration coefficients are valuable for the lens designer to understand. The first is that lower-order aberration coefficients affect similar high-order aberration coefficients. An alternative way to express this behavior is that higher-order aberration coefficients do not affect the value of lower-order aberration coefficients; that is, adjustment of say τ_1 does not change the third- and fifth-order contributions. The second characteristic is that higher-order aberration coefficients move or change their values slowly with changes in constructional parameters (radii, thickness, etc.) compared to the movement of lower-order aberration coefficients. In short, this means that higher-order aberrations, be they astigmatic or comatic, are far more stable than lower-order aberrations.

4.3 ABERRATION DETERMINATION USING RAY TRACE DATA

The elements comprising the total ray aberration can be computed directly from specific ray trace data. How this is done and the relationship to the aberration coefficients are presented in this section. It should be noted that the method discussed decouples the defocus element from the astigmatic elements thereby enhancing the utility of these elements in the optical design process. Each of the following aberrations is briefly introduced and will be discussed in detail in subsequent chapters.

4.3.1 Defocus

Defocus can be used as a first-order aberration that is measured from the paraxial image plane. It depends only on entrance pupil coordinates, not on the object height or field angle. Defocus impacts imagery uniformly over the entire field of view. Often defocus can be used to balance or improve symmetric (astigmatic) aberrations, while having no effect on asymmetric (comatic) aberrations. Figure 4.6 shows the upper and lower marginal rays exiting the optical system, focusing at the paraxial image plane, and forming a blur at the image plane located a longitudinal distance ξ from the paraxial image plane. Defocus can be expressed as

$$
\begin{aligned}
DF(\rho, \xi) &= -\xi \tan v'_a \\
&= -\frac{\rho}{f} \xi
\end{aligned}
\tag{4-4}
$$

where v'_a is the angle of the marginal paraxial ray in image space and f is the focal length.

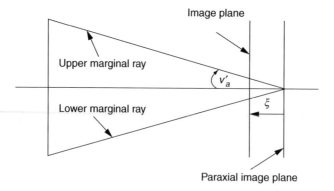

Figure 4.6 Defocus.

For finite conjugate systems, f should be replaced by the paraxial image distance. The intersection height of the marginal ray with the image plane is the transverse defocus aberration, and the defocus blur is

$$\left|\frac{2\xi\rho}{f}\right|$$

The ray fan plot of ε_x or ε_y versus ρ is simply a straight line. A plot of $\vec{\varepsilon}$ versus θ for a fixed value of ρ shows a circle because of the radial symmetry about the optical axis. As will be discussed in subsequent chapters defocus can be used as a means to improve the image quality when astigmatic errors are present; however, defocus has no effect on comatic aberrations.

PROBLEM: Show that $DF(\rho, \xi)$ is independent of the image height and is therefore a field-independent aberration.

4.3.2 Spherical Aberration

Spherical aberration can be defined as a variation with aperture of the image distance or focal length in the case of infinite conjugates. Figure 4.7 shows a positive lens that suffers undercorrected or negative spherical aberration, which is typical of such lenses.[13] A close-up view of the image region of Figure 4.7a is shown in Figure 4.7b. The paraxial rays come to a focus at the paraxial focal plane while, in this case, meridional rays farther from the optical axis progressively intersect this axis farther from the paraxial image plane and closer to the lens. This is referred to as longitudinal spherical aberration and is referenced to the marginal ray intercept as shown in the figure. In a similar manner, these rays intercept the paraxial image plane below the optical axis and are referred to as transverse spherical aberration.

Figure 4.8a presents the meridional ray fan plot, which more clearly presents the transverse ray error ε_y as a function of entrance pupil radius ρ. Figure 4.8b shows the longitudinal spherical aberration as a plot of the axial intercept location as a function of the entrance pupil radius ρ. An alternative presentation of the ray error is the spherical aberration contribution to the wavefront error as a function of the entrance pupil radius ρ as illustrated in Figure 4.8c. As will be explained, the longitudinal, transverse, and wave presentations of spherical aberration are related to each by simple multiplicative factors. Each form of spherical aberration has utility and none has general superiority.

The transverse spherical aberration at the paraxial image plane is given by the displacement of a ray having coordinates $(\rho, 0^0, 0, 0)$ from the optical axis, which can be expressed as

$$
\begin{aligned}
SPH(\rho, 0^0, 0) &= Y(\rho, 0^0, 0, 0) \\
&= \sigma_1 \rho^3 + \mu_1 \rho^5 + \tau_1 \rho^7 + \dots
\end{aligned}
\tag{4-5}
$$

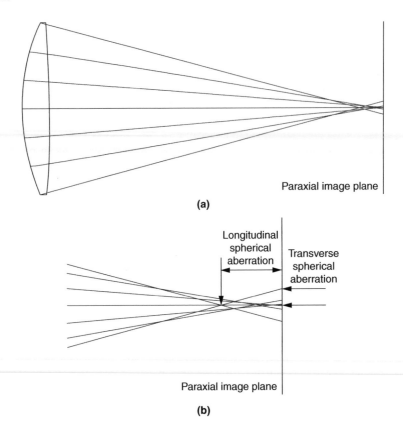

Figure 4.7 (a) Positive lens that suffers undercorrected or negative spherical aberration. (b) Close-up view of the image region.

where $Y(\rho, 0^0, 0, 0)$ is the real ray value in the polynomial expansion as also shown.

Figure 4.9 illustrates the general behavior of the third-, fifth- and seventh-order spherical aberration terms. In this particular case, σ_1, μ_1, and τ_1 are all given a value of unity. It should be noted that the higher the order of the terms, the flatter the plots are until progressively larger values of ρ are reached, at which point the curves increase rapidly. The distance from the paraxial image plane to the intersection point of the ray with the optical axis is called longitudinal spherical aberration. Assuming that the ray slope is negative, then the longitudinal spherical aberration is considered positive, or overcorrected, if the intersection point is beyond the paraxial image plane; and is considered negative, or undercorrected, if the intersection point precedes the paraxial image plane.

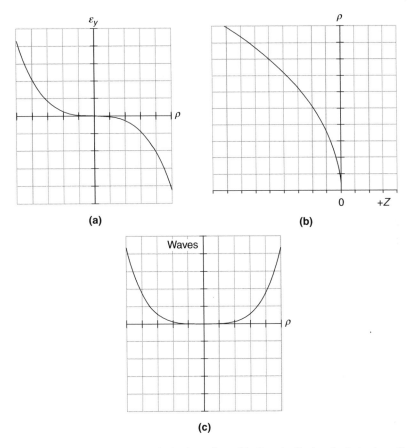

Figure 4.8 (a) Transverse spherical aberration. (b) Longitudinal spherical aberration. (c) Spherical aberration contribution to the wavefront error.

4.3.3 Tangential and Sagittal Astigmatism

Field-dependent astigmatism and curvature of field are inherently related to displace the image from the paraxial image plane. As is illustrated in Figure 4.10, the meridional rays come to a focus some distance from the paraxial image plane, forming a line lying in the sagittal plane whose length is determined by the width of the sagittal fan of rays at that point. In a like manner, the sagittal focus is determined by where the sagittal fan focuses in the tangential plane and has a length determined by the width of the tangential fan at that point. The tangential astigmatism for a given value of ρ and H can be determined exactly by tracing three rays, namely the corresponding upper and lower off-axis rays,

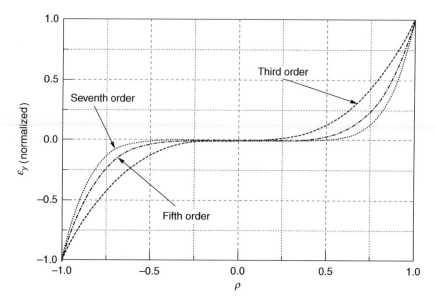

Figure 4.9 Orders of spherical aberration.

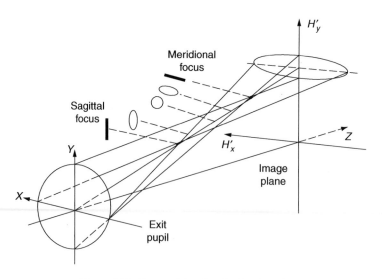

Figure 4.10 Tangential and sagittal astigmatism.

and the marginal ray. Consequently, the tangential astigmatism is computed using the ray data in the following equation.

$$
\begin{aligned}
TAST(\rho, H) &= Y(\rho, 0^0, H, \xi) - Y(\rho, 180^0, H, \xi) - 2Y(\rho, 0^0, 0, \xi) \\
&= 2AST_y(\rho, 0^0, H) \\
&= 2[(3\sigma_3 + \sigma_4)H^2 + \mu_{10}H^4 + \tau_{18}H^6 + \ldots]\rho \\
&\quad + 2[(\mu_4 + \mu_6)H^2 + (\tau_{11} + \tau_{12})H^4 + \ldots]\rho^3 \\
&\quad + 2[(\tau_4 + \tau_6)H^2 + \ldots]\rho^5 + \ldots
\end{aligned}
\tag{4-6}
$$

In addition, the resulting aberration coefficients are also shown. Notice that the polynomial expansion is expanded in odd orders of ρ. The importance of this will be explained presently. The purpose of including the marginal ray in the above calculation is to remove the field-independent components from the upper and lower rays, that is, defocus and spherical aberration. The portion of the expansion that is linear with ρ is known as linear tangential astigmatism and has a ray fan plot similar to the plot for defocus. Figure 4.11 illustrates the behavior of tangential astigmatism for ρ, ρ^3, and ρ^5 for a particular value of H. Notice that these plots have the same form as defocus, and third- and fifth-order spherical aberration; however, $TAST(\rho, H)$ varies with H.

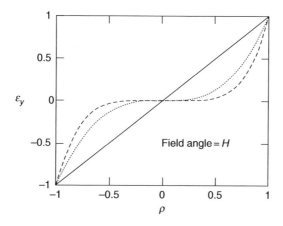

Figure 4.11 Tangential ray fan plot.

In a similar manner, sagittal astigmatism is computed using the ray data in the following equation.

$$\begin{aligned}
SAST(\rho, H) &= 2[X(\rho, 90^0, H, \xi) - Y(\rho, 0^0, 0, \xi)] \\
&= 2AST_x(\rho, 90^0, H) \\
&= 2[(\sigma_3 + \sigma_4)H^2 + \mu_{11}H^4 + \tau_{19}H^6 + \ldots]\rho \qquad (4\text{-}7) \\
&\quad + 2[\mu_5 H^2 + \tau_{13}H^4 + \ldots]\rho^3 \\
&\quad + 2[\tau_5 H^2 + \ldots]\rho^5 + \ldots
\end{aligned}$$

Subtraction of the meridional ray from the x-component of the sagittal ray, having coordinates $(\rho, 90^0, H, \xi)$, removes the field-independent astigmatic contributions since the axial meridional and sagittal rays $(H = 0)$ contain the same values. The y-component of this sagittal ray is used to compute sagittal coma. The σ_3 coefficient represents third-order astigmatism, while σ_4 represents Petzval. Assuming all other aberration coefficients are zero with the exception of σ_4, it is easily shown that the image formed on the resulting Petzval surface is stigmatic.

> **PROBLEM:** Determine an equation that expresses the longitudinal image displacement from the paraxial image plane when all aberration coefficients are zero except σ_4.

4.3.4 Tangential and Sagittal Coma

Coma can be viewed as a variation in magnification from one zone to another zone in the entrance pupil. Figure 4.12 shows the basic geometry for computing tangential coma. The upper and lower rim rays are shown and intersect some distance behind the paraxial image plane. The principal ray also intersects the

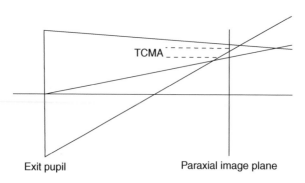

Figure 4.12   Tangential coma.

paraxial image plane as illustrated. Tangential coma is determined by subtracting the principal-ray height from the average value of the upper and lower rim ray intercept heights. Figure 4.12 depicts the computation in the paraxial image plane. It is also common to compute the value of tangential coma in the plane where the upper and lower rim rays intersect and subtract from that height the height of the principal ray in this plane. The value computed will be the same because comatic aberrations are unaffected by defocus. The defining equation for tangential coma and its polynomial expansion are as follows:

$$
\begin{aligned}
TCMA(\rho, H) &= \left[\frac{Y(\rho, 0^0, H, \xi) + Y(\rho, 180^0, H, \xi)}{2} \right] - Y(0, 0^0, H, \xi) \\
&= CMA_y(\rho, 0^0, H) \\
&= [3\sigma_2\rho^2 + (\mu_2 + \mu_3)\rho^4 + (\tau_2 + \tau_3)\rho^6 + \ldots]H \\
&\quad + [(\mu_7 + \mu_8)\rho^2 + (\tau_7 + \tau_8 + \tau_{10})\rho^4 + \ldots]H^3 \\
&\quad + [(\tau_{15} + \tau_{16})\rho^2 + \ldots]H^5 + \ldots
\end{aligned}
\tag{4-8}
$$

Figure 4.13 shows plots of second-, fourth-, and sixth-order tangential coma as a function of ρ for a specific field angle. These correspond to the third-, fifth-, and seventh-order aberration coefficients for linear coma. Figure 4.14 illustrates the general functional relationship between the various orders of tangential coma $(H + H^3 + H^5 + \ldots)$ versus field angle/image height H. Examination of these two figures clearly illustrates that linear coma is dominant for small field angles as are the ρ^2 aberration coefficients at a specific value of H.

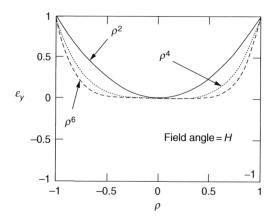

Figure 4.13 Tangential coma as a function of ρ for constant field angle.

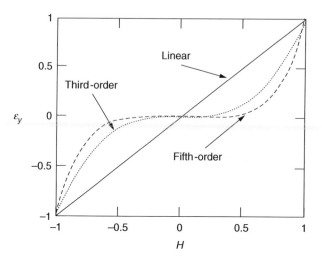

Figure 4.14 Tangential coma as a function of H.

The exact sagittal coma is computed using the y-component of the sagittal ray having coordinates $(\rho, 90^0, H, \xi)$ and the principal ray. Notice that this is measured in the meridional plane as is the tangential coma. The defining equation for sagittal coma and its polynomial expansion are as follows:

$$SCMA(\rho, H) = Y(\rho, 90^0, H, \xi) - Y(0, 0^0, H, \xi)$$
$$= CMA_y(\rho, 90^0, H)$$
$$= [\sigma_2 \rho^2 + (\mu_2 - \mu_3)\rho^4 + (\tau_2 - \tau_3)\rho^6 + \ldots]H \qquad (4\text{-}9)$$
$$+ [(\mu_7 - \mu_8)\rho^2 + (\tau_7 - \tau_8 + \tau_{10})\rho^4 + \ldots]H^3$$
$$+ [(\tau_{15} - \tau_{16})\rho^2 + \ldots]H^5 + \ldots$$

A point-source image formed by an optical system suffering coma spreads out the light into a comet-shaped flare. Coma is a rather annoying aberration since its flare is non-symmetrical and makes it quite difficult to make accurate determination of image position in contrast to a symmetric or circular blur made, for example, by spherical aberration. Since approximately half of the energy in the coma patch is located in the region near the head of the coma patch, sagittal coma provides a more reasonable estimate of the image blur than does tangential coma. When the coma tail lies between the optical axis and the Gaussian image, it is referred to as negative or undercorrected coma. If the coma tail is farther from the optical axis than the Gaussian image, it is referred to as positive or overcorrected coma.

Also, the *offense against a sine condition* or *OSC* is given by

$$\frac{SCMA(\rho, H)}{Y(0, 0°, H, 0)}$$

which is the sagittal coma divided by the principal-ray height for regions near the optical axis. The *OSC* for the doublet objective lens described in Sections 2.5 and 5.2 is -0.000173 for a marginal ray of height 2.0 at the vertex of the first surface. This should be compared to the *OSC* for this lens computed by two alternative methods in Section 9.3.2. For field angles up to a bit over 1°, the agreement is essentially exact for the above method. It can also be shown that, to the third order, the $OSC \propto \sigma_2$, and for small obliquities, $TCMA = 3 \cdot SCMA$.

The flare for an image suffering third-order linear coma is confined between a pair of lines intersecting at the Gaussian image height and having a 60° included angle. The image patch or flare for higher-order linear coma is flatter and results in a wider comet tail appearance. The included angle between the bounding lines increases to 84° for fifth-order linear coma and 97° for seventh-order linear coma. Consequently, the presence of the higher-order linear coma should be visually evident by inspection of the spot diagram.

PROBLEM: Show that, to the third order, the $OSC \propto \sigma_2$, and for small obliquities, $TCMA = 3 \cdot SCMA$.

PROBLEM: Assuming that all aberration coefficients are zero except for third-order linear coma, show that coma appears as a family of circles as ρ and varies where the circles have radii of $|\sigma_2 \rho^2 H|$. Where is the center of each circle located with respect to the Gaussian image height? Show that these circles are confined between a pair of lines intersecting at the Gaussian image height and having a 60° included angle.

4.3.5 Distortion

As previously discussed, one of the requirements for an optical system to produce ideal imagery is that the image it forms must be geometrically similar to the object, that is, the image dimensions are a linear factor of the object dimensions. Consider now that the image formation is also stigmatic. The image height is determined by the intersection of the principal ray with the paraxial image plane. In general, the geometrical similarity of the image to the object is *not* a linear relationship with the object height and is referred to as distortion of the image. Just as the Gaussian image height is aperture independent, so is the distortion, which is the aperture-independent comatic aberration.

In Figure 4.5, the Gaussian image height is shown in a dashed box as the linear portion of the aperture independent comatic aberration. This should make sense in that comatic aberrations are considered associated with variation in magnification with respect to object height and entrance pupil. It also fills out the H expansion sequence in the aperture independent comatic aberrations. Hence, distortion of the image can be considered as the aberration of the principal ray and is defined by the following equation.

$$DIST(H) = Y(0, 0^0, H, \xi) - GIH(H, \xi)$$
$$= \sigma_5 H^3 + \mu_{12} H^5 + \tau_{20} H^7$$

(4-10)

where the Gaussian image height is $GIH(H, \xi) = GIH(H) + DF_y(\rho, 0^0, \xi)$.

Distortion is considered negative when the actual image is closer to the axis than the ideal image, and positive distortion is the converse. This physically means that the image of a square suffering negative distortion will take on a barrel-like appearance and is referred to as barrel distortion. In the case of positive distortion, the image takes on a pincushion-like appearance and is referred to as pincushion distortion. To reiterate, distortion is an aperture independent comatic aberration. For most lenses, distortion beyond the third-order term is minimal.

4.3.6 Selection of Rays for Aberration Computation

Table 4.1 presents the five rays necessary to compute the astigmatic and comatic aberrations for a particular set of (ρ, H). It should be noticed that the first three rays all contained in the meridional plane. The remaining two rays are skew rays. The sagittal astigmatism is the only aberration to utilize x-coordinate ray data.

Table 4.1

Table of Rays Required to Compute the Astigmatic and Comatic Aberrations

Ray Coordinates	ρ $0°$ 0	0 $0°$ H	ρ $0°$ H	ρ $90°$ H	ρ $180°$ H
SPH	Y				
TAST	Y		Y		Y
SAST	Y			X	
TCMA		Y	Y		Y
SCMA		Y		Y	
DIST		Y			

4.3.7 Zonal Aberrations

Due to the great labor and tedious nature of tracing rays, early lens designers carefully chose the rays to trace, which were primarily meridional rays and lesser sagittal rays. Even if the meridional rays and sagittal rays came to perfect focus, these designers understood the importance of tracing a few general skew rays and the significant increase in computational labor required to do so. Why was this important? At the time, the designers had little in-depth theoretical knowledge to reach this conclusion; however, their experience was that such a ray trace was necessary to assure the quality of the design.

Looking at the 37 optical aberration coefficients through the seventh order, it can be shown that not all of the aberration coefficients are accounted for by the determination of spherical aberration, coma, astigmatism, and distortion. The "missing" aberration coefficients are μ_9, τ_9, τ_{14}, and τ_{17}. To account for these coefficients, several additional defect definitions are added to those already discussed.[14] These are denoted as tangential and sagittal zonal astigmatism, and tangential and sagittal zonal coma, which use evaluation-plane ray intercept data from the two rays having coordinates of $(\rho, 45^\circ, H, \xi)$ and $(\rho, 135^\circ, H, \xi)$ in addition to intercept data from the marginal ray.

4.3.8 Tangential and Sagittal Zonal Astigmatism

The defining equations for tangential zonal astigmatism and sagittal zonal astigmatism and their polynomial expansion are as follows:

$$
\begin{aligned}
TZAST(\rho, H) &= Y(\rho, 45^\circ, H, \xi) - Y(\rho, 135^\circ, H, \xi) - \sqrt{2}\,Y(\rho, 0^\circ, 0, \xi) \\
&= 2AST_y(\rho, 45^\circ, H) \\
&= \sqrt{2} \left\{
\begin{array}{l}
[(3\sigma_3 + \sigma_4)H^2 + \mu_{10}H^4 + \tau_{18}H^6 + \ldots]\rho \\
+ \left[\left(\mu_4 + \frac{\mu_6}{2}\right)H^2 + \left(\tau_{11} + \frac{\tau_{12}}{2}\right)H^4 + \ldots\right]\rho^3 \\
+ \left[\left(\tau_4 + \frac{\tau_6}{2}\right)H^2 + \ldots\right]\rho^5 + \ldots
\end{array}
\right\}
\end{aligned} \tag{4-11}
$$

$$
\begin{aligned}
SZAST(\rho, H) &= X(\rho, 45^\circ, H, \xi) + X(\rho, 135^\circ, H, \xi) - \sqrt{2}\,Y(\rho, 0^\circ, 0, \xi) \\
&= 2AST_x(\rho, 45^\circ, H) \\
&= \sqrt{2} \left\{
\begin{array}{l}
[(\sigma_3 + \sigma_4)H^2 + \mu_{11}H^4 + \tau_{19}H^6 + \ldots]\rho \\
+ \left[\left(\mu_5 + \frac{\mu_6}{2}\right)H^2 + \left(\tau_{13} + \frac{\tau_{14}}{2}\right)H^4 + \ldots\right]\rho^3 \\
+ \left[\left(\tau_5 + \frac{\tau_6}{2}\right)H^2 + \ldots\right]\rho^5 + \ldots
\end{array}
\right\}
\end{aligned} \tag{4-12}
$$

As explained previously, astigmatic aberrations computed using these equations inherently have the defocus contribution subtracted thereby yielding aberrations terms not dependent on the image plane location. Notice that only one of the four missing aberration coefficients, τ_{14}, appears in the sagittal zonal astigmatism, and none in the tangential zonal astigmatism. This particular aberration is formally known as seventh order, third-degree astigmatism.

4.3.9 Tangential and Sagittal Zonal Coma

The defining equations for tangential zonal coma and sagittal zonal coma and their polynomial expansion are as follows:

$$
\begin{aligned}
TZCMA(\rho, H) &= \left[\frac{Y(\rho, 45^\circ, H, \xi) + Y(\rho, 135^\circ, H, \xi)}{2}\right] - Y(0, 0^\circ, H, \xi) \\
&= CMA_y(\rho, 45^\circ, H) \\
&= [2\sigma_2\rho^2 + \mu_2\rho^4 + \tau_2\rho^6 + \ldots]H \\
&\quad + [\mu_7\rho^2 + (\tau_7 - \tau_{10})\rho^4 + \ldots]H^3 \\
&\quad + [\tau_{15}\rho^2 + \ldots]H^5 + \ldots
\end{aligned}
\tag{4-13}
$$

$$
\begin{aligned}
SZCMA(\rho, H) &= \left[\frac{X(\rho, 45^\circ, H, \xi) + X(\rho, 135^0, H, \xi)}{2}\right] \\
&= CMA_x(\rho, 45^\circ, H) \\
&= [\sigma_2\rho^2 + \mu_3\rho^4 + \tau_3\rho^6 + \ldots]H \\
&\quad + [\mu_9\rho^2 + \tau_9\rho^4 + \ldots]H^3 \\
&\quad + [\tau_{17}\rho^2 + \ldots]H^5 + \ldots
\end{aligned}
\tag{4-14}
$$

As explained previously, comatic aberrations computed using these equations are not dependent on the image plane location. Notice that three of the four missing aberration coefficients—μ_9, τ_9, and τ_{17}—appear in the sagittal zonal coma, and that none are in the tangential zonal coma. These aberrations are formally called fifth-order, third-degree coma; seventh-order, third-degree coma; and seventh-order, fifth-degree coma. It should be observed that these astigmatic and comatic terms are of a reasonably high order and degree, and consequently are difficult in general to control during the design process.

4.3.10 Higher-Order Contributions

It should be evident at this point that the computation of aberrations using real ray data is not an approximation of the aberrations, but is accurate. The reason for this is that all of the aberration coefficients are incorporated within the preceding aberration definitions. Just as in any design process, the designer needs to appropriately select object heights and entrance pupil coordinates for the design task at hand. In addition, the astigmatic aberrations were formulated to remove dependence on defocusing of the image plane, with respect to the Gaussian image plane. The comatic aberrations are inherently independent of image plane location. It is also helpful to have an estimation of the higher-order contributions to the aberration coefficients. Conrady was perhaps the first to derive equations expressing the higher-order astigmatic and comatic aberrations.

Referring to Figure 4.5, the field-dependent astigmatic aberrations are divided into linear and nonlinear terms with respect to entrance pupil radius. The tangential astigmatism previously defined contains both linear and nonlinear terms. The nonlinear term is typically referred to as oblique spherical aberration and is a particularly onerous aberration. It is actually rather simple to compute oblique spherical aberration by subtracting the linear term of tangential astigmatism from the total tangential astigmatism term. The linear term is determined by computing the tangential astigmatism for an entrance pupil radius ρ_0 much smaller than the radius ρ being used to calculate the tangential astigmatism itself. The linear term is appropriately scaled and subtracted from the tangential astigmatism to obtain the tangential oblique spherical aberration. This is expressed by the following equation.

$$TOSPH(\rho, H) = TAST(\rho, H) - \frac{\rho}{\rho_0} TAST(\rho_0, H)$$

(4-15)

where $\rho_0 \ll \rho$.

In a like manner, the sagittal oblique spherical aberration is computed using the following equation.

$$SOSPH(\rho, H) = SAST(\rho, H) - \frac{\rho}{\rho_0} SAST(\rho_0, H)$$

(4-16)

where $\rho_0 \ll \rho$.

Obviously the linear terms

$$\frac{\rho}{\rho_0} TAST(\rho_0, H) \quad \text{and} \quad \frac{\rho}{\rho_0} SAST(\rho_0, H)$$

for the tangential and sagittal astigmatism, respectively, can be utilized in the design process.

Similarly, the comatic aberration has an aperture dependent set of aberrations, namely, linear coma and nonlinear coma. The nonlinear tangential coma is found by subtracting the appropriately scaled linear tangential coma from the tangential coma. The linear tangential coma is determined by computing the tangential coma for a comparatively small object height, that is, $H_0 \ll H$. The equation for the nonlinear tangential coma is given by

$$NLTCMA(\rho, H) = TCMA(\rho, H) - \frac{H}{H_0} TCMA(\rho, H_0) \qquad (4\text{-}17)$$

where $H_0 \ll H$. In a like manner, the nonlinear sagittal coma is given by

$$NLSCMA(\rho, H) = SCMA(\rho, H) - \frac{H}{H_0} SCMA(\rho, H_0) \qquad (4\text{-}18)$$

where $H_0 \ll H$. The linear terms

$$\frac{H}{H_0} TCMA(\rho_0, H) \text{ and } \frac{H}{H_0} SCMA(\rho_0, H)$$

for the tangential and sagittal coma, respectively, can be utilized in the design process.

4.4 CALCULATION OF SEIDEL ABERRATION COEFFICIENTS

In 1856, Philip Ludwig von Seidel published his work on a systematic method for computing third-order aberrations and provided explicit formulas. These aberrations are commonly referred to as the Seidel aberrations and are denoted in order of spherical, coma, astigmatism, Petzval (field curvature), and distortion by a variety of symbols in different books and papers such as (a) σ_1 though σ_5; (b) SC, CC, AC, PC, and DC; (c) $S_I, S_{II}, \ldots,$ and S_V; (d) B, F, C, P, and E; (e) $_0a_{40}, _1a_{31}, _2a_{22}, _2a_{20},$ and $_3a_{11}$; and others. When using any computation scheme to determine the Seidel aberrations, care should be taken to understand if the values are coefficients only, transverse aberrations, longitudinal aberrations, or wave aberrations. In the following, a method will be presented for computing σ_1 though σ_5 aberration coefficients from simply marginal and principal paraxial ray data. By multiplying these coefficients by the appropriate factor, transverse, longitudinal, and wave aberrations can be obtained although it is often common that the symbols σ_1 though σ_5 be used after the transformation to transverse, longitudinal, or wave aberrations.

There are a variety of approaches to derive equations to compute the Seidel aberration coefficients. The method followed here is after Buchdahl, but only the general approach is presented as the details can be easily worked out. By tracing a marginal paraxial ray and a principal paraxial ray, using Eq. (3-2), at a surface, it can be shown that

$$ yn\bar{u} - \bar{y}nu = yn'\bar{u}' - \bar{y}n'u' $$

where $\bar{y}$ and $\bar{u}$ represent the principal-ray values. This implies that $yn\bar{u} - \bar{y}nu$ is a constant across any surface. Using Eq. (3-3), it can be shown that $(yn\bar{u} - \bar{y}nu)_i$ at the ith surface is equal to $(yn\bar{u} - \bar{y}nu)_{i+1}$ at the $(i+1)^{\text{th}}$ surface, which means that the term is also constant within the space between the surfaces. This term is called the *optical invariant*.

> **PROBLEM:** Using Eqs. (3-2) and (3-3), show that $yn\bar{u} - \bar{y}nu$ is invariant across surfaces and in the space between surfaces.

Consider now an object located in the meridional plane having height $H_y = h$ and $H_x = 0$ since the object is aberration free. For a stigmatic optical system, the paraxial and real ray image heights must be identical and related to the object height by the magnification, that is, $h' = mh = mH_y$. As stated previously, an imperfect system will suffer some ray aberration and the transverse ray aberration is given by $\varepsilon_y \equiv H'_y - h'$ and $\varepsilon_x \equiv H'_x$. Now trace two rays from the object with one starting at the base of the object and the other at the object's head. Using a subscript o to designate the object, it is evident that $\lambda = -hn_ou_o$ and is called the *Lagrange invariant*. So it follows that for the ith surface,

$$ \lambda_i = y_i n_i \bar{u}_i - \bar{y}_i n_i u_i = hn_o u_o $$

If the image is located at the kth surface, then $y_k = 0$ and $\lambda_k = -h'n_k u_k$. As shown in the prior chapter, the lateral system magnification is given by

$$ m = \frac{h'}{h} = \frac{n_o u_o}{n_k u_k}. $$

Using the Lagrange invariant, the image height can be expressed in terms of the axial ray final slope angle and the Lagrange invariant. This is simply

$$ h' = \frac{\lambda}{n_k u_k}. $$

It is evident that the Lagrange invariant can be used to form intermediate images by each surface comprising the system. In other words, the image formed by the first surface of the object becomes the object for the second surface to form an image, and so on until the final image is reached.

Buchdahl recognized that imaging could be achieved by propagating the image surface by a surface utilizing the Lagrange invariant for an astigmatic optical system.[3] He then defined the Buchdahl quasi-invariant defined as

$$ \Lambda \equiv Hnu $$

where H is the image height of a real ray in contrast to a paraxial ray. In the paraxial limit, Λ reduces to λ. Since Λ is based on the real ray height at each intermediate image, the aberration at each surface causes the real-ray intermediate image heights to differ from the corresponding paraxial image heights, which is why Buchdahl called Λ the quasi-invariant. Now, because image height for the ith surface is the same as the object height for the $(i+1)^{\text{th}}$ surface,

$$H'_i = H_{i+1}$$

and it is apparent that

$$\Lambda'_i = \Lambda_{i+1}.$$

Consequently, it follows that at the final system image located at the kth surface (image plane),

$$\Lambda'_k = \Lambda_1 + \sum_{i=1}^{k} \Delta\Lambda_i$$

where Δ represents the difference between Λ before and after refraction/reflection at a surface.

So $\Delta\Lambda_i \equiv \Lambda'_i - \Lambda_i$. Using the above definition for Λ, we obtain

$$\sum_{i=1}^{k} \Delta\Lambda_i = H'n_k u_k - H n_o u_o.$$

Recalling that $H'_y = h' + \varepsilon_y$ and the lateral system magnification definition, it follows that

$$\sum_{i=1}^{k} \Delta\Lambda_i = \varepsilon_y n_k u_k.$$

For ε_x, the Lagrange invariant is zero. The ray aberration can now be defined as follows,

$$\varepsilon_x = \frac{\sum_{i=1}^{k} \Delta\Lambda_{x_i}}{n_k u_k} \text{ and } \varepsilon_y = \frac{\sum_{i=1}^{k} \Delta\Lambda_{y_i}}{n_k u_k}.$$

The total ray aberration is the sum of the individual surface contributions. It is important to understand that the surface contributions are related to the final image rather than the intermediate images. Although it is possible to compute the transverse aberration of the intermediate images by using the local marginal ray slope angle $n_i u_i$ rather than $n_k u_k$, these aberrations are not additive, that is, they may *not* be added together to get the final image aberration. Computing the transverse aberration at the intermediate images has no practical utility or meaning.

A general skew ray can be specified at the ith surface by its spatial coordinates (X_i, Y_i, Z_i) and direction cosines (K_i, L_i, M_i). The paraxial ray coordinates (y, nu) can be generalized in the following manner. In the prior chapter, it was shown that the paraxial ray height at a surface is actually the height at the surface tangent plane. In addition, nu is properly interpreted as $n \tan u$. For a meridional ray, the real ray coordinates can be written in a form similar to the paraxial ray coordinates as (Y, U_y) where

$$U_y \equiv \frac{L}{M} = \tan U$$

Buchdahl referred to the (Y, U_y) coordinates as canonical coordinates and they can be used for ray tracing as well; however, the prime object is to determine $\Delta\Lambda$ for each surface. Although the derivation of $\Delta\Lambda$ is tedious, it is straightforward to show that

$$\Delta\Lambda = yn(U + cY)\left(\frac{M}{M'} - 1\right) + niZ\Delta U \qquad (4\text{-}19)$$

where $\Delta U = U_{i+1} - U_i$. The change in the Buchdahl quasi-invariant across a surface boundary is given exactly by Eq. (4-19).

The canonical coordinates (Y_i, U_i) are nonlinear functions of the object ray coordinates (Y_1, U_1). Consequently, the coordinate values needed to solve Eq. (4-19) are unknown. The solution is to perform a series expansion of $\Delta\Lambda$ in terms of the canonical coordinates. It can be shown that $\Delta\Lambda$ can be expanded as an odd-order polynomial, namely

$$\Delta\Lambda = \Delta \overset{1}{\Lambda} + \Delta \overset{3}{\Lambda} + \Delta \overset{5}{\Lambda} + \dots$$

where $\overset{\chi}{\Lambda}$ represents the χth-order of the polynomial expansion of $\Delta\Lambda$. Since $\overset{1}{\Lambda} = \lambda$, then $\Delta \overset{1}{\Lambda} = \Delta\lambda = 0$ and $\Delta\Lambda = \Delta \overset{3}{\Lambda} + \Delta \overset{5}{\Lambda} + \dots$. This is consistent with the premise that first-order or paraxial optics is aberration free. Now, because the ray aberrations are linearly related to $\Delta\Lambda$, we can write

$$\varepsilon = \overset{3}{\varepsilon} + \overset{5}{\varepsilon} + \overset{7}{\varepsilon} + \dots.$$

which is a statement that the ray aberrations can be expressed as a summation of third, fifth, seventh, and higher orders. Once the expansion is completed, it is observed that the third-order term of $\Delta\Lambda$ depends only on the *linear* part of the approximations of Y and U while the nonlinear parts of these approximations give rise to fifth- and higher-order aberrations. Seidel and others realized that the third-order aberrations can be computed using data from only two paraxial rays (marginal and principal).

An orderly iterative process for computing the higher-order aberration terms was achieved by Buchdahl somewhat less than a hundred years after Seidel published his work. As mentioned previously, from Buchdahl's work and that of others, it became understood that aberration coefficients comprise intrinsic and extrinsic contributions.[15] Extrinsic contributions of, say, the ith surface affect the aberration coefficient values of subsequent surfaces while the intrinsic contributions remain local to that surface. Third-order aberration coefficients do not have extrinsic contributions which means these coefficients are decoupled from one another unlike the higher-order aberration coefficients. The nonlinear parts of the approximations of Y and U, and the existence of the extrinsic contributions are reasons the general lens design problem is quite nonlinear and often difficult to optimize.

In actual practice, the lens designer observes that the higher the order of the aberration, the more stable the aberration is with respect to changes in constructional parameters such as curvature and thickness. For example, the values of the third-order aberrations will change much more rapidly, in general, than the fifth-order aberrations if a curvature is changed. It is generally understood by lens designers that if a lens suffers from higher-order aberrations, some significant change to the current optical configuration will be necessary.

With further algebraic effort, $\Delta\Lambda$ is transformed into the third-order form of ε_x and ε_y which can be written in terms of paraxial entering ray coordinates, (ρ, θ, H), namely,

$$\varepsilon_x = \underbrace{\sigma_1\rho^3 \sin(\theta)}_{\text{SPHERICAL}} + \underbrace{\sigma_2\rho^2 H \sin(2\theta)}_{\text{LINEAR COMA}} + \underbrace{(\sigma_3 + \sigma_4)\rho H^2 \sin(\theta)}_{\text{LINEAR ASTIGMATISM}}$$

$$\varepsilon_y = \underbrace{\sigma_1\rho^3 \cos(\theta)}_{\text{SPHERICAL}} + \underbrace{\sigma_2\rho^2 H(2 + \cos(2\theta))}_{\text{LINEAR COMA}} + \underbrace{(3\sigma_3 + \sigma_4)\rho H^2 \cos(\theta)}_{\text{LINEAR ASTIGMATISM}} + \underbrace{\sigma_5 H^3}_{\text{DISTORTION}}$$

The third-order aberration coefficients, σ_1 through σ_5, for a given optical system can be calculated using the ray data obtained by tracing the marginal and principal paraxial rays using the following equations. The coefficient form with the presubscript is used to denote the aberration contribution of the ith surface. It is important to understand that these coefficients can be used to compute transverse, longitudinal, and wave aberrations, which are related by scaling factors.

$$\bar{i}_i = c_i\bar{y} + \frac{n_{i-1}\bar{u}_{i-1}}{n_{i-1}}$$

$$i_i = c_i y_i + \frac{n_{i-1}u_{i-1}}{n_{i-1}}$$

$$q_i = \frac{\bar{i}_i}{i_i}$$

$$i_i + u_i = i_i + \frac{n_i u_i}{n_i}$$

$${}_i\sigma_1 = \frac{n_{i-1} y_i i_i^2 (n_{i-1} - n_i)(i_i + u_i)}{n_i}$$

$${}_i\sigma_2 = q_i \, {}_i\sigma_1$$

$${}_i\sigma_3 = q_i^2 \, {}_i\sigma_1$$

$${}_i\sigma_4 = \frac{c_i(n_{i-1} - n_i)(yn_{-1}\bar{u}_{-1} - \bar{y}n_{-1}u_{-1})^2}{n_{i-1} n_i}$$

$${}_i\sigma_5 = q_i(q_i^2 \, {}_i\sigma_1 + {}_i\sigma_4)$$

The transverse third-order aberration coefficients are determined by summation of the surface contributions and then multiplying by the factor

$$\frac{-1}{2n_k u_k}$$

Notice that the Petzval term σ_4 is also multiplied by the square of the Lagrange invariant, $yn_{-1}\bar{u}_{-1} - \bar{y}n_{-1}u_{-1}$.

$$\sigma_1 = \frac{-1}{2n_k u_k} \sum_{i=1}^{k} {}_i\sigma_1 \text{ Spherical Aberration}$$

$$\sigma_2 = \frac{-1}{2n_k u_k} \sum_{i=1}^{k} {}_i\sigma_2 \text{ Coma}$$

$$\sigma_3 = \frac{-1}{2n_k u_k} \sum_{i=1}^{k} {}_i\sigma_3 \text{ Astigmatism} \qquad (4\text{-}20)$$

$$\sigma_4 = \frac{-(yn_{-1}\bar{u}_{-1} - \bar{y}n_{-1}u_{-1})^2}{2n_k u_k} \sum_{i=1}^{k} {}_i\sigma_4 \text{ Petzval}$$

$$\sigma_5 = \frac{-1}{2n_k u_k} \sum_{i=1}^{k} {}_i\sigma_5 \text{ Distortion}$$

To convert these values into longitudinal aberrations, the $\frac{-1}{2n_k u_k}$ factor is replaced by $\frac{1}{2n_k u_k^2}$. Transverse and longitudinal aberrations are in lens units.

Conversion to wave aberrations requires that the $\frac{-1}{2n_k u_k}$ factor be replaced as follows:

$$\sigma_1 = \frac{1}{8\lambda} \sum_{i=1}^{k} {}_i\sigma_1 \text{ Spherical aberration}$$

$$\sigma_2 = \frac{1}{2\lambda} \sum_{i=1}^{k} {}_i\sigma_2 \text{ Coma}$$

$$\sigma_3 = \frac{1}{2\lambda} \sum_{i=1}^{k} {}_i\sigma_3 \text{ Astigmatism} \qquad\qquad (4\text{-}21)$$

$$\sigma_4 = \frac{\left(yn_{-1}\bar{u}_{-1} - \bar{y}n_{-1}u_{-1}\right)^2}{4\lambda} \sum_{i=1}^{k} {}_i\sigma_4 \text{ Petzval}$$

$$\sigma_5 = \frac{1}{2\lambda} \sum_{i=1}^{k} {}_i\sigma_5 \text{ Distortion}$$

where λ is the wavelength and the wave aberrations are measured at the edge of the exit pupil in units of wavelength.

ENDNOTES

[1] H. A. Buchdahl, *Optical Aberration Coefficients*, Dover Publications, New York (1968).

[2] Historically, lens designers used a left-hand Cartesian coordinate system with positive slopes of rays bending downwards. This was done for computational convenience and error mitigation when doing manual computations. Most current optical design and analysis software packages use the right-hand Cartesian coordinate system.

[3] It should be understood that *astigmatic* means *not stigmatic*. This should not be confused with astigmatism or more specific astigmatic aberrations, which will be discussed later. In a like manner, the term *anastigmatic lens* means a highly corrected lens having sensibly perfect imagery in contrast to meaning a *stigmatic lens* (not not stigmatic).

[4] The entrance pupil is the image of the aperture stop formed by all of the optical elements preceding the aperture stop. The exit pupil is the image of the aperture stop formed by all of the optical elements following the aperture stop.

[5] Note that changing the sign of ρ is the same as changing the signs of both X and Y, or the angle θ by π.

[6] The value of ρ can have values of either sign. Consequently, a ray having entrance pupil coordinates of $(-\rho, 0^\circ)$ is equivalent to having entrance pupil coordinates of $(\rho, 180^\circ)$.

[7] The number of independent aberration coefficients for the nth-order is given by

$$\frac{(n+3)(n+5)}{8} - 1$$

For $n = 1$, or the first-order, there are two independent coefficients, namely magnification and defocus.

[8] G. C. Steward, *The Symmetrical Optical System*, Cambridge University Press (1928).

[9] A. E. Conrady, *Applied Optics and Optical Design*, Dover Publications, New York; Part I (1957), Part II (1960).

[10] Andrew Rakich and Raymond Wilson, "Evidence supporting the primacy of Joseph Petzval in the discovery of aberration coefficients and their application to lens design," *SPIE Proc.* 6668:66680B (2007).

[11] A. E. Conrady, p. 289–290.

[12] R. Barry Johnson, "A Historical perspective on the understanding optical aberrations," *SPIE Proc.*, CR41:18–29 (1992).

[13] A negative singlet lens has overcorrected or positive spherical aberration.

[14] If the primary coefficients negligible, then the identity $\mu_9 = \mu_7 - \mu_8$ is reasonably valid if the system is well corrected. See previous Buchdahl Eq. (31.8).

[15] Extrinsic contributions are also referred to as transfer contributions.

Chapter 5

Chromatic Aberration

5.1 INTRODUCTION

In 1661, Huygens created the two-lens compound negative eyepiece which generally corrected lateral color, that is, yielding an image of a white object which subtends the same angle for all colors (see Chapter 16). This was a remarkable achievement and won him acclaim at the scientific conferences, since other eyepieces of the day yielded poor performance and contained often 5, 8, and even 19 lenses. An interesting point is that Huygens had no concept of achromatizing his eyepieces or any other kind of optical system for that matter; nevertheless, it worked better than other eyepieces of the day. The reason for Huygens' lack of understanding was that no one understood the dispersive properties of glass.

About two years later, Newton began to study the dispersion of glass, in part, to understand why, the Huygens compound eyepiece was corrected for lateral color. Newton was the first, it should be noted, to develop the concept of the dispersion of glass. Remarkably, Newton failed to recognize one important property of glass—different glasses have different dispersions. In contrast, he put forth the concept that all glasses have the same dispersion; consequently, he asserted that one could not achieve an achromatic system. Newton was also the first to differentiate between the aberrations of spherical and color by assigning the spherical aberration to the surface and the color to the materials. He was the first to explain that spherical aberration varied with the cube of the aperture, and published the results in his book *OPTICKS*.[1] Also, Newton presented a detailed description of chromatic aberrations.

After the work by Newton, there was a lull in the development of optical aberrations of about 60 years. Then in 1729, Chester Hall discovered, rather accidentally it is noted, that achromatic lenses could be constructed by cementing positive and negative lenses together when the lenses were made of different glasses. By achromatic, it was meant only in the context that the chromatic aberrations of the lenses were not corrected, but notably reduced. Hall's discovery gave rise to renewed research into understanding optical materials and recognizing that the dispersion of glasses can vary from type to type. John Dolland, a London optician, in 1757 began to design and fabricate a variety of achromatic lenses after he empirically determined by experimenting with a variety of positive and negative lens combinations that longitudinal chromatic aberration could be mitigated by combining a convex crown-glass lens with a weaker concave flint-glass lens. According to Conrady, John Dolland produced the first achromatic telescope objective and was the first person to patent the achromatic doublet.[2]

The Swedish mathematician Klingenstierna, in 1760, was the first to develop a mathematical theory of achromatic lenses and, what was called at that time, the aplanatic lens. Part of Klingenstierna's work was based on John Dolland's initial understanding of achromatic lenses. The next year, Clairaut was the first to explain the concept of secondary spectrum (see Section 5.5) and he also observed that certain crown and flint glasses had different partial dispersions. He further deduced theorems for pairing glasses, not unlike those found in modern books. Also that same year, John Dolland made an effort to correct secondary spectrum by the use of a third glass. In 1764, D'Alembert described a triple glass objective in which he also distinguished between longitudinal and transverse features of spherical aberration and chromatic aberration.[3]

A discussion was presented in Chapter 2 about the refractive index of glass and other optical materials changing with wavelength. From this behavior of optical materials, it follows that every property of a lens depending on its refractive index will also change with wavelength. This includes the focal length, the back focus, the spherical aberration, field curvature, and all of the other aberrations. In this chapter, we explore field-independent chromatic aberrations[4] while field-dependent chromatic aberrations (including lateral color) are discussed in Chapter 11.

Figure 5.1 depicts the chromatic aberration of a single positive lens having "white" light incident upon the lens. As will be mentioned in Section 5.9.1, it is common to select F (blue), d (yellow), and C (red) spectral lines for design and analysis of visual systems.[5] As seen in the figure, the focus for F light is

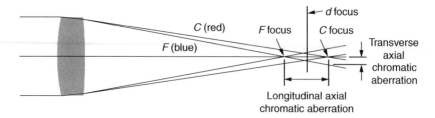

Figure 5.1   Undercorrected chromatic aberration of a simple lens.

inside the paraxial focus for d light while the C light focus lies to the outside. This should be evident since the refractive index is progressively greater for C, d, and F light thereby increasing the optical power of the lens ($\phi_\lambda = (n_\lambda - 1)(c_1 - c_2)$). The longitudinal axial chromatic aberration is given by $L'_{ch} = L'_F - L'_C$ (see 5.2.3) and transverse axial chromatic aberration[6] is given by $L'_{ch} \tan u'$. A simple converging lens that is uncorrected for aberrations, as shown in Figure 5.1, is said to have *undercorrected* aberrations. If the sign of an aberration of the optical system is opposite to that of a simple converging lens, the lens system is said to be *overcorrected*. When a specific aberration is made zero or less than some desired tolerance, the lens system is said to be *corrected*.

5.2 SPHEROCHROMATISM OF A CEMENTED DOUBLET

Consider a cemented doublet objective lens, as illustrated in Figure 5.2. The prescription of this lens, repeated from Section 2.5, is as follows:

$$r_1 = 7.3895 \qquad c_1 = 0.135327$$
$$d_1 = 1.05 \qquad n_1 = 1.517$$
$$r_2 = -5.1784 \qquad c_2 = -0.19311$$
$$d_2 = 0.40 \qquad n_2 = 1.649$$
$$r_3 = -16.2225 \qquad c_3 = -0.06164$$

If we now trace through it a marginal, zonal, and paraxial ray in each of five wavelengths, we obtain Table 5.1, which shows image distances expressed relative to the paraxial focus in D light.

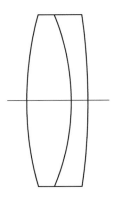

Figure 5.2 A cemented doublet objective.

Table 5.1

Image Distance versus Wavelength Relative to the Paraxial Focus

Wavelength	A' (0.7665)	C (0.6563)	D (0.5893)	F (0.4861)	g (0.4358)
Crown index	1.51179	1.51461	1.517	1.52262	1.52690
Flint index	1.63754	1.64355	1.649	1.66275	1.67408
Marginal $Y = 2$	0.0203	0.0100	0.0081	0.0265	0.0588
Zonal $Y = 1.4$	0.0059	−0.0101	−0.0176	−0.0153	0.0025
Paraxial	0.0327	0.0121	0	−0.0101	−0.0033

These data may be plotted in two ways. First we can plot the longitudinal spherical aberration against aperture, separately in each wavelength (Figure 5.3a); and second, we can plot aberration against wavelength for each zone (Figure 5.3b). The first set of curves represents the chromatic variation of spherical aberration, or "spherochromatism," and the second set represents the chromatic aberration curves for the three zones. On these curves we notice several specific aberrations.

5.2.1 Spherical Aberration (LA')

This is given by $L'_{\text{marginal}} - l'_{\text{paraxial}}$ in brightest (D) light. It has the value 0.0081 in this example, and is slightly overcorrected.

5.2.2 Zonal Aberration (LZA')

This is given by $L'_{\text{zonal}} - l'_{\text{paraxial}}$ in D light. It has the value −0.0175, and is undercorrected. The best compromise between marginal and zonal aberration for photographic objectives is generally to secure that $LA' + LZA' = 0$, but for visual systems it is better to have $LA' = 0$.

5.2.3 Chromatic Aberration (L'_{ch})

This is given by $L'_F - L'_C$, and its magnitude varies from zone to zone (Figure 5.4) as shown in Table 5.2.

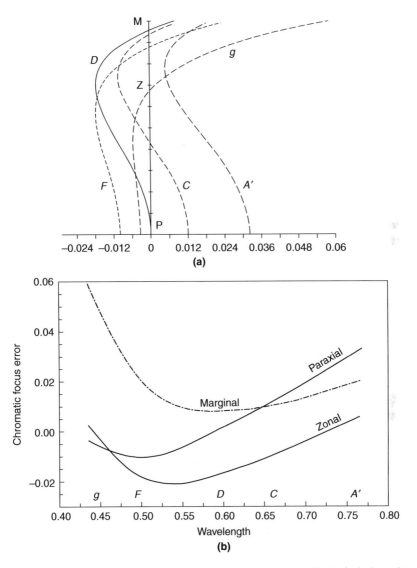

Figure 5.3 Spherochromatism ($f = 12$). (a) Chromatic variation of spherical aberration; (b) chromatic aberration for three zones.

If no zone is specified, we generally refer to the 0.7 zonal chromatic aberration because zero zonal chromatic aberration is the best compromise for a visual system. Photographic lenses, on the other hand, are generally stopped down somewhat in use, and it is often better to unite the extreme colored foci for about the 0.4 zone instead of the 0.7 zone suggested here.

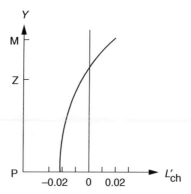

Figure 5.4 Variation of chromatic aberration with aperture.

Table 5.2

Chromatic Aberration for Three Zones in the Aperture

Zone	$L'_{ch} = L'_F - L'_C$
Marginal	+0.0165
0.7 Zonal	−0.0052
Paraxial	−0.0222

Chromatic aberration can be represented as a power series of the ray height Y:

$$\text{chromatic aberration} = L'_{ch} = a + bY^2 + cY^4 + \ldots$$

The constant term a is the paraxial or "primary" chromatic aberration. The secondary term bY^2 and the tertiary term cY^4 represent the variation of chromatic aberration with aperture as shown in Figure 5.4.

5.2.4 Secondary Spectrum

Secondary spectrum is generally expressed as the distance of D focus from the combined $C - F$ focus, taken at the height Y at which the C and F curves intersect. In the example shown later in this section, the C and F curves intersect at about $Y = 1.6$, and at that height the other wavelengths depart from the combined C and F focus by

Spectrum line	A'	C	D	F	g
Departure of focus	0.005	0	−0.016	0	0.012

In the absence of secondary spectrum the curves in Figure 5.3b would all be straight lines. The fact that achromatizing a lens for two colors fails to unite the other colors is known as secondary spectrum; it should not be confused with the secondary chromatic aberration mentioned in Section 5.2.3.

5.2.5 Spherochromatism

This is the chromatic variation of spherical aberration and is expressed as the difference between the marginal spherical aberration in F and C light:

$$
\begin{aligned}
\text{spherochromatism} &= (L' - l')_F - (L' - l')_C \\
&= (L'_F - L'_C)(l'_F - l'_C) \\
&= \text{marginal chromatic aberration} \\
&\quad - \text{paraxial chromatic aberration} \\
&= 0.0165 + 0.0222 = 0.0386
\end{aligned}
$$

5.3 CONTRIBUTION OF A SINGLE SURFACE TO THE PRIMARY CHROMATIC ABERRATION

To determine the contribution of a single spherical surface to the paraxial chromatic aberration of a lens, we recall from Section 3.1.5 that

$$
\frac{n'}{l'} - \frac{n}{l} = \frac{n' - n}{r}
$$

and write it in F and C light as

$$
\frac{n'_F}{l'_F} - \frac{n_F}{l_F} = \frac{n'_F - n_F}{r} \quad \text{and} \quad \frac{n'_C}{l'_C} - \frac{n_C}{l_C} = \frac{n'_C - n_C}{r}.
$$

Subtracting F from C gives

$$
\frac{n'_C}{l'_C} - \frac{n'_F}{l'_F} - \frac{n_F}{l_C} + \frac{n_F}{l_F} = \frac{(n'_C - n'_F) - (n_C - n_F)}{r}
$$

We now write $(n_F - n_C) = \Delta n$; hence $n_F = n_C + \Delta n$ and $n'_F = n'_C + \Delta n'$. Since for all optical glasses the difference between n_F and n_C is a small fraction of n_d and the d line is not very far from being midway between the F and C lines, only a small inaccuracy is introduced by replacing both n_F and n_C with $n_d = n$, and similarly for the primed terms. If in the denominator, we also replace l'_F and l'_C with $l'_d = l'$, and similarly for the unprimed terms, we have that

$$
\frac{n'}{l'_2}(l'_F - l'_C) - \frac{n}{l^2}(l_F - l_C) = \Delta n\left(\frac{1}{r} - \frac{1}{l}\right) - \Delta n'\left(\frac{1}{r} - \frac{1}{l'}\right).
$$

We next multiply through by y^2, noting that $(1/r - 1/l) = i/y$. Then

$$n'u'^2 L'_{ch} - nu^2 L_{ch} = yi\Delta n - yi'\Delta n' = yni(\Delta n/n - \Delta n'/n')$$

We write this expression for every surface and add. Much cancellation occurs because of the identities $n'_1 \equiv n_2, u'_1 \equiv u_2$, and $L'_{ch1} \equiv L_{ch2}$. Hence, if there are k surfaces, we get

$$(n'u'^2 L'_{ch})_k - (nu^2 L_{ch})_1 = \sum yni(\Delta n/n - \Delta n'/n')$$

and dividing through by $(n'u'^2)_k$ gives

$$L'_{ch_k} = L_{ch_1} \left(\frac{n_1 u_1^2}{n'_k u'^2_k} \right) + \sum \frac{yni}{n'_k u'^2_k} \left(\frac{\Delta n}{n} - \frac{\Delta n'}{n'} \right) \qquad (5\text{-}1a)$$

The quantity within the summation sign is a surface contribution to the longitudinal paraxial chromatic aberration and the first term is a chromatic aberration of the object. Thus we can write that the resultant longitudinal paraxial chromatic aberration is

$$L'_{ch}C = \frac{yni}{n'_k u'^2_k} \left(\frac{\Delta n}{n} - \frac{\Delta n'}{n'} \right). \qquad (5\text{-}1b)$$

The chromatic aberration of the object, if any, is transferred to the image by the ordinary longitudinal magnification rule (see Section 3.2.2) and added to the aberration arising at the surfaces of the lens.

In Table 5.3 we have used these formulas to calculate the paraxial chromatic aberration contributions of the three surfaces of the cemented doublet already used several times. The sum of the contributions is –0.022255. For comparison, we note from the data in Table 5.1 that $l'_F - l'_C = -0.022178$ (shown with more significant digits than in Table 5.1). The agreement between this contribution formula and actual paraxial ray tracing is extremely close (about 0.35%) in spite of the various small approximations that we made in deriving the formula.

Table 5.3

Primary Chromatic Aberration Contributions

y		2		1.903148		1.880973
n		1		1.517		1.649
i		0.270654		–0.459757		–0.171386
$1/u'^2_k$		36		36		36
$n_F - n_C = \Delta n$	0		0.00801		0.01920		0
$\Delta n/n$	0		0.005280		0.011643		0
$(\Delta n/n - \Delta n'/n')$		–0.005280		–0.006363		0.011643	
$L'_{ch}C$		–0.105746		0.312485		–0.228994	$\sum = -0.022255$

5.4 CONTRIBUTION OF A THIN ELEMENT IN A SYSTEM TO THE PARAXIAL CHROMATIC ABERRATION

The classical relation between object and image distances for a thin lens is

$$\frac{1}{l'} = \frac{1}{l} + (n-1)c$$

where $c = c_1 - c_2 = \Delta c$ and is known as the *total curvature* or *element curvature*. We write this imaging equation in F and C light and subtract F from C. This gives

$$\frac{l'_C - l'_F}{l'^2} - \frac{l_C - l_F}{l^2} = (n_F - n_C)c = \frac{1}{fV} \tag{5-2}$$

Multiplying by $(-y^2)$ gives

$$L'_{ch}\left(\frac{y^2}{l'^2}\right) - L_{ch}\left(\frac{y^2}{l^2}\right) = -\frac{y^2}{fV} \quad \text{or} \quad L'_{ch}u'^2 - L_{ch}u^2 = -\frac{y^2}{fV}$$

We write this expression for each thin element in the system and add up. After much cancellation, and assuming that there are k elements in the system, we get

$$L'_{chk}u'^2_k - L_{ch1}u^2_1 = -\sum \frac{y^2}{fV}$$

Finally, dividing through by u'^2_k gives an expression for the chromatic aberration of the image as

$$L'_{chk} = L_{ch1}\left(\frac{u_1}{u'_k}\right)^2 - \frac{1}{u'^2_k}\sum \frac{y^2}{fV} \tag{5-3}$$

In these expressions, f refers to the focal length of each individual thin element, and V refers to its *Abbe number* or reciprocal dispersive power,

$$V = \frac{n_d - 1}{n_F - n_C}$$

The magnitude of V varies from 25 for a very dense flint to about 75 for an extra light crown. Every type of optical glass can thus be represented by a point on a chart connecting the mean refractive index n_d with the V number (Figure 5.5). The vertical line at $V = 50$ divides the so-called crown (*kron* in German) and flint types, although these names have long lost any significance. However, we still use the terms loosely to represent glasses having relatively low and high dispersive powers. This diagram was also shown in the first edition of *Lens Design Fundamentals* using Schott's 1973 catalog and is similar today although some specific glasses have been deleted and others added.

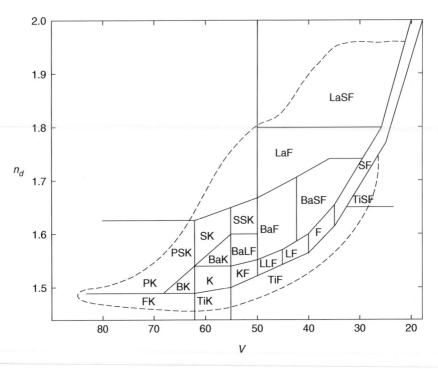

Figure 5.5   Glass chart.

The narrow band of crowns, light flints, flints, and dense flints in Figure 5.5 contains all the older soda-lime–silica glasses having a progressively increasing lead content. Above this band comes, first, the barium glasses and then (since 1938) a wide range of lanthanum or rare-earth glasses. In the early 1970s, some titanium flints were introduced, which fall below the old crown–flint line. At the far left are found some fluor and phosphate crowns, some of which have extreme properties, introduced about the same time. Because optical glasses vary enormously in price, from a few dollars to over $500 a pound, the lens designer must watch the price catalog very carefully when selecting glasses to be used in any particular lens.

In the ensuing years, a number of changes have been made to the formulation and availability of glasses by the various glass manufacturers to achieve more ecologically acceptable glasses by, for example, removing lead, arsenic, and/or radioactive materials from many of them. In the early 1990s, the interesting deep crown TiK (alkali alumoborosilicate), titanium flint TiF, and titanium short-flint TiSF (titanium alkali alumoborosilicate) were deleted by Schott. However, Ohara S-FTM16 and Hoya FF5 are presently offered as a substitute for Schott TiFN5.

The reason for using lead, arsenic, and other materials in glasses considered non-ecologically acceptable is that the optical properties achievable could often be most helpful in realizing higher performance optical designs with fewer glasses. Starting in the 1970s with the growth of pollution reduction, glass manufacturers began exploring ways to remove toxic materials such as cadmium and later arsenic, lead, and so on. The challenge these glass companies had, and still have, was to develop new glass compositions that are ecologically acceptable while still providing adequate richness of properties for lens designers to utilize in their designs.

Significant success has been realized by the manufacturers, but development of new compositions of glass continues to meet optical performance, manufacturability, and cost objectives. Nikon is an excellent example of an integrated corporation that makes its own glasses, manufactures its optical components, and produces a wide variety of optical products. In about 1990, approximately 100 types of its optical glasses contained arsenic or lead. By 1999, the company was using new ecologically acceptable glasses throughout its optical design department. In 2000, new optical designs of Nikon consumer products (cameras, binoculars, etc.) utilized essentially none of the new glasses while in 2008, the use of the new glasses had risen to 100%.

Returning to Eq. (5-3) we see that the paraxial chromatic aberration of an isolated single thin lens in air is given by

$$L'_{ch} = -\frac{y^2}{fV}\left(\frac{l'^2}{y^2}\right) = -\frac{l'^2}{fV}$$

and if the object is very distant, this becomes merely

$$L'_{ch} = -f/V$$

The chromatic aberration of a single thin lens with a distant object is therefore equal to the focal length of the lens divided by the V number of the glass. It thus falls between 1/25 and 1/75 of the focal length, depending on the type of glass used in its construction.

For a thin system of lenses in close contact (see Section 3.4.6), we can write Eq. (5-2) for each element and then add. This gives

$$\left[\frac{L'_{ch}}{l'^2} - \frac{L_{ch}}{l^2}\right] = -\sum c\,\Delta n = -\sum \frac{\phi}{V}$$

The quantity on the left-hand side we call the *chromatic residual R*, which is zero for an achromatic lens with a real object. If the total power of the thin-lens system is Φ, then

$$\Phi = \sum \phi = \sum (Vc\,\Delta n) \text{ and } R = -\sum (c\,\Delta n)$$

For the very common case of a thin doublet, these equations become

$$1/F' = \Phi = V_a(c\,\Delta n)_a + V_b(c\,\Delta n)_b$$
$$-R = (c\,\Delta n)_a + (c\,\Delta n)_b$$

Solving for c_a and c_b gives the important relationships

$$c_a = \frac{1}{F'(V_a - V_b)\Delta n_a} + \frac{RV_b}{(V_a - V_b)\Delta n_a}$$

$$c_b = \frac{1}{F'(V_b - V_a)\Delta n_b} + \frac{RV_b}{(V_b - V_a)\Delta n_b}$$
(5-4)

These are the so-called (c_a, c_b) equations which are used to start the design of any *thin achromatic doublet.*

In most practical cases the chromatic residual R is zero and then only the first terms need be considered. The condition for achromatism is then independent of the object distance, and we say that achromatism of a thin system is "stable" with regard to object distance. Notice also that c_a and c_b do not depend explicitly upon the refractive index of each material.

Since for a thin lens $f' = 1/c(n - 1)$, we can convert the (c_a, c_b) formulas into the corresponding focal-length formulas for $R = 0$:

$$f'_a = F'\left(\frac{V_a - V_b}{V_a}\right) \quad \text{and} \quad f'_b = F'\left(\frac{V_b - V_a}{V_b}\right)$$
(5-5)

For an ordinary crown glass with $V_a = 60$ and an ordinary flint with $V_b = 36$, we have $V_a - V_b = 24$, and the power of the crown element is seen to be 2.5 times the power of the combination, while the power of the flint is -1.5 times as strong as the doublet. Hence, to achromatize a thin lens requires the use of a crown element 2.5 times as strong as the element itself (Figure 5.6).

Consequently, although a single lens of aperture $f/1$ is not excessively strong, it is virtually impossible to make an achromat of aperture much over $f/1.5$.

It is important to note that *chromatic aberration depends only on lens powers and not at all on bendings or surface configuration.* Attempts to modify the chromatic correction of a lens by hand rubbing on one of the surfaces are generally quite unsuccessful, because it requires a very large change in the lens to produce a noticeable change in the chromatic aberration.

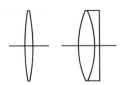

Figure 5.6   An $f/3.5$ single lens and an achromat of the same focal length.

5.5 PARAXIAL SECONDARY SPECTRUM

We have so far regarded an achromatic lens as one in which the C and F foci are coincident. However, as we have seen, in such a case the d (yellow) focus falls short and the g (blue) focus of the same zone falls long. To determine the magnitude of the paraxial secondary spectrum of a lens in which the paraxial C and F foci coincide, we write the chromatic aberration contribution of a single thin element, for two wavelengths λ and F, as

$$L'_{\text{ch}}C(\text{for } \lambda \text{ to } F) = -\frac{y^2 c}{u'^2_k}(n_\lambda - n_F) = L'_{\text{ch}}C\left(\frac{n_\lambda - n_F}{n_F - n_C}\right)$$

The quantity in parentheses is another intrinsic property of the glass, known as the *partial dispersion ratio* from λ to F. It is generally written $P_{\lambda F}$. Hence for any succession of thin elements

$$l'_\lambda - l'_F = \sum P_{\lambda F}(L'_{\text{ch}}C) = -\frac{1}{u'^2_k}\sum \frac{Py^2}{f'V} \qquad (5\text{-}6)$$

For the case of a thin achromatic doublet, y is the same for both elements, and Eq. (5-5) shows that $f'_a V_a = -f'_b V_b = F'(V_a - V_b)$; hence

$$l'_\lambda - l'_F = -F'\left(\frac{P_a - P_b}{V_a - V_b}\right) \qquad (5\text{-}7)$$

For any particular pair of wavelengths, say F and g, we can plot the available types of glass on a graph connecting P_{gF} with V, as in Figure 5.7. All the common types of glass lie on a straight line that rises slightly for the very dense flints. Below this line come the "short" glasses, which exhibit an unusually short blue end to the spectrum; these are mostly lanthanum crowns and so-called short flints (KzF and KzFS types). Above the line are a few "long" crowns with an unusually stretched blue spectrum (this region also contains some plastics and crystals such as fluorite). The titanium flints also fall above the line, as can be seen.

If we join the points belonging to the two glasses used to make an achromatic doublet, the slope of the line is given by

$$\tan \psi = \frac{P_a - P_b}{V_a - V_b}$$

and clearly the secondary spectrum is given by $F' \tan \psi$. The fact that most of the ordinary glasses lie on a straight line indicates that the secondary spectrum will be about the same for any reasonable selection of glass types. For example,

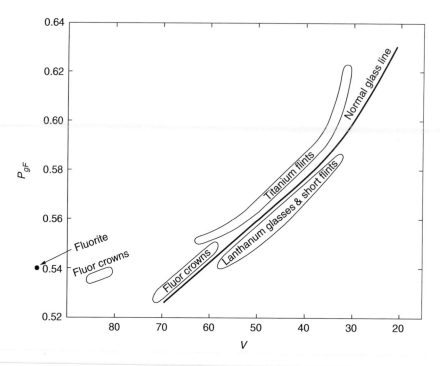

Figure 5.7 Partial dispersion ratio versus dispersive power of optical glasses.

if we choose Schott's N-K5 and N-F2, we find that the secondary spectrum for a number of wavelengths, assuming a focal length of 10, is

		$r - F$	$d - F$	$g - F$	$h - F$
N-K5	$V_a = 59.48$	$P_a = -1.17372$	-0.69558	0.54417	0.99499
N-F2	$V_b = 36.43$	$P_b = -1.16275$	-0.70682	0.58813	1.10340
		$l_\lambda' - l_F' = 0.00476$	-0.00488	0.01907	0.047033

We can reduce the secondary spectrum by choosing a long crown, such as fluorite,[7] with a matching dense barium crown glass as the flint element[8]:

		$r - F$	$d - F$	$g - F$	$h - F$
Fluorite[9]	$V_a = 95.23$	$P_a = -1.17428$	-0.69579	0.53775	0.98112
N-SK5	$V_b = 61.27$	$P_b = -1.17512$	-0.69468	0.53973	0.98690
		$l_\lambda' - l_F' = -0.00025$	0.00033	0.00058	0.00170

This amount of secondary spectrum is obviously vastly smaller than we found using ordinary glasses. On the other hand, we shall increase the secondary spectrum if we use a normal crown with a dense flint such as N-SF15 glass:

		$d - F$	$g - F$
N-K5	$V_a = 59.48$	$P_a = -0.69558$	0.54417
N-SF15	$V_b = 30.20$	$P_b = -0.71040$	0.60366
		$l'_\lambda - l'_F = -0.00506$	-0.02032

These residuals are about 1.5 times as large as for the normal glasses listed here.

In view of the apparent inevitability of secondary spectrum, we may wonder why it is necessary to achromatize a lens at all. This question will be immediately answered by a glance at Figure 5.8, where we have plotted to the same scale the paraxial secondary spectrum curve of the example in Figure 5.3b and the corresponding curve for a simple lens of crown glass, N-K5, both with $f' = 10$.

If a lens has a small residual of primary chromatic aberration, the secondary spectrum curve will become tilted. The three curves sketched in Figure 5.9 show what happens in this case. It will be noticed that when the chromatic aberration is undercorrected the wavelength of the minimum focus moves toward the blue; for a $C - F$ achromat it falls in the yellow-green; and for an overcorrected lens

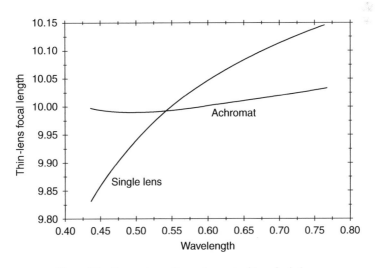

Figure 5.8 Comparison of an achromat with a single lens.

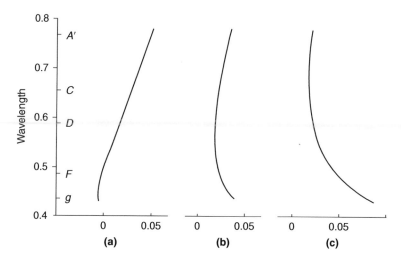

Figure 5.9 Effect of chromatic residual in a cemented doublet ($f' = 10$): (a) Undercorrected by -0.03, (b) achromat, and (c) overcorrected by $+0.03$.

it rises up toward the red. Lenses for use in the near infrared are often decidedly overcorrected, whereas lenses intended for use with color-blind film or bromide paper should be chromatically undercorrected.

In any achromat of high aperture, the spherochromatism and other residual aberrations are likely to be so much greater than the secondary spectrum that the latter can often be completely ignored. However, in a low-aperture lens of long focal length, such as an astronomical telescope objective in which the other aberration residuals are either corrected in the design or removed by hand figuring, the secondary spectrum may well be the only outstanding residual, and it is then important to consider the possibility of removing it by a suitable choice of special types of glass. For example, fluorite is commonly used in microscope objectives for this purpose.

5.6 PREDESIGN OF A THIN THREE-LENS APOCHROMAT

As there are many practical objections to the use of fluorite as a means for reducing secondary spectrum, it is often preferred to unite three wavelengths at a common focus by the use of three different types of glass.

For a thin system with a very distant object, which is achromatized and also corrected for secondary spectrum, we have the three relationships

$$\sum (Vc\,\Delta n) = \Phi \quad \text{(power)}$$

$$\sum (c\,\Delta n) = 0 \quad \text{(achromatism)}$$

$$\sum (Pc\,\Delta n) = 0 \quad \text{(secondary spectrum)}$$

For a thin three-lens apochromat, these equations can be extended to

$$V_a(c_a\,\Delta n_a) + V_b(c_b\,\Delta n_b) + V_c(c_c\,\Delta n_c) = \Phi$$

$$(c_a\,\Delta n_a) + (c_b\,\Delta n_b) + (c_c\,\Delta n_c) = 0$$

$$P_a(c_a\,\Delta n_a) + P_b(c_b\,\Delta n_b) + P_c(c_c\,\Delta n_c) = 0$$

These can be solved for the three curvatures as follows:

$$c_a = \frac{1}{F'E(V_a - V_c)}\left(\frac{P_b - P_c}{\Delta n_a}\right)$$

$$c_b = \frac{1}{F'E(V_a - V_c)}\left(\frac{P_c - P_a}{\Delta n_b}\right)$$

$$c_c = \frac{1}{F'E(V_a - V_c)}\left(\frac{P_a - P_b}{\Delta n_c}\right)$$

Note the cyclic order of the terms, and that the coefficient in front of the parentheses is the same in each case.

The meaning of E is as follows: If we plot the three chosen glasses on the $P - V$ graph shown in Figure 5.10 and then join the three points to form a triangle, E is the vertical distance of the middle glass from the line joining the two outer glasses, E being considered negative if the middle glass falls below the line. Algebraically E is computed by

$$E = \frac{V_a(P_b - P_c) + V_b(P_c - P_a) + V_c(P_a - P_b)}{V_a - V_c}$$

$$= (P_c - P_a)\left(\frac{V_b - V_c}{V_a - V_c}\right) - (P_c - P_b)$$

Since E appears in the denominator of all three c expressions, it is clear that the lenses will become infinitely strong if all three glasses fall on a straight line, and conversely, all the elements will become as weak as possible if we select glass types having a large E value. The most usual choice is some kind of crown for

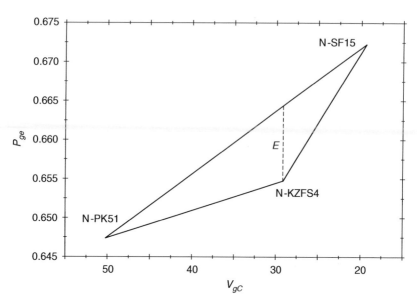

Figure 5.10 $P - V$ graph of the glasses used in a three-lens apochromat.

lens a, a very dense flint for lens c, and a short flint or lanthanum crown for the intermediate lens b. Once the three curvatures have been calculated, the actual lenses can be assembled in any order.

As an example in the use of these formulas, we will select three glasses forming a wide triangle on the graph of Figure 5.10, namely, Schott's N-PK51, N-KZFS4, and N-SF15. Since the calculated curvatures depend on the differences between the P numbers, it is necessary to know these to many decimal places, requiring a knowledge of the individual refractive indices to about seven decimals, which is beyond the capability of any measurement procedure. We therefore use the interpolation formula given in the current Schott catalog to calculate refractive indices to the required precision. Failure to do this will result in such scattered points that it is impossible to plot a smooth chromatic spectrum focus curve for the completed design.

To unite the C, e, and g lines at a common focus, we use Table 5.4. In this case, $V_{gC} = \frac{n_e - 1}{n_g - n_C}$, analogous to the commonly used Abbe number formula for C, d, and F lines, and $P_{ge} = \frac{n_g - n_e}{n_e - n_C}$. Using these somewhat artificially accurate numbers, we calculate the value of E as -0.009744, and assume that the focal length is $F' = 10$ mm, which gives $c_a = 0.5461855$, $c_b = -0.3830219$, and $c_c = 0.0661602$. As illustrated in Figure 5.10, a negative value of E means that the glass having the intermediate dispersion lies below the line connecting the other two glasses.

Table 5.4

Depression and Partial Depression for Glasses Used in a Three-Lens Apochromatic

Lens	Glass	n_e	$\Delta n = n_g - n_C$	$n_g - n_e$	P_{ge}	V_{gC}
a	N-PK51	1.5301922	0.0105790	0.0068488	0.6473933	50.117231
b	N-KZFS4	1.6166360	0.0214990	0.0140786	0.6544848	28.682091
c	N-SF15	1.7043784	0.0371291	0.0249650	0.6723844	18.971081

Using other refractive indices, also calculated by the interpolation formula, we can plot the chromatic spectrum focus curve that duly passes through the points for *C*, *e*, and *g* as required (Figure 5.11). It can be seen that there is a very small residual of tertiary spectrum, the foci for the *d* and *F* lines being slightly back and forward, respectively, while the two ends of the curve move rapidly inward toward the lens. In this particular configuration, a fourth crossing at 0.39 μm occurs. The peak residual tertiary chromatic aberration for this apochromat is

$$\frac{0.00015}{10} \times 100\% = 0.0015\%$$

which is insignificant. By comparison, the peak residual tertiary chromatic aberration for an ordinary doublet such as shown in Figure 5.8 is about 0.2% or more than a 100 times greater than for this apochromat.

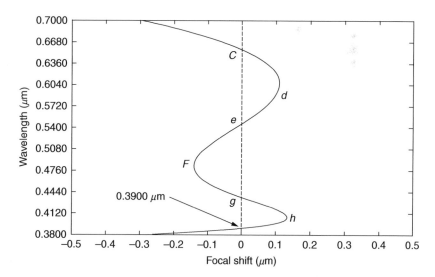

Figure 5.11 Tertiary spectrum of a 10mm focal-length thin three-lens apochromat with the *C*, *e*, and *g* lines brought to a common focus. Focal shift is with respect to the paraxial focus of the *e* line.

Depending on the choice of glasses, the chromatic aberration curve for an apochromat can take on other shapes such as shown in Figure 5.11. In this case, both the foci for the *d* and *h* lines are slightly back from the *e* line focus. Note that if the image plane is shifted slightly away from the lens, a common focus for four wavelengths is again obtained and the residual chromatic aberration is reduced.

This system should be called a "superachromat," since the three glasses satisfy Herzberger's condition[10] for the union of four wavelengths at a common focus. Failure to meet this condition generally ends up with three united wavelengths in the visible spectrum with the fourth wavelength falling far out into the infrared. In Section 7.4, the design of the apochromat will be completed after inserting suitable thicknesses and choosing such a shape that the spherical aberration is also corrected.

DESIGNER NOTE

As has been explained, longitudinal or axial chromatic aberration is a first-order aberration. When using a computer program to aid in the design of a lens, the designer can effectively use the axial color operand to define an optimization defect term that is the axial image distance between two selected wavelengths. For an achromat, the designer might select *C* and *F* lines to unite their foci. In the case of an apochromat, defect terms might be formed that measure the axial image distance between, say, *g* and *e*, *g*, and *C*, and *e* and *C* to unite foci for the *g*, *e*, and *C* lines. Always remember the great importance of proper selection of the glasses as the bendings of the lenses have essentially no impact upon the first-order chromatic aberration.

5.7 THE SEPARATED THIN-LENS ACHROMAT (DIALYTE)

In the early 1800s, the fabrication of a large achromatic doublet for astronomical objectives was problematic due to the difficulty of obtaining large disks of flint glass. An elegantly simple solution was introduced at that time to mitigate this difficulty, which no longer exists today. This solution was known as "*dialyte* objectives," which comprises a positive crown lens with a smaller negative flint lens separated by some modest distance. This is in effect a telephoto lens, since the track length[11] of the lens is significantly smaller than the effective focal length. When used as a telescope objective, the dialyte lens has the advantage that both sides of the objective lens are exposed to the atmosphere, thereby allowing quicker and more uniform thermal tracking to maintain sharp imagery

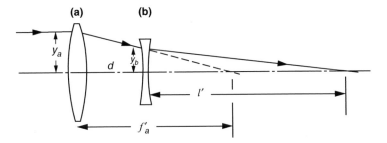

Figure 5.12 The dialyte lens with positive lens (a) and negative lens (b).

and spatial stability of the image. Gauss recognized that by choosing the proper separation between the two lenses, it is possible to correct the spherical aberration for two different colors.[12]

In Figure 5.12 we show the dialyte lens having the two components of a thin achromatic doublet separated by a finite distance d, where we find that the flint element particularly must be considerably strengthened. It is convenient to express d as a fraction k of the focal length of the crown lens, that is, $k = d/f_a'$. Since the chromatic aberration contributions of the two elements in an achromat must add up to zero, we see that

$$\frac{y_a^2}{f_a V_a} + \frac{y_b^2}{f_b V_b} = 0$$

but by Figure 5.12 it is clear that $y_b = y_a(f_a' - d)/f_a'$, or $y_b = y_a(1 - k)$. Combining these gives

$$f_b V_b = -f_a V_a (1-k)^2$$

Since the system must have a specified focal length F', we have

$$\frac{1}{F'} = \frac{1}{f_a'} + \frac{1}{f_b'} - \frac{d}{f_a' f_b'} = \frac{1}{f_a'} + \frac{1-k}{f_b'}$$

Combining the last two relationships gives the focal lengths of the two components as

$$f_a' = F'\left[1 - \frac{V_b}{V_a(1-k)}\right], \quad f_b' = F'(1-k)\left[1 - \frac{V_a(1-k)}{V_b}\right] \tag{5-8}$$

Since it is evident that $\dfrac{y_a}{F'} = \dfrac{y_b}{l'}$, then the back focal length l' is given by $l' = \dfrac{y_b}{y_a} F'$.

As an example, let us assume that $V_a = 60$ and $V_b = 36$. The two focal lengths are then related to the value of k as follows:

k	0	0.1	0.2	0.3
f'_a	$0.4F'$	$0.333F'$	$0.25F'$	$0.143F'$
f'_b	$-0.667F'$	$-0.45F'$	$-0.267F'$	$-0.117F'$
d	0	$0.033F'$	$0.05F'$	$0.043F'$

As k is increased the powers of both lenses become greater, but the power of the negative lens increases more rapidly than that of the positive lens. For this example, the two powers become identical at $k = 0.225$, at which point both the focal lengths are $0.225F'$. This property of an achromatic dialyte is employed with great effect in the predesign of a dialyte-type four-element photographic objective (see Section 13.2). The limiting value of k occurs when $V_a(1 - k) = V_b$, that is, when both elements become infinitely strong. In the present example this is when $k = 0.4$.

A more general solution to the dialyte problem has been provided by Conrady.[13] In this case, he developed the equations for the object being at a finite distance rather than at infinity. It is observed that when the distance between the two lenses is adjusted to produce achromatism for a particular object distance, the dialyte lens will display chromatic aberration when the object is located at any other distance.

5.7.1 Secondary Spectrum of a Dialyte

In Eq. (5-6) we saw that for a succession of thin elements

$$l'_\lambda - l'_F = -\frac{1}{u'^2_k} \sum \frac{Py^2}{Vf'}$$

Substituting in this the values of f'_a, f'_b, and y_b for a dialyte gives

$$l'_\lambda - l'_F = -\frac{F'(1 - k)}{V_a(1 - k) - V_b}(P_a - P_b) \qquad (5-9)$$

When $k = 0$ for a cemented lens this, of course, degenerates to the equation for a thin achromatic doublet (see Eq. (5-7)).

Actually, neither the achromatism relation, Eq. (5-8), nor the secondary spectrum expression, Eq. (5-9), is strictly correct because in their derivation we assumed that $y_b = y_a(1 - k)$ for all wavelengths. Because of the dispersion of the front element and the finite separation between the elements, it turns out that y_b is a little smaller in blue than in red light. Thus, a dialyte made in

Table 5.5

Glasses for Thin-Lens Dialyte

Glass	n_C	n_e	n_F	$\Delta n = n_F - n_C$	$V_e = \frac{n_e - 1}{n_F - n_C}$
Crown	1.51554	1.52031	1.52433	0.00879	59.193
Flint	1.61164	1.62058	1.62848	0.01684	36.852
					$V_a - V_b = 22.341$

accordance with Eq. (5-8) turns out to be slightly overcorrected chromatically, requiring a slight decrease in the power of the flint element to achromatize. For the same reason the secondary spectrum turns out to be slightly less than the amount given by Eq. (5-9).

To illustrate, suppose we design a thin-lens dialyte using the glasses shown in Table 5.5. Using formulas in Eq. (5-8) we find that, for $F' = 10$ and $k = 0.2$

$$f'_a = 2.21783, \quad \text{where } c_a = 0.866581 \quad [\text{because } c = 1/f'(n-1)]$$
$$f'_b = -2.27991, \quad \text{where } c_b = -0.706781$$
$$d = 0.443566$$

Tracing paraxial rays in C, e, and F through this system using the ordinary thin-lens $(y - u)$ method gives

$$l'_C = 8.008133, \quad l'_e = 8.0, \quad l'_F = 8.008431$$

There is thus a small residual of paraxial chromatic aberration of magnitude 0.000298 in the overcorrected sense. To remove this, we must weaken the flint element slightly, to $c_b = -0.706449$, which gives

$$l'_C = 7.994955, \quad l'_e = 7.986857, \quad l'_F = 7.994962$$

The $F - C$ aberration is now corrected, and the e image lies closer to the lens by an amount of secondary spectrum equal to -0.008103. A thin cemented achromat of the same focal length made of the same glasses has a $D - F$ secondary spectrum of -0.004820, only about half that of the dialyte (see Figure 5.13).

5.7.2 A One-Glass Achromat

It has been known for a long time that it is actually possible to design an air-spaced achromat using only one kind of glass.[14] If we write $V_a = V_b$ in Eq. (5-8) we obtain the focal lengths of the two elements of a one-glass achromat as follows:

$$f'_a = \frac{kF'}{k-1}, \quad f'_b = -kF'(k-1), \quad d = kf'_a, \quad l' = -F'(k-1) \quad (5\text{-}10)$$

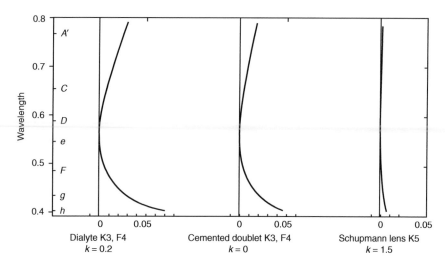

Figure 5.13   Three secondary spectrum curves.

once again assuming a very distant object and thin lenses. Since the air-space d must always be positive, we see that k must have the same sign as f_a', and $k-1$ must have the same sign as F'.

For a positive lens, k must be greater than 1.0, which makes for a very long system (see Figure 5.14). This is known as a *Schupmann* lens, but it is seldom used because the image is inside (between the lenses); however, it can be used for an eyepiece or as part of a more complex optical system (see Chapters 15 and 16).

For a negative system, $k-1$ must be negative, so that k must be less than 1.0. If the front element is positive, k must be positive and thus must lie between 0 and 1. This gives a compact system (see Figure 5.15a). If the front element is negative, k must be negative but may have any value. If k is small the system is compact, but if k is large the system becomes very long (Figures 5.15b and 5.15c). A negative one-glass achromat could, for instance, be used in the rear member of a telephoto lens.

When designing a Schupmann dialyte, the colored rays become separated at the rear component because of the long air space, and so the simple formulas

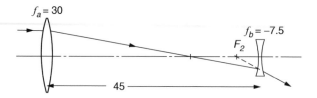

Figure 5.14   A Schupmann lens ($f = 10$).

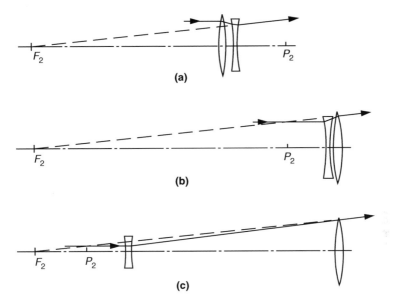

Figure 5.15 Negative one-glass achromatic dialytes ($f' = -10$). (a) $k = 0.2$, $f_a = 2.5$, $f_b = -1.6$, $d = 05$; (b) $k = -0.2$, $f_a = -1.66$, $f_b = 2.4$, $d = 0.33$; (c) $k = -5.0$, $f_a = -8.33$, $f_b = 300$, $d = 41.7$.

fail to give a perfect achromat and we must readjust the rear lens power for achromatism. Similarly, since both elements have the same dispersion, we might expect the secondary spectrum to be zero, but it is actually slightly undercorrected.

As an example, we will design a Schupmann dialyte of focal length 10.0 using the same crown glass for both elements. We take $k = 1.5$, and Eq. (5-10) tells us that

$$f_a' = 30, \quad f_b' = -7.5, \quad d = 45, \quad l' = -5$$

The refractive indices of this crown glass are

$$n_C = 1.51981, \quad n_D = 1.52240, \quad n_r = 1.52857$$

Therefore $c_a = 0.063808$ and $c_b = -0.255232$. Tracing paraxial rays in these wavelengths through the thin-lens solution gives

$$l_F' = -4.998653 \quad \text{and} \quad l_C' = -4.999746$$

leaving a residual of chromatic aberration of $+0.001093$. To remove this we weaken the flint component to $c_b = -0.250217$, which gives the back foci shown in Table 5.6 for a number of wavelengths.

These data are plotted in Figure 5.13 in comparison with the corresponding secondary spectrum curves for a cemented achromat and a dialyte made with ordinary glasses.

Table 5.6

Residual Chromatic Aberration for the Example Schupmann Dialyte

	Wavelength	Back focus	Departure from D
A'	0.7682	−5.06442	+0.00194
C	0.6563	−5.06577	+0.00059
D	0.5893	−5.06636	0
e	0.5461	−5.06647	−0.00011
F	0.4861	−5.06578	+0.00058
g	0.4359	−5.06353	+0.00283
h	0.4047	−5.06058	+0.00578

Another form of a single-glass achromat is a thick singlet lens where the lens thickness and glass type are used to correct the paraxial chromatic aberration.[15] The first surface is convex, having a curvature c_1, and the second surface is flat. The thickness is given by

$$t = \frac{n_F n_C}{c_1(n_F - 1)(n_C - 1)}.$$

When collimated light enters the lens, it forms a real focus suffering from chromatic aberration as explained in Section 5.1. The light then diverges and forms a virtual focus within the glass that is free of axial color, and interestingly, the secondary spectrum is quite small. Since both foci are located inside the lens, this lens could perhaps be used as the secondary element of a beam expander or telescope with a concave mirror serving as the primary element. The virtual focus of the lens and the focus of the primary element should be coincident. See Section 15.4.8 for a related discussion.

5.8 CHROMATIC ABERRATION TOLERANCES

5.8.1 A Single Lens

In the seventeenth century astronomers used simple lenses of very long focal length as telescope objectives. In this way they managed to make the chromatic aberration insignificant. The logic behind this procedure is that the chromatic aberration of a simple lens is equal to f/V, while the focal range based on diffraction theory is equal to $\lambda/\sin^2 U' = 4\lambda f^2/D^2$, where D is the diameter of the lens. Assuming that because of the drop in sensitivity of the eye at the deep red and blue we may let the chromatic aberration reach twice the focal range, we have

$$f/V = 8\lambda f^2/D^2$$

When $\lambda = 0.58$ μm approximately, and if $V = 60$, we then find our formula tells us that the shortest possible focal length to meet this relation is roughly equal to 40 times the square of the lens diameter in centimeters (or 100 times the square of the lens diameter in inches). Thus an objective of 10-cm aperture will have an insignificant amount of chromatic aberration if its focal length is greater than about 40 m.

5.8.2 An Achromat

By a similar logic, we can determine the minimum focal length of an achromatic telescope objective for the secondary spectrum to be invisible to the observer. Now we equate the secondary spectrum in d light to the whole focal range, or

$$f/2200 = 4\lambda f^2/D^2$$

where $f = 2D^2$ if in centimeters or $f = 5D^2$ if in inches approximately. Consequently a 10-cm aperture achromatic objective will have an insignificant amount of secondary spectrum if its focal length is greater than about 2 m (or 80 inches). The enormous gain resulting from the process of achromatizing is clearly evident.

5.9 CHROMATIC ABERRATION AT FINITE APERTURE

It is clear from the graphs in Figure 5.3 that the chromatic aberration of a lens, expressed as $L'_F - L'_C$, varies across the aperture, and a graph of chromatic aberration against incidence height Y appears in Figure 5.4. Thus a normal achromat has some degree of chromatic undercorrection for the paraxial rays and a corresponding degree of chromatic overcorrection for the marginal rays, it being well-corrected for the 0.7 zonal rays. To achromatize a finite-aperture lens therefore requires the tracing of zonal rays in the two wavelengths that are to be united at a common focus, and experimentally varying one of the radii until these two foci become coincident.

5.9.1 Conrady's $D - d$ Method of Achromatization

Although this method is not frequently used today because of the availability of powerful lens design programs that can operate on desktop computers, the student of lens design will obtain additional valuable knowledge of optical design by understanding this very useful and simple procedure for achromatizing a lens which Conrady[16] introduced in 1904. The method he suggested depends on the fact that in an achromat

$$\sum (D-d)\,\Delta n = 0$$

where D is the distance measured along the traced marginal ray in brightest light from one surface to the next, and d is the axial separation of those surfaces. Δn is the index difference between the two wavelengths that are to be united at a common focus for the material occupying the space between the two lens surfaces under consideration. Since Δn for air is zero, we need consider only glass lenses in making this calculation. The argument used in deriving this relation is as follows.

Suppose we have a series of rays in one wavelength originating at an axial object point and passing through a lens. Each point in the wavefront will travel along the ray and will eventually emerge from the rear of the lens, the moving wavefront being always orthogonal to the rays (Malus' theorem).

Since the emerging wavefront has the property that light takes the same time to go from the source to every point on the wavefront, we see (Figure 5.16) that time $= \sum (D/v)$, where v is the velocity of light in each section of the ray path of length D. Hence time $= \sum (D/c)(c/v)$ where c is the velocity of light in air. Thus time $= (1/c) \sum (Dn)$ since the refractive index n is equal to the ratio of the velocity of light in air to its velocity in the glass. The $\sum (Dn)$ is the length of the optical path along the traced ray, from the original object point to the emerging wavefront, and all points on a given wavefront have the same value of $\sum (Dn)$.

Conrady then proceeded to assume that in a lens having some residual of spherical aberration and spherochromatism, as most lenses do, the best possible state of achromatism occurs when the emerging wavefronts in C and F light (red and blue) cross each other on the axis and at the margin of the lens aperture, as indicated in Figure 5.17. Since the C and F wavefronts will then be parallel to each other at about the 0.7 zone, the C and F rays through that zone will lie

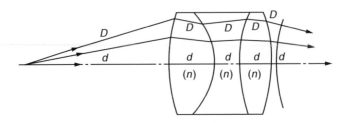

Figure 5.16   The emerging wavefronts from a lens.

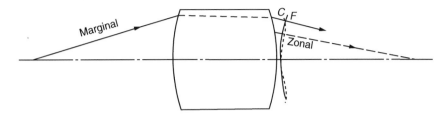

Figure 5.17 The emerging wavefronts from an achromat.

together and cross the axis at the same point. Under these circumstances

$$\sum (Dn)_C = \sum (Dn)_F$$

along the marginal ray. However, since all points on a wavefront have the same value of $\sum (Dn)$, it is clear that in an achromat

$$\sum (D-d)n_C = \sum (D-d)n_F \quad \text{or} \quad \sum (D-d)(n_F-n_C)= 0 \qquad (5\text{-}11)$$

This is Conrady's condition for the best possible state of achromatism in a lens that suffers from other residuals of aberration. The presence of spherochromatism, for example, causes the two emerging wavefronts in C and F light to separate between the axis and margin of the aperture, while the presence of spherical aberration causes the wavefronts to assume a noncircular shape.

In stating this condition, we are tacitly assuming that the values of D within all the lens elements are equal for C and F light. This is certainly not true, but we shall make only a very small error if we trace the marginal ray in brightest light, which is usually d or e for $C - F$ achromatism, and calculate the distances D along that ray. The argument breaks down if there is a long air space between unachromatized or only partially achromatized separated components, but in most cases it is surprisingly accurate.

The $D - d$ relation would be impossibly difficult to use if we had to calculate every D value from the original object right up to the emerging wavefront, but the method is saved by the fact that the dispersion $\Delta n = n_F - n_C$ of air is zero. For this reason we must calculate $D - d$ only for those sections of the marginal ray that lie in glass. The length D is found by the usual relation

$$D = (d + Z_2-X_1)/\cos U_1'$$

where $Z = r[1 - \cos(I - U)]$ as explained in Section 2.3. The choice of dispersion values depends on the region of the spectrum in which achromatism is desired. For ordinary visual achromatism, we trace the ray in d or e light and use $\Delta n = n_F - n_C$; for photographic achromatism, we may prefer to trace the marginal ray in F light and use $\Delta n = n_g - n_D$ for the dispersion. The process for interpolating dispersions suggested in Chapter 1 is of value here; however, all modern lens design programs include data tables for optical materials and the appropriate interpolating dispersion equation for each.

5.9.2 Achromatization by Adjusting the Last Radius of the Lens

To achromatize a lens then, we must make the sum $\sum (D - d)\,\Delta n$ equal to zero by some means or other. Commonly we calculate that value of the last radius of the lens that will accomplish this. Alternatively we may design the lens using any suitable refractive indices, and then at the end search the glass catalog for glass types with dispersion values that will make the $(D - d)\,\Delta n$ sum zero. To use the first method, suppose that the value of the $D - d$ sum for all the lens elements prior to the last element is $\sum_0$; then for the last element we must have

$$\sum (D-d)\Delta n = -\sum_0$$

We now calculate the Z and Y at the next-to-last surface, and knowing the desired value of D in the last element to achieve achromatism, we calculate

$$Z_2 = D \cos U'_1 + Z_1 - d \quad \text{and} \quad Y_2 = Y_1 - D \sin U'_1$$

(Here the indices 1 and 2 refer to the first and second surfaces of the last element.) The radius of curvature of the last surface is given by

$$r = \frac{(Z^2 + Y^2)}{2Z}$$

and the problem is solved. As a check on our work, we may wish to trace zonal rays in F and C light through the whole lens; if everything is correct, these rays should cross the axis at the same point in the image space.

5.9.3 Tolerance for the $D - d$ Sum

Conrady[17] suggests that in a visual system the tolerance for the $D - d$ sum is about half a wavelength. However, there is no point in achieving perfect achromatism for the 0.7 zonal rays, which the $D - d$ method does, if there is considerable spherochromatism in the lens since this will swamp the excellent color correction. Therefore we have found that a more reasonable tolerance is about 1% of the contribution of either the crown or the flint element in the lens. If these contributions are small, it indicates that the spherochromatism will be small and a tight tolerance for the $D - d$ sum is sensible.

Example

As an example in the use of the $D - d$ method, we return to the cemented doublet lens used as a ray-tracing example in Section 2.5, and compute the $(D - d)$ Δn sum along the traced marginal ray (Table 5.7). It will be seen that there is a small residual of the sum, amounting to –0.0000578, which is about 1% of the separate contributions of the crown and flint elements. We must therefore regard this lens as noticeably undercorrected for chromatic aberration. That is the reason why the C and F curves in Figure 5.3a (see page 141) cross somewhat above the 0.7 zone of the aperture.

If we wish to achromatize this lens perfectly, we can solve for the last radius by the method described in Section 5.9.1. This tells us that a last radius of –16.6527 would make the $D - d$ sum exactly zero. As this radius is decidedly different from the given radius of –16.2225, we see once again that it is necessary to change a lens drastically if we wish to affect the chromatic correction.

As an alternative method for achromatizing, we could calculate what value of Δn for either the crown or the flint glass would be required to eliminate the $D - d$ sum. The numbers shown in Table 5.6 (see page 162) tell us that we could achromatize with the given crown if we had a flint with $\Delta n = 0.01941$; this represents a V number of 33.43 instead of the given 33.80. Or we could retain the given flint and seek a crown with $\Delta n = 0.00792$; this represents a V number of 65.26 instead of the given 64.54. In both cases the required change in V number is only slightly larger than the normal factory variation in successive glass melts, indicating that the small residual of chromatic aberration in this lens is really almost insignificant.

Table 5.7
Calculation of the $(D - d)$ Δn Sum

C	0.1353271	–0.1931098		–0.0616427	
Z	0.2758011	–0.3865582		–0.1149137	
D		1.05		0.4	
$\cos U$		0.9955195		0.9985902	
D		0.3893853		0.6725927	
$D - d$		–0.6606147		0.2725927	
Δn		0.00801		0.01920	
Prod.		–0.0052916		0.0052338	$\sum = -0.0000578$

5.9.4 Relation between the $D - d$ Sum and the Ordinary Chromatic Aberration

D. P. Feder[18] has shown that, for any zone of a lens, the vertical displacement in the paraxial focal plane between marginal rays in F and C light is given closely by

$$H'_F - H'_C = \frac{\partial \Sigma}{\partial(\sin\ U')}$$

where Σ is the sum $\sum (D - d)\ \Delta n$ calculated along the zonal ray in question, and $\sin U'$ is the emerging slope of the same ray. Thus if we can express $\sum$ as a polynomial of the form

$$\sum = a \sin^2 U' + b \sin^4 U' + c \sin^6 U' \tag{5-12}$$

then

$$(H'_F - H'_C) = 2a \sin U' + 4b \sin^3 U' + 6c \sin^5 U'$$

By calculating $\sum$ for three zones of a lens, we can solve for the three coefficients a, b, and c, and we shall see this is in excellent agreement with Eq. (5-13).

A more convenient but only approximate relation between $(H'_F - H'_C)$ and $\sum$ can be found by neglecting the $\sin^6 U'$ term in Eq. (5-12). When this is done, we can relate the 0.7 zonal chromatic aberration with the marginal $\sum$ in the following way:

Writing

$$S = a \sin^2 U' + b \sin^4 U'$$

for the $D - d$ sum along any zonal ray, we see that if the angle between the C and F rays at any zone is α, then (Figure 5.18) computing the derivative of S with respect to $\sin(U')$ we find that

$$L'_\alpha = \frac{dS}{d \sin(U')} = 2a \sin U' + 4b \sin^3(U').$$

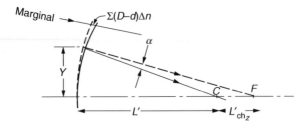

Figure 5.18   Relation between the $D - d$ sum and the zonal chromatic aberration.

The longitudinal chromatic aberration for this zone is given approximately by

$$L'_{ch} = L'_a/\sin U' = 2a + 4b \sin^2 U'$$

and hence the 0.7 zonal chromatic aberration will be given by

$$(L'_{ch})_z = 2(a + 2b \sin^2 U'_z)$$

But $\sin U'_z = \sin U'_m/\sqrt{2}$ approximately, and the calculated marginal $\sum (D - d)$ Δn sum is

$$\sum = S_m = a \sin^2 U'_m + b \sin^4 U'_m$$

Hence

$$(L'_{ch})_z = 2 \sum /\sin^2 U'_m \qquad (5\text{-}13)$$

As a check on this result, we recall that the residual of $\sum$ in our cemented telescope doublet was -0.0000578 and $\sin U'$ was 0.16659. Therefore, we should expect the zonal chromatic aberration to be -0.00417. By actual ray tracing we find

$$\text{zonal } L'_F = 11.27022$$

$$\text{zonal } L'_C = \underline{11.27523}$$

$$\therefore F - C = -0.00501$$

The small discrepancy is due to our having neglected the $\sin^6 U'$ term in the expression for S.

5.9.5 Paraxial $D - d$ for a Thin Element

We can readily reduce the $D - d$ expression to its paraxial form for a single thin lens element. In this case the length D in the paraxial region becomes

$$D = d + Z_2 - Z_1 = d + \frac{Y^2}{2r_2} - \frac{Y^2}{2r_1}$$

Hence

$$(D - d) = \frac{Y^2}{2} \left(\frac{1}{r_2} - \frac{1}{r_1} \right) = -\frac{Y^2}{2f'(n - 1)}$$

$$(D - d) \Delta n = -\frac{Y^2}{2f'} \left(\frac{\Delta n}{n - 1} \right) = -\frac{Y^2}{2f'V}$$

Now, by Eq. (5-13),

$$\text{paraxial chromatic aberration} = \frac{2\Sigma}{u'^2} = -\frac{Y^2}{f'Vu'^2}$$

which is in exact agreement with Eq. (5-3).

DESIGNER NOTE

Current optical design programs allow the lens designer to specify an operand that measures the optical path difference (OPD) with respect to the principal ray in a user-specified wavelength. In determining the chromatic aberration at a finite aperture, the lens designer can select OPD operands for F and C for the axial object and (relative) pupil coordinates of $\rho = 1$ and $\theta = 0$, then subtract OPD_C from OPD_F. By changing $\rho = 0.707$, the zonal chromatic aberration can be computed. When using the Conrady $D - d$ method for achromatizing, the goal is to make $\sum (D - d)\Delta n = 0$ which, it should be recalled, is computed in d light.[19] As mentioned previously, this can lead to some error, but the calculation is typically adequate. The aforementioned OPD method is of course accurate and the tolerances explained in Section 5.9.3 are applicable.

ENDNOTES

[1] Sir Isaac Newton, *OPTICKS*, Reprint of Fourth Edition (1730), Dover Publications, New York (1959).

[2] A. E. Conrady, p. 143.

[3] The introductory material is extracted in part from R. Barry Johnson, "A historical perspective on understanding optical aberrations," *Lens Design*, Warren J. Smith (Ed.); "Critical reviews of optical science and technology," *Proc. SPIE*, CR41:18–29 (1992).

[4] Axial chromatic aberration is a first-order aberration similar to defocus in that it is a field-independent aberration, as discussed in Chapter 4.

[5] The spectral lines associated with the helium d and sodium D are 0.5876 μm and 0.5893 μm. In this book, both are used; however, it is noted that glass manufacturers presently use the d value for the base refractive index and the Abbe number when stating the glass code. For example, Schott N-FK5 is 487704 ($n_d = 1.487$ and $v_d = 70.4$). See Chapter 2 for additional discussion.

[6] The term *transverse chromatic aberration* is often used synonymously with *lateral color*. In this book, "transverse axial color" is used in the context shown in Figure 5.1 and "lateral color" is used in the context discussed in Chapter 16.

[7] I. H. Malitson, "A redetermination of some optical properties of calcium fluoride," *Appl. Opt.*, 2:1103 (1963).

[8] Although the N-SK5 is actually classified as a crown glass, it is serving the function of a flint glass since the fluorite has a much higher Abbe number.

[9] Schott LITHOTEC-CAF2.

[10] M. Herzberger, "Colour correction in optical systems and a new dispersion formula," *Opt. Acta (London)*, 6:197 (1959).

[11] Track length is the distance from the front element of the lens system to the image plane.

[12] A. E. Conrady, p. 177.

[13] A. E. Conrady, pp. 175–183.

[14] Prof. Ludwig Schupmann (German, b. 1851, d. 1920), "Optical Correcting Device for Refracting Telescopes, U.S. Patent No. 620,978, March 14 (1899).

[15] William Swantner, private communications (2009).

[16] A. E. Conrady, p. 641. See also A. E. Conrady, "On the chromatic correction of object glasses" (first paper), *M. N. Roy. Astron. Soc.*, 64:182 (1904).

[17] A. E. Conrady, p. 647.

[18] D. P. Feder, "Conrady's chromatic condition" (research paper 2471), *J. Res. Nat. Bur. Std.*, 52:47 (1954).

[19] Note that d and D are used in several different contexts.

Chapter 6

Spherical Aberration

It was discussed in Chapter 4 and illustrated in Figure 4.5 that the field-independent astigmatic aberration comprises defocus and spherical aberration. This means that they are not functions of field angle and are constant aberrations over the entire field of view, and that collectively they are a function of the odd powers of the entrance pupil radius ρ for the transverse aberration form; that is,

$$\underbrace{\rho}_{\text{Defocus}} + \underbrace{\rho^3 + \rho^5 + \rho^7 + \ldots}_{\text{Spherical Aberration}}$$

In this chapter, we will consider both defocus and spherical aberration. In Section 4.3.1, the transverse defocus in an image plane located ξ from the paraxial image plane was shown to be expressed as

$$
\begin{aligned}
DF(\rho, \xi) &= -\xi \tan v'_a \\
&= -\frac{\rho D_{ent.pupil}}{2f} \xi \ \text{(when the object is located at infinity)} \\
&= -\frac{\rho}{2\,f\text{-number}} \xi
\end{aligned}
\tag{6-1}
$$

where v'_a is the angle of the marginal paraxial ray in image space, ρ is the normalized entrance pupil radius, and f is the focal length. Figure 6.1 shows a typical ray plot of a fan of axial meridional rays. The ray intercepts with the defocused image plane and forms a line that is rotated from the abscises. Now, the marginal ray intercept $\varepsilon_y = (DF(1, \xi))$ times $2\,f$-number equals the defocus ξ.

The direct calculation of spherical aberration is a simple matter. A meridional ray is traced from object to image, passing through the desired zone of a lens, and the image distance L' is found. This is compared directly with the l' of a corresponding paraxial ray from the same object point. Then

$$\text{longitudinal spherical aberration} = LA' = L' - l'. \tag{6-2}$$

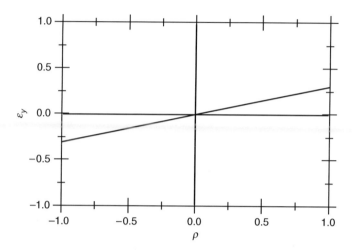

Figure 6.1   Transverse defocus error in the paraxial image plane.

Historically, longitudinal spherical aberration was used for several reasons. First, the intersection of the meridional ray with the optical axis directly provided the length L'. Second, the calculation was reasonably easy using hand computing methods. And finally, the longitudinal spherical aberration is inherently independent of defocus.[1]

Figure 5.3a illustrated a typical presentation of longitudinal spherical aberration for various colors of light, which was mentioned already, spherochromatism. In the following discussions, consider only the D line. The transverse spherical aberration for a desired zone of a lens is measured as the intersection height of a meridional ray with the paraxial image plane ($\varepsilon_y = -LA' \tan U'$); however, the actual image plane may be displaced from the paraxial image plane by ξ so that the ray intersection height comprises both spherical and defocus components. Figure 6.2 shows a ray plot when the image plane is inside the paraxial focus and for positive primary spherical aberration. Notice that the composite aberration (sum of defocus and spherical aberration) ray plot is rotated because of the presence of defocus.

As was explained in Chapter 4, defocus can influence the blur caused by the other astigmatic aberrations, but does nothing to the comatic aberrations. Consider now the problem of locating the position of the defocused image plane to achieve the minimum blur diameter in the presence of primary spherical aberration. Figure 6.3 shows the ray plot for third-order spherical aberration in the paraxial image plane. The aberration is expressed by $\varepsilon_y(\rho, 0, 0, 0) = 0.6\rho^3$ where $\sigma_1 = 0.6$. It is evident that the blur diameter is 1.2 (lens units). The central dashed line in Figure 6.3 represents the defocus aberration. If one then views

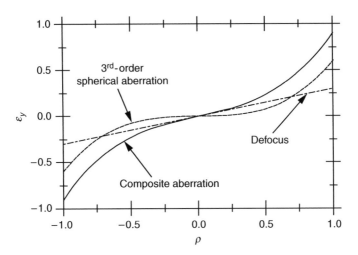

Figure 6.2 Transverse ray plot when image plane is inside the paraxial focus and for positive primary spherical aberration.

this line as the x-axis in the image plane, the primary spherical aberration curve is positioned to give the minimum blur diameter. The parallel lines bound the primary spherical aberration curve and it is easily shown that the minimum defocused blur diameter is $\frac{1}{4}$ the amount in the paraxial image plane, that is, $\frac{1}{2}\sigma_1$. Using Eq. (6-1), the amount of defocus needed is $\frac{3}{4}LA'_{marginal}$. In the case shown in Figure 6.3, the best focus image plane lies outside of the paraxial image plane.

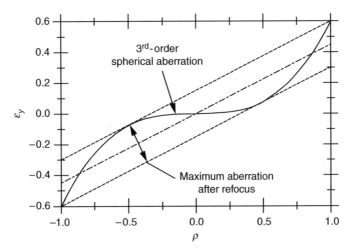

Figure 6.3 Ray plot for third-order spherical aberration in the paraxial image plane.

6.1 SURFACE CONTRIBUTION FORMULAS

The simple relationship given by Eq. (6-2) is often inadequate, both because it gives the aberration as a small difference between two large numbers, and also because it gives no clue as to where the aberration arises. It is therefore much more useful to compute the aberration as the sum of a series of surface contributions. A convenient formula has been given by Delano[2]; the derivation follows from Figure 6.4. Note that these surface contributions are for all orders of spherical aberration, not just for the primary term. In this diagram, entering marginal and paraxial rays are shown at a spherical surface. The length S is the perpendicular drawn from the paraxial object point P onto the marginal ray. The marginal ray is defined by its Q and U, the paraxial ray by its y and u. Then

$$S = Q - l \sin U, \quad \text{hence,} \quad Su = Qu - y \sin U$$

We now replace u on the right by $i - yc$ and $\sin U$ by $\sin I - Qc$, where c is the surface curvature as usual. Multiplying through by n gives

$$Snu = Qni - yn \sin I$$

Doing the same thing for the refracted ray and subtracting plain from prime gives

$$S'n'u' - Snu = (Q' - Q)ni$$

We write this for every surface and add. After extensive cancellation because $(S'n'u')_1 = (Snu)_2$, we get for k surfaces

$$(S'n'u')_k - (Snu)_1 = \sum (Q' - Q)ni \tag{6-3}$$

Inspection of Figure 6.4 shows that

$$LA = -S/\sin U \quad \text{and} \quad LA' = -S'/\sin U'$$

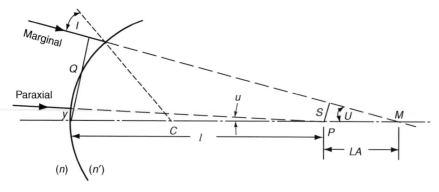

Figure 6.4 Spherical aberration contribution.

Hence,

$$LA' = LA\left(\frac{n_1 u_1 \sin U_1}{n'_k u'_k \sin U'_k}\right) - \sum \frac{(Q' - Q)ni}{n'_k u'_k \sin U'_k} \tag{6-4}$$

The quantity under the summation sign is the contribution of each surface to the spherical aberration of this particular ray (see Section 4.4), and the first term is the transfer of the object aberration across the lens to the image space. It may be thought of as the contribution of the object to the final aberration.

As an example of the use of this formula, we will take the lens used in Section 2.5. A marginal ray and a corresponding paraxial ray entering this lens from infinity at height 2.0 has been traced; additional data required for use of Delano's formulas are given in Table 6.1. It will be noted that the sum of the aberration contributions agrees closely with the $L' - l'$ aberration obtained directly from ray tracing:

$$L' = 11.29390$$
$$l' = 11.28586$$
$$LA' = L' - l' = 0.00804$$

However, the L' and l' values are good only to about 1 in the fifth decimal place, when using tables for manual ray tracing whereas the contributions are good to 1 in the seventh place. The contribution method is clearly the more precise of the pair.

Note, too, that the first and third surfaces of this lens contribute undercorrected aberration, the third giving twice as much as the first in spite of its flat

Table 6.1

Surface Contributions to Spherical Aberration

c	0.1353271		−0.1931098		−0.0616427
d		1.05		0.40	
n		1.517		1.649	
		Paraxial ray data			
u	0	−0.0922401		−0.0554372	−0.1666664
$yc + u = i$	0.2706542		−0.4597566		−0.1713855
		Marginal ray data			
Q	2.0		1.9178334		1.9186619
Q'	2.0171179		1.9398944		1.8814033
$Q' - Q$	0.0171179		0.0220610		−0.0372586
ni	0.2706542		−0.6974508		−0.2826147
$- n'_k u'_k \sin U'_k$	−0.0277643		−0.0277643		−0.0277643
Spherical contribution	−0.1668701		0.5541815		−0.3792578 $\sum = 0.0080536$

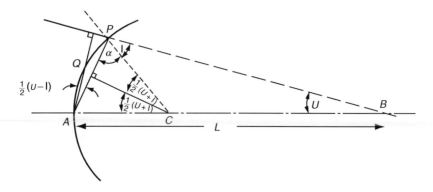

Figure 6.5   Diagram showing that $Q = PA \cos \frac{1}{2}(-U - I)$.

curvature; the second surface contributes more overcorrection than the total undercorrection of the two outer surfaces in spite of the small index difference between the media on each side of it.

An alternative representation of the contribution formula is sometimes useful. Its derivation depends on the relation between Q and the chord PA (Figure 6.5). In triangle APB we have

$$\frac{PA}{\sin U} = \frac{-L}{\sin(\alpha + I)}$$

Therefore,

$$PA = \frac{-L \sin U}{\sin(\alpha + I)} = \frac{-Q}{\sin(\alpha + I)}$$

However,

$$\alpha = 90° - \tfrac{1}{2}(I - U)$$

Therefore,

$$\alpha + I = 90° - \tfrac{1}{2}(I - U) \quad \text{and} \quad Q = PA \cos\tfrac{1}{2}(I - U)$$

Hence,

$$(Q - Q') = PA[\cos\tfrac{1}{2}(I - U) - \cos\tfrac{1}{2}(I' - U')]$$
$$= PA[-2\sin(\tfrac{1}{2}\text{sum})\sin(\tfrac{1}{2}\text{diff})]$$
$$= 2PA \sin\tfrac{1}{2}(I' + U') \sin\tfrac{1}{2}(I' - I)$$

The spherical aberration contribution formula can therefore be written

$$LA' = LA\left(\frac{n_1 u_1 \sin U_1}{n'_k u'_k \sin U'_k}\right) + \sum \frac{2PA \sin\frac{1}{2}(I' - I)\sin\frac{1}{2}(I' + U)ni}{n'_k u'_k \sin U'_k} \tag{6-5}$$

where

$$I' - I = U' - U, \text{ and } I' + U = I + U'$$

6.1.1 The Three Cases of Zero Aberration at a Surface

In Eq. (6-5), the quantity under the summation sign becomes zero in the following special cases:

(a) if $PA = 0$,
(b) if $I' = I$,
(c) if $i = 0$,
(d) if $I' = -U$

In case (a) the object and image are both at the vertex of the surface. In case (b) the marginal ray suffers no refraction at the surface; this could occur because the object is at the center of curvature of the surface, as also in case (c), but it could occur trivially if the refractive index were the same on both sides of the surface. Case (d) arises if $I' = -U$ or if $I = -U'$. This very important case must be considered further.

By Eq. (2-1) we see that in this case

$$\sin I = Qc + \sin U = \sin U - \left(\frac{L \sin U}{r}\right) = \left(1 - \frac{L}{r}\right) \sin U$$

But $\sin I = (n'/n) \sin I'$, and since, in this special case, $I' = -U$, we find that $(L/r) - 1 = n'/n$, where

$$L = r(n + n')/n$$

and similarly

$$L' = r(n + n')/n'$$

It can also be shown that, for this particular pair of conjugates,

$$Q = Q', \quad nL = n'L', \quad 1/L + 1/L' = 1/r$$

We can understand case (b) better with a numerical example. Consider the aplanatic hemispherical magnifier shown in Figure 6.6, which has a convex surface with air on the right and glass of index 1.5 on the left. We find that

$$L = 2.5r \text{ and } L' = 1.6667r$$

The image is free of all orders of spherical aberration, third-order coma, and axial color.

The aplanatic points B and B' are shown in Figure 6.7. All rays in the object space directed toward B will pass through B' after refraction, no matter at what angle they enter the surface. This pair of conjugates is known as the *aplanatic points* of the surface. Note that the distances of these points from the center of curvature of the surface are respectively equal to $B = r(n'/n)$ and

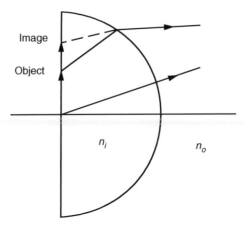

Figure 6.6 Aplanatic hemispherical magnifier with the object and image located at the center of curvature of the spherical surface. This type of magnifier has a magnification of n'/n and can be used as a contact magnifier or as an immersion lens.

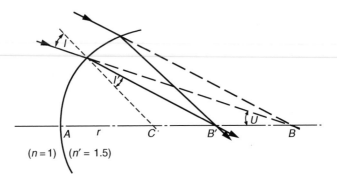

Figure 6.7 The aplanatic points of a surface.

$B' = r(n/n')$. Aplanatic surfaces of this type are used in many types of lens, particularly high-power microscope objectives and immersion lenses which make detectors appear larger.

A similar magnifier can be constructed by using a hyperhemispherical surface and a plano surface as depicted in Figure 6.8. The lateral magnification is $(n'/n)^2$. This lens, called an *Amici lens,* is based on the fourth aplanatic case. The image is free from all orders of spherical aberration, third-order coma, and third-order astigmatism. These magnifiers are often used as desktop magnifiers, having a magnification of about 2.5.

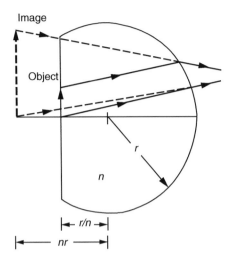

Figure 6.8 Aplanatic hyperhemispherical magnifier or Amici lens has the object located at an aplanatic point.

DESIGNER NOTE

It must be borne in mind that an aplanatic surface is capable only of increasing the convergence of converging light or increasing the divergence of diverging light. The greater the convergence or divergence, the greater will be the effect of an aplanatic surface. For parallel entering light, the aplanatic surface is a plane and produces no change in convergence.

6.1.2 An Aplanatic Single Element

It is possible to make an aplanatic single-element lens for use in a converging light beam by making the front face aplanatic and the rear face perpendicular to the marginal ray. Such a lens increases the convergence of a converging beam, which is useful in certain situations. In parallel light an aplanatic lens is merely a parallel plate. In a diverging beam an aplanatic lens element is a negative meniscus that increases the divergence of the beam without, of course, introducing any spherical aberration.

6.1.3 Effect of Object Distance on the Spherical Aberration Arising at a Surface

We have seen that the contribution of a single surface to the spherical aberration is zero if the (virtual) object is at *A, C,* or *B,* as in Figure 6.7. We may

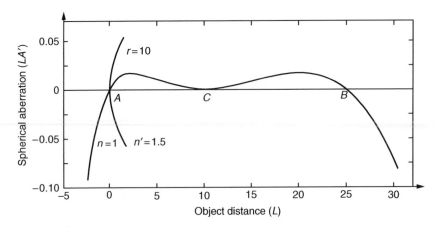

Figure 6.9   Effect of object distance on spherical aberration.

now inquire what will happen if the object lies in any of the regions between these points, the light entering from the left in all cases. As an example, consider the case of a surface of radius 10 with air on the left and glass of index 1.5 to the right. We will let a ray enter this surface at a fixed slope angle of 11.5°, and we calculate the spherical aberration in the image as the object moves along the axis. This is shown in Figure 6.9.

If the object lies between the surface *A* and the aplanatic point *B*, a collective surface such as we are considering here contributes *overcorrected* spherical aberration, which is decidedly unexpected and can be quite useful. The peak value of this overcorrection occurs, in our case, when the object distance is about twice the surface radius. As can be seen in Figure 6.9 the peak of overcorrection is close to the aplanatic point, and to achieve it we must use a surface somewhat flatter than the aplanatic radius. There is also a second but much less useful peak close to the surface itself, the value of *L* in this case being about 0.2 times the surface radius. As a general rule, using ordinary glasses, the maximum overcorrection will be obtained if *r* is set at a value about 1.2 times the aplanatic radius of $L(1 + n')$, assuming that there is air to the left of the surface as shown in Figure 6.9.

6.1.4 Effect of Lens Bending

One of the best methods for changing the spherical aberration of a lens is to *bend* it (see Section 3.4.5). If the lens is thin, changing all the surface curvatures by the same amount has the effect of changing the lens shape while leaving the focal length and the chromatic aberration unchanged. Generally spherical aberration varies with bending in a parabolic fashion when plotted against some reasonable shape parameter such as c_1. At extreme bendings either to the left or right, a thin positive lens is decidedly undercorrected, and the aberration reaches a mathematical maximum at some intermediate bending. The aberration of a single thin lens with a distant object is never zero, but in a positive achromat the aberration exhibits a region of overcorrection at and close to the maximum. To bend a thick lens, it is customary to change all the surface curvatures except the last by a chosen value of Δc, the last radius being then solved by the ordinary angle solve procedure (Section 3.1.4) to maintain the paraxial focal length.

DESIGNER NOTE

This procedure, of course, slightly affects the chromatic aberration but it alters the spherical aberration far more. It should be noted, however, that if the aberration is at the maximum, then quite a significant bending will have little or no effect on the spherical aberration. When this condition exists, bending can be used as an effective design tool to vary other aberrations such as coma or field curvature while minimally impacting the spherical aberration.

6.1.5. A Single Lens Having Minimum Spherical Aberration

A single positive lens can be made to have minimum spherical aberration at one wavelength by taking a series of bendings, in each case solving for the last radius to hold focal length. When this is done, it is found that in the minimum-aberration lens each surface contributes about the same amount of aberration, with the front surface (in parallel light) contributing slightly more than the rear surface.

As an example, suppose we wish to design such a lens of focal length 10 and aperture $f/4$, using glass of index 1.523. A suitable thickness is 0.25. The two surface contributions become equal at $c_1 = 0.1648$, for which bending the total aberration is found to be -0.15893. A careful plot shows that the true minimum occurs at $c_1 = 0.1670$ with an aberration of -0.15883, but of course the difference between these two values of the aberration is utterly insignificant. We shall therefore make no noticeable error by aiming at equal contributions for the minimum bending. The error may become much greater, however, in lenses made from high-index materials for the infrared.

The results of ray tracing with $Y_1 = 1.25$ are as shown in the following:

c_1	0.15	0.16	0.17	0.18
Solved c_2	−0.041742	−0.031639	−0.021519	−0.011380
Spherical aberration contribution (1)	−0.05977	−0.07267	−0.08730	−0.10376
Spherical aberration contribution (2)	−0.10434	−0.08705	−0.07174	−0.05828
Total	−0.16411	−0.15972	−0.15904	−0.16205

6.1.6 A Two-Lens Minimum Aberration System

A considerable reduction in spherical aberration can be achieved by taking two identical lenses of twice the desired focal length and mounting them close together. In our case this procedure, after scaling to a focal length of 10, gave a spherical aberration of −0.0788, about half that of the original single element. However, a much greater improvement can be made by bending the second lens so that each of the four refracting surfaces contributes an identical amount of aberration. The required condition is that each surface should have the same value of $(Q' - Q)ni$, since it is this product that determines the aberration contribution of the surface. When computed manually, the curvature of each surface is determined by a few trials, and then, if the resulting focal length is not correct, the whole lens is scaled up or down until it is.

As an example, suppose we add another element to the single minimum-aberration lens of Section 6.1.5. Finding c_3 and c_4 by trial to make all four contributions equal gives the lens shown in Figure 6.10 that has the following prescription and spherical aberration contributions:

c	d	n	Spherical aberration contribution for $Y_1 = 1.25$
0.1648			−0.01703
	0.25	1.523	
−0.02678			−0.01702
	0.05	(air)	
0.3434			−0.01703
	0.25	1.523	
0.1216			−0.01700

The focal length is now 4.6155 and the aperture is $f/1.85$. Scaling the lens to a focal length of 10.0 and tracing a ray at $Y_1 = 1.25$ ($f/4$) gives the total spherical aberration as −0.0310, about one-fifth that of the single element. It is worth noting that the focal lengths of the two elements are now not equal, being respectively 21.7 and 18.4.

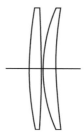

Figure 6.10   A two-lens minimum aberration system.

There is a common misconception regarding this type of two-element lens—namely, that to secure minimum spherical aberration, the marginal ray must be deviated equally at each of the four surfaces. To see how far this is from the truth, these are the surface ray deviations for the last example:

Surface	Angle U (deg)	Angle U' (deg)	Deviation $U' - U$ (deg)
1	0	1.877	1.877
2	1.877	3.318	1.441
3	3.318	6.019	2.701
4	6.019	7.195	1.176

The reason the third surface does so much refracting "work" without the introduction of excessive aberration is its close proximity to the aplanatic condition.

It should be noted that when designing a two-element infrared lens with a material having a refractive index higher than about 2.5, such as silicon or germanium, it will be found that if r_3 is chosen to give the maximum possible overcorrection, it may actually overcompensate the undercorrection of the front minimum-aberrations lens, making it possible to correct the spherical aberration completely. The last radius is then chosen to have its center of curvature at the final image to eliminate any aberration there.

As an example, we will design an $f/1$ lens made of silicon having a refractive index of 3.4. Following the suggested procedure, we come up with the prescription that follows Figure 6.11. This figure shows a longitudinal section of this lens. The strong rear element is highly meniscus, as can be seen. High-index materials such as silicon and germanium appear to behave quite oddly to anyone only familiar with the properties of ordinary glass lenses.

Figure 6.11   An *f*/1 silicon lens.

c	d	n	Spherical aberration contribution	
0.02790			−0.006017	$f' = 10.283$
	0.25	3.4		
0.01572			−0.006004	$l' = 9.717$
	0.05	(air)		
0.12632			+0.012009	aperture $= 10$ (*f*/1)
	0.50	3.4		
0.10291			0	

Focal length of front component, 33.99; of rear component, 14.88

6.1.7 A Four-Lens Monochromat Objective

As was stated in Section 6.1.2, a single aplanatic lens element for use in parallel light is nothing but a planoparallel plate and not a lens at all. However, by making use of the small overcorrection that can be obtained from a convex surface slightly weaker than a true aplanat, it is possible to construct an aplanatic system for use with a distant object by placing a minimum-aberration lens first, and following this by a series of overcorrected menisci in the converging beam produced by the first lens.

As an example we may take the single minimum-aberration *f*/4 lens in Section 6.1.5, and follow it with three menisci, the front face of each being chosen to give the maximum of overcorrection, while the rear faces are perpendicular to the marginal ray. Nothing is gained by departing from the strict perpendicular condition for the rear surfaces of the menisci because, being dispersive surfaces, any departure from perpendicularity in either direction would yield spherical undercorrection, which is just what we are trying to avoid.

After several trials to obtain the greatest possible amount of overcorrection, and finally scaling to $f' = 10.0$ with an aperture of *f*/2, we obtain the following system:

c	d	n	Spherical aberration contribution at $f/2$	
0.066014^a			-0.020622	⎫
	0.3	1.523		⎬ -0.041232
-0.0103636^a			-0.020610	⎭
	0.05	(air)		
0.082192			$+0.002463$	
	0.3	1.523		
0.055672			0	
	0.05	(air)		
0.113932			$+0.005962$	
	0.3	1.523		
0.077543			0	
	0.05	(air)		
0.158867			$+0.014476$	
	0.3	1.523		
0.109134			0	
		Total	-0.018331	

^{a}*Crossed lens* in parallel light (see Section 6.3.2).

The focal length of the first lens alone is now 24.969. It is clear that even with three menisci it is not possible to compensate for the undercorrection of the first lens.

However, we can do much better by starting with the two-lens minimum aberration form given in Section 6.1.6, and following this with only two menisci. By this procedure we can design a four-lens spherically corrected system for use in parallel light with an aperture as high as $f/2$. Scaled to $f' = 10$ this becomes

c	d	n	Spherical aberration contribution at $f/2$
0.041520			-0.005090
	0.3	1.523	
-0.006726			-0.005098
	0.05		
0.084883			-0.005106
	0.3	1.523	
0.029164			-0.005098
	0.05		
0.113764			$+0.005966$
	0.3	1.523	
0.077891			0
	0.05		
0.159353			$+0.014387$
	0.3	1.523	
0.109941			-0.000016
		Total	-0.000068

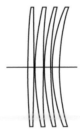

Figure 6.12 A four-lens *f*/2 aplanatic objective.

This lens is shown in Figure 6.12. The focal length of the first two lenses is now 18.380. This system has been used in monochromat microscope objectives made of quartz for use at a single wavelength in the ultraviolet. The design has been discussed by Fulcher.[3]

6.1.8 An Aspheric Planoconvex Lens Free from Spherical Aberration

Two cases arise, the first when the curved aspheric surface faces the distant object, and the other when the plane surface faces the object. In 1637, Descartes described and explained the general properties of utilizing concave and convex lenses, both singularly and in combination. He was the first to create a mathematical formulation to explain spherical aberration. Descartes also made a detailed study of elliptical and hyperbolic surfaces, particularly the plano-hyperbolic lens. Descartes and his colleagues spent substantial money and effort trying to fabricate such a lens since it would be free of spherical aberration.

For all their efforts, not one lens could be fabricated with the tools and methods then available. Fortunately, technology has advanced to allow fabrication of both elliptical and hyperbolic surfaces of high quality. The topic of perfect imaging from one point to another point was first addressed by utilizing Fermat's principle, which states that perfect conjugate imaging occurs when all the rays passing through the conjugate points have the same optical path length. Using Fermat's principle, Luneburg[4] showed in 1944 that the surface can be represented by a fourth-order curve, which is also known as the Cartesian Oval after René Descartes.

Convex to the Front

The left side of the ellipse shown in Figure 6.13a is the portion of the ellipse used as the surface contour for the ellipsoid-plano lens with the image being formed at the rear surface of the lens, which is planar. This lens has no spherical

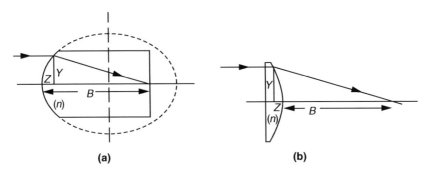

(a)   **(b)**

Figure 6.13   Aspheric single lenses corrected for spherical aberration.

aberration at the design wavelength for collimated light input and does suffer
from spherochromatism. The coordinate system used for ray tracing has the
coordinate system origin located at the left surface vertex of the ellipsoid. The
z-axis is the major axis of ellipsoid and the x-axis and y-axis define the vertex
tangent plane. The sag or z-coordinate displacement from the tangent plane at
the vertex of the surface can be determined from a commonly used mathematical representation of conical surfaces given by

$$z = \frac{y^2 c}{1 + \sqrt{1 - (1 + \kappa)c^2 y^2}}$$

where y is the coordinate of the ray intercept on the surface, c is the radius of
curvature (reciprocal of the radius) at the surface vertex, and κ is the conic constant (see Eq. (2-7)). This equation form is typically used in optical design programs with c (or r) and κ as an input description of the surface.

Figure 6.14 presents the basic parameters describing an ellipse.[5] These
include the lengths of its major and minor axes, and foci. The distance d can

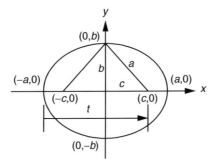

Figure 6.14   Geometrical parameters of an ellipse.

be considered the focal length of the elliptical lens and is the sum of the major axis semi-length a and the foci distance c from the ellipse center.

A conic section can also be represented by

$$\frac{\left(x - d\dfrac{n_1}{n_0 + n_1}\right)^2}{d^2\left(\dfrac{n_1}{n_0 + n_1}\right)^2} + \frac{z^2}{d^2\dfrac{n_1 - n_0}{n_0 + n_1}} = 1$$

and is an ellipse if $n_1 > n_0$ and a hyperbola if $n_0 > n_1$. This surface is called a rotationally symmetric surface. Examination of this equation indicates that the major a and minor b axes' semi-lengths can be computed from

$$a^2 = d^2\left(\frac{n_1}{n_0 + n_1}\right)^2$$

and

$$b^2 = d^2\frac{n_1 - n_0}{n_0 + n_1}.$$

By basic geometry, the foci distances are computed by

$$c^2 = a^2 - b^2$$

and by substituting in the terms for a^2 and b^2, we obtain

$$c^2 = d^2\left(\frac{n_0}{n_0 + n_1}\right)^2.$$

From geometry, the eccentricity $\varepsilon = \frac{c}{a}$ which is determined by using the preceding equations to be $\varepsilon = \frac{n_0}{n_1}$. The conic constant κ is defined as

$$\kappa = -\varepsilon^2 = \frac{-n_0^2}{n_1^2}.$$

Again from geometry (semi-latus rectum), the vertex radius r is given by

$$r = \pm\frac{b^2}{a} = d\left(\frac{n_1 - n_0}{n_1}\right).$$

It should be evident that the same relationship between r, t, n_0, and n_1 can be determined using paraxial optics and a purely spherical surface. This result should be expected when the aperture is very small. The inclusion of the conic constant to transform the surface into an ellipsoid does not change the first-order properties of this lens, but does mitigate the inherent spherical aberration of a spherical surface.

Consider an example where $n_0 = 1$ and $n_1 = 1.5$, and $d = 20$ mm. It follows that

$$\kappa = \frac{-n_0^2}{n_1^2} = \frac{-1}{1.5^2} = -0.44444$$

$$r = d\left(\frac{n_1 - n_0}{n_1}\right) = 20\left(\frac{0.5}{1.5}\right) = 6.66666$$

A surface of this kind has long been used on highway reflector "buttons."[6] The same surface profile can be used to form a cylindrical lens, which has various applications. An array of such lenses can be created to form a lenticular array commonly used in the printing industry to make prints providing 3D photographic projections or the display of different images as the print is tilted.[7]

Consider a situation when an ellipsoid-plano lens is bonded to one or more materials. An example would be a detector array affixed to the plano side of the lens using an optical glue which can have different refractive index from the lens. The lens thickness would need to be reduced to account for the thickness of the glue. From basic aberration theory, the different refractive indices of the lens and substrate materials can introduce additional spherical aberration since the light beam is converging (see Section 6.4).

To compensate for the different refractive indices, the equation relating r, n, and d can be modified as presented in the following equation to determine the radius r of the lens when the composite optical element comprises two or more different materials.

$$r = (n_1 - 1)\left(\frac{d_1}{n_1} + \frac{d_2}{n_2} + \ldots \frac{d_n}{n_n}\right)$$

This equation is readily derived using paraxial ray tracing. Note that the substrates are each assumed to be planar. The summation shown in the parentheses is the effective optical thickness of the assemblage. If the substrate thicknesses are small compared to the lens thickness d_1, then the conic constant can be estimated to be given by $-n_1^{-2}$.

In the event that the substrate thicknesses are a significant fraction of the lens thickness, the conic constant estimation is somewhat more complex in that an estimate of the effective refractive index is required. Consider that a paraxial ray having $n_0 u_0 = 0$ is incident on the refracting surface at a height y_1 where u_0 is the angle the ray makes with the optical axis. As this ray propagates through the assemblage, it intercepts the substrates at heights $y_2, y_3, \ldots, y_n$.

The effective refractive index n_{eff} can then be expressed, following the mean value theorem of integral calculus, as

$$n_{\text{eff}} = \frac{\sum\limits_{i=1}^{n} y_i t_i n_i}{\sum\limits_{i=1}^{n} y_i t_i}$$

and the corresponding conic constant is $-n_{\text{eff}}^{-2}$.

Figure 6.15 illustrates a lens system comprising a spherical refracting surface and two substrates while Figure 6.16 shows the same lens system with the effective conic constant. The prescriptions of these lens systems are provided in Table 6.2 where the only difference between them is the conic constant. As it is evident by examining Figure 6.15, the lens system with spherical refracting surface suffers from significant spherical aberration. The inclusion of the effective conic constant effectively mitigates the spherical aberration illustrated in Figure 6.16.

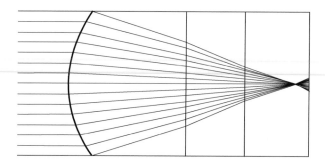

Figure 6.15 Lens system with multiple substrates and spherical refractive surface.

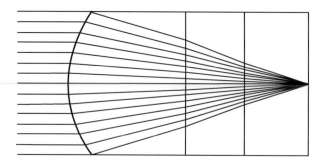

Figure 6.16 Lens system with multiple substrates and elliptical refractive surface.

Table 6.2

Prescription of Lens System Depicted in Figures 6.15 and 6.16

Surface	Radius	Thickness	Refractive index	Marginal ray height	Conic constant
Object	∞	∞	1		
1	10	10	1.8	6	−0.34542
2	∞	5	1.4	3.33333	
3	∞	5.396825	1.6	1.61905	
Image	∞			0	

A solid-optics (containing no air gaps) lens system for making an afocal telescope can be designed by using the foregoing information. Two ellipsoid-plano lenses would be placed with their plano sides facing one another and some optical bonding material of finite thickness placed between them after having accounted for the several thicknesses so that the focal point of the two lenses coincide. The lenses can be of different materials and radii, and the angular magnification is simply the ratio of their effective focal lengths.

Plane Surface in Front

Equating optical paths in the air behind the lens shown in Figure 6.13b gives

$$B + nZ = [Y^2 + (B + Z)^2]^{1/2}$$

where

$$\frac{\{Z + [B/(n+1)]\}^2}{[Bn/(n+1)]^2} - \frac{Y^2}{B^2(n-1)/(n+1)} = 1$$

There is a clear resemblance between these two cases. The plane-in-front lens has a hyperbolic surface with semimajor axis equal to $B/(n+1)$, and semiminor axis equal to $B[(n-1)/(n+1)]^{1/2}$ as before (Figure 6.13b), the eccentricity now being equal to the refractive index n and the conic constant being $\kappa = -n^2$.

Using a y-nu ray trace, it is trivial to show that the focal length of a plano-hyperboloid lens is $\frac{-r}{n-1}$ which is shown as B in Figure 6.13b. To create a solid optical element for use at finite magnification, a hyperbolic surface can be applied on both faces of a biconvex lens with the ratio of focal lengths being the ratio of the radii. The light is of course collimated between the faces so there is no fundamental restriction of the lens thickness. The axial image is free of spherical aberration for up to as high an aperture as required; however, the field of this lens is restricted by coma.

6.2 ZONAL SPHERICAL ABERRATION

As we have seen, it is possible by the use of opposing positive and negative elements to design a lens such that the focus of the marginal ray coincides with the paraxial image point. We say that this lens has zero spherical aberration. However, it generally happens that the foci of rays passing through the intermediate zones of the lens fall closer to the lens than the paraxial image-point, and occasionally but rarely fall further from it. Thus we can plot a graph connecting entrance height Y with the spherical aberration, as shown in Figure 6.17. This zonal residual is generally known as *zonal aberration*. It can be expressed as a power series containing only even powers of Y, as

$$LA' = aY^2 + bY^4 + cY^6 + \ldots$$

The successive terms of this series have been called primary, secondary, tertiary ... aberration, but of course they have no separate existence, and the actual aberration of the lens is the sum of all these terms. However, we can plot them separately to see how they vary (Figure 6.17). If Y is small, the secondary and higher terms are very small or negligible, and the primary term represents

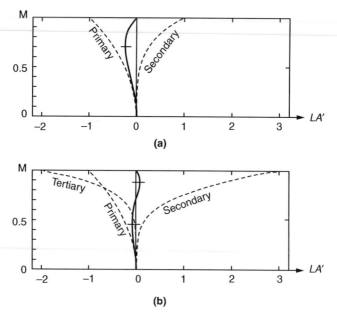

Figure 6.17 Interaction of various orders of spherical aberration: (a) Primary and secondary only; (b) primary, secondary, and tertiary.

the whole aberration. Then, at increasing values of Y, first the secondary and then the tertiary terms begin to increase and finally dominate the situation.

In the example shown in Figure 6.17a, the primary term is negative and the secondary term is positive, and they have equal and opposite values for the marginal ray. Consider now when only the first two terms are present and the marginal spherical aberration is zero, that is, $LA' = a\rho^2 + b\rho^4 = 0$ when $\rho = 1$. This implies that $a = -b$. Now the peak residual can be found by solving $\frac{dLA'}{d\rho} = 2a\rho + 4b\rho^3 = 0$. Now dividing by ρ and substituting $a = -b$, it follows that $2a + 4b\rho^2 = -2b + 4b\rho^2$ which yields that $\rho = 1/\sqrt{2} = 0.707$. In an actual lens system, the peak zonal residual occurs when ρ is equal to the marginal ρ_m multiplied by 0.7071. The magnitude of the zonal residual, in the case of suffering only third- and fifth-order spherical aberration, equals one-quarter of the primary term at the marginal zone of the lens, that is, $LA'_{0.707} = a/4$.

Because tertiary aberration is not greatly different from secondary, it may be positive and add to the secondary; in this case the maximum zonal residual falls higher than the 0.7 zone, and the marginal aberration increases very rapidly. On the other hand, if the tertiary aberration is negative, it tends to oppose the secondary, and it is then possible to eliminate both the marginal and the zonal aberrations, as indicated in Figure 6.17b. It will be noticed that the secondary and tertiary aberrations are now much larger than in the simple case of Figure 6.17a, but the resulting aberration curve is nearly flat, having small equal and opposite residuals above and below the 0.7 zone. An analysis of the situation reveals that the maximum and minimum residuals fall at values of ρ given by

$$\frac{\rho}{\rho_m} = \sqrt{\frac{1 \pm 1/\sqrt{3}}{2}} = 0.8881 \text{ and } 0.4597$$

The locations of the maximum residuals are indicated by short horizontal lines on these diagrams.

DESIGNER NOTE

As a consequence of the nature of the expansion of spherical aberration being in even orders (for longitudinal form), it is virtually always true that the signs of the coefficients a, b, and c must alternate to achieve correction for the marginal ray. This is seen in Figure 6.17a for primary and secondary aberration only and Figure 6.17b when the tertiary aberration is present. Examination of the ray plot for a lens can tell the lens designer what orders of spherical aberration are present and if they have the correct signs to achieve correction.

The effect of refocusing when only primary spherical aberration is present was shown in Figure 6.3. Consider now the effect of refocusing when both primary and secondary spherical aberration are present, which will be seen to be more complicated. When only primary spherical aberration is present, it is fairly evident what the optimum refocus should be. With the presence of both primary and secondary aberration, the optimum refocus is not so simple.

Consider Figure 6.18 that shows the transverse ray errors versus the normalized entrance pupil radius for the case where the marginal spherical aberration is zero when $\rho = 1$ (curve A) and the case where the marginal spherical aberration is zero when $\rho = 1.12$ (curve B). The first case represents what lens designers often attempt to achieve, that is, having the marginal-ray error equal zero in the paraxial image plane $\left(\varepsilon_y = \sigma_1\rho^3 + \mu_1\rho^5 = 0 \text{ when } \rho = 1\right)$. The refocus boundaries are shown and represent the blur diameter that contains 100% of the energy. If a different refocus is used (slope of the boundary lines indicates the amount of refocus) in this case, a brighter core can be obtained. This is illustrated in Figure 6.18. Since these boundary lines intersect curve A at about $\rho = \pm0.9$, this bright core region contains about 80% of the energy and would provide an improvement in resolution.

The remaining 20% of the energy would be spread around this bright core to form a dim flare having a diameter about five times larger than the core. Now consider the second case which is illustrated as curve B in Figure 6.18. It can be shown that if $\varepsilon_y = \sigma_1\rho^3 + \mu_1\rho^5 = 0$ when $\rho = 1.12$, then the smallest 100% blur diameter is obtained when refocused and it is about 50% larger than the bright core for

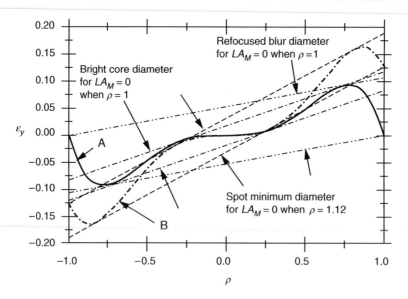

Figure 6.18 Geometric blur for third- and fifth-order spherical aberration for zero marginal spherical aberration when (curve A) $\rho = 1$ and (curve B) $\rho = 1.12$.

curve A. Which of the preceding cases and amount of refocus are best for a given application must be determined by the lens designer. It is important to note that achieving a marginal-ray error equal to zero in the paraxial image plane is not always appropriate.

6.3 PRIMARY SPHERICAL ABERRATION

6.3.1 At a Single Surface

To isolate the primary term, we would have to make Y of infinitesimal magnitude, and then we cannot use the formula in Eq. (6-4) to compute the aberration for the same reason that we cannot trace a paraxial ray by the ordinary ray-tracing formulas. However, the primary term can be determined as a limit:

$$LA'_{\text{primary}} = \lim_{y \to 0}(LA'_y)$$

To find this limit, we use paraxial ray data to fill in the numbers in the accurate version of Eq. (6-5). Making this substitution gives the primary aberration equation:

$$LA'_p = LA_p \left(\frac{n_1 u_1^2}{n'_k u'^2_k}\right) + \sum \frac{2y \cdot \frac{1}{2}(i' - i) \cdot \frac{1}{2}(i' + u)ni}{n'_k u'^2_k}$$

Here LA_p is the primary aberration of the object, if any; it is transferred to the final image by the ordinary longitudinal magnification rule. The quantity under the summation sign is the primary aberration arising at each surface.

These *surface contributions* (*SC*) can be written

$$SC = yni(u' - u)(i + u')/2n'_k u'^2_k \qquad (6-6)$$

Only paraxial ray data are required to evaluate this formula. To interpret it, we note that for pure primary aberration,

$$LA'_p = aY^2$$

and that the radius of curvature of the spherical-aberration graphs in Figure 6.17, at the point where the graph crosses the axis, has the value

$$\rho = Y^2/2LA'_p = 1/2a$$

Therefore, the coefficient of primary aberration a is an inverse measure of twice the radius of curvature of the spherical-aberration graph at the point where it crosses the lens axis. Hence, by tracing one paraxial ray, we not only discover the location of the image point, but we also ascertain the shape of the aberration curve as it crosses the axis at that point. It is remarkable how much information can be obtained from so very little ray-tracing effort.

Table 6.3

Calculation of Primary Spherical Aberration

y	2		1.9031479		1.8809730	
n	1		1.517		1.649	
$yc + u = i$	0.2706542		−0.4597566		−0.1713855	
u	0	−0.0922401		−0.0554372		−0.1666664
y	2		1.9031479		1.8809730	
ni	0.2706542		−0.6974508		−0.2826147	
$u' - u$	−0.0922401		0.0368029		−0.1112292	
$i + u'$	0.1784141		−0.5151938		−0.3380519	
$1/2u_k'^2$	18		18		18	
Product $= SC$	−0.160349		0.453014		−0.359792	$\Sigma = -0.067127$

As an example of the use of this formula, we will calculate the primary spherical aberration contributions of the three surfaces of the cemented doublet shown in Section 2.5 that we have already used several times (see Table 6.3). It is interesting to compare these primary aberration contributions with the exact contributions given in Section 6.1. The contributions are as follows:

Surface	1	2	3	Sum
Exact contribution	−0.16687	+0.55418	−0.37926	0.00805
Primary contribution	−0.16035	+0.45301	−0.35979	−0.06713
Difference (contribution of higher orders)	−0.00652	+0.10117	−0.01947	

At each surface the true and primary contributions are similar in magnitude and have the same sign, but the cemented interface shows the greatest difference. This is due to the presence of a significant amount of secondary and higher-order aberrations there, while the outer surfaces show very little sign of higher-order aberrations. It is the presence of the considerable amount of higher-order aberrations at the cemented interface that is the cause of the large zonal aberration of this lens. Examination of curve D in Figure 5.3a indicates that the spherical aberration is comprised of almost totally primary and secondary contributions.

6.3.2 Primary Spherical Aberration of a Thin Lens

By combining the SC values for the two surfaces of a thin lens element, we find that a thin lens, or a thin group of lenses in close contact, within a system contributes the following amount to the primary spherical aberration at the final image:

$$SC = -\frac{y^4}{n_0' u_0'^2} \sum (G_1 c^3 - G_2 c^2 c_1 + G_3 c^2 v_1 + G_4 c c_1^2 - G_5 c c_1 v_1 + G_6 c v_1^2) \quad (6\text{-}7)$$

where the terms with suffix 0 refer to the final image, the other terms applying to each single element. Here c and c_1 have their usual meanings, namely, $c_1 = 1/r_1$ and $c = 1/f'(n-1)$. The symbol v_1 is the reciprocal of the object distance of the element, and the Gs are functions of the refractive index, namely,

$$G_1 = \tfrac{1}{2} n^2 (n-1), \qquad G_2 = \tfrac{1}{2}(2n+1)(n-1),$$

$$G_3 = \tfrac{1}{2}(3n+1)(n-1), \quad G_4 = (n+2)(n-1)/n,$$

$$G_5 = 2(n^2-1)/n, \qquad G_6 = \tfrac{1}{2}(3n+2)(n-1)/n$$

The details of the derivation of this formula have been given by Conrady.[8] The summation sign in Eq. (6-7) is used only if the thin lens contains more than one element, for example, if it is a thin doublet or triplet; otherwise it may be omitted. If there is more than one element we must assume a very thin layer of air to exist between the elements in place of cement, c_1 being the curvature of the first surface of each element and v_1 being the reciprocal of the object distance in air. Thus for the second lens of a cemented doublet we take

$$(c_1)_b = (c_1)_a - c_a \quad \text{and} \quad (v_1)_b = (v_1)_a + c_a(n_a - 1)$$

In the case of an isolated thin element or thin system in air, not forming part of a more complex system, $n_0' = 1$ and $u_0' = y/l'$. Also the aberration of the object (if any) must be transferred to the image and added to the new aberration arising at the lens. Thus in such a case we have

$$LA_p' = LA_p \left(\frac{l'}{l}\right)^2 - y^2 l'^2 \sum (G \text{ sum})$$

the (G sum) referring to the six-term expression in parentheses in Eq. (6-7).

By use of the G-sum formula we can plot a graph showing how the primary spherical aberration of a thin lens varies with the bending (Figure 6.19). For a single thin positive element, this graph is a vertical parabola, the vertex of which nearly but not quite reaches the zero aberration line.

The thin single lens having the minimum primary spherical aberration is called a *crossed lens*. Its shape can be found by differentiating the G-sum expression with respect to c_1 yielding

$$c_1 = \frac{\tfrac{1}{2} n(2n+1)c + 2(n+1)v_1}{n+2}$$

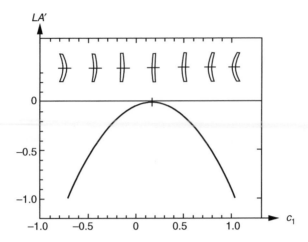

Figure 6.19 Effect of bending on spherical aberration.

For the special case of a very-distant object, $v_1 = 0$, we find that $c_1/c = n(2n + 1)/2(n + 2)$ and $c_2/c_1 = (2n^2 - n - 4)/n(2n + 1) = r_1/r_2$. For glass having an index of 1.6861, the crossed lens is exactly planoconvex; however, for other glass indices, the departure from the planconvex form is slight. The very high-refractive indices of infrared materials cause the crossed lens to be a deeply curved thin meniscus.

It is well-known that the spherical aberration of a lens is a function of its shape factor or bending. Although several definitions for the shape factor have been suggested (see Section 3.4.5), a useful formulation is

$$\chi = \frac{c_1}{c_1 - c_2} \tag{6-8}$$

where c_1 and c_2 are the curvatures of the lens, with the first surface facing the object. By adjusting the lens bending, the spherical aberration can be seen to have a minimum value.

The power of a thin lens or the reciprocal of its focal length is given by

$$\phi = \frac{(n - 1)c_1}{\chi} \tag{6-9}$$

When the object is located at infinity, the shape factor for minimum spherical aberration can be represented by

$$\chi = \frac{n(2n + 1)}{2(n + 2)} \tag{6-10}$$

and

$$\frac{c_2}{c_1} = \frac{2n^2 - n - 4}{n(2n + 1)}.$$

The resultant third-order spherical aberration of the marginal ray in angular units is

$$SA3 = \frac{n^2 - (2n+1)k + (1+2/n)\chi^2}{16(n-1)^2(f\text{-number})^3} \tag{6-11}$$

or after some algebraic manipulations,

$$SA3 = \frac{n(4n-1)}{64(n+2)(n-1)^2(f\text{-number})^3} \tag{6-12}$$

When the object is located at a finite distance s_0, the equations for the shape factor and residual spherical aberration are more complex. Recalling that the magnification m is the ratio of the object distance to the image distance and that the object distance is negative if the object lies to the left of the lens, the relationship between the object distance and the magnification is

$$\frac{1}{s_0\phi} = \frac{m}{1-m} \tag{6-13}$$

where m is negative if the object distance and the lens power have opposite signs. The term $1/s_0\phi$ represents the reduced or ϕ-normalized reciprocal object distance v_1; that is, s_0 is measured in units of focal length, ϕ^{-1}. The shape factor for minimum spherical aberration is given by

$$\chi = \frac{n(2n+1)}{2(n+2)} + \frac{2(n^2-1)}{n+2}\left(\frac{m}{1-m}\right) \tag{6-14}$$

and the resultant third-order spherical aberration of the marginal ray in angular units is

$$SA3 = \frac{1}{16(n-1)^2(f\text{-number})^3}\left[n^2 - (2n+1)\chi + \frac{n+2}{n}\chi^2 + (3n+1)(n-1)\left(\frac{m}{1-m}\right)\right.$$
$$\left. -\frac{4(n^2-1)}{n}\left(\frac{m}{1-m}\right)\chi + \frac{(3n+2)(n-1)^2}{n}\left(\frac{m}{1-m}\right)^2\right] \tag{6-15}$$

and

$$c_1 = \frac{\frac{1}{2}n(2n+1)c + 2(n+1)v_1}{n+2}. \tag{6-16}$$

When the object is located at infinity, the magnification becomes zero and the above equations reduce to those previously given.

Figure 6.20 illustrates the variation in shape factor as a function of reciprocal object distance for refractive indices of 1.5 to 4 for an f-number $= 1$.[9] Notice that lenses have a shape factor of 0.5 regardless of the refractive index when

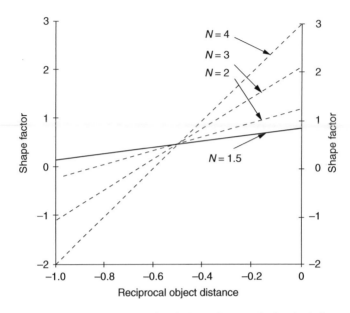

Figure 6.20 The shape factor for a single lens is shown for several refractive indexes as a function of reciprocal object distance v_1 where the distance is measured in units of focal length.

the magnification is -1 or $v_1 = -0.5$. For this shape factor, all lenses have biconvex surfaces with equal radii. When the object is at infinity, a lens having a refractive index of 1.5 has a somewhat biconvex shape with the second surface having a radius about six times greater than the first surface radius.

Since the minimum-spherical lens shape is selected for a specific magnification, the spherical aberration will vary as the object-image conjugates are adjusted. For example, a lens having a refractive index of 1.5 and configured for $m = 0$ ($v_1 = 0$ and image at f) exhibits a substantial increase in spherical aberration when the lens is used at a magnification of -1. Figure 6.21 illustrates the variation in the angular spherical aberration as both a function of refractive index and reciprocal object distance when the lens bending is for minimum spherical aberration with the object located at infinity. As can be observed from Figure 6.21, the ratio of the spherical aberration, when $m = -0.5$ and $m = 0$, increases as n increases.

Figure 6.22 shows the variation in angular spherical aberration when the lens bending is for minimum spherical aberration at a magnification of -1. In a like manner, Figure 6.23 presents the variation in angular spherical aberration for a convex-plano lens with the plano side facing the image. The figure can also be used when the lens is reversed by simply replacing the object distance with the image distance. For these plots, the actual aberration value is determined by dividing the aberration value shown by $(f\text{-number})^3$.

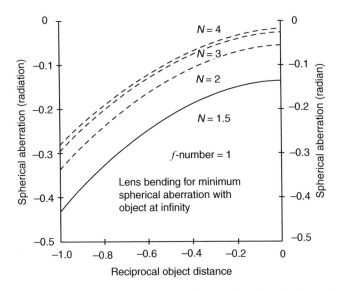

Figure 6.21 Variation of angular spherical aberration as a function of reciprocal object distance v_1 for various refractive indices when the lens is shaped for minimum spherical aberration with the object at infinity. Spherical aberration for a specific *f*-number is determined by dividing the aberration value shown by $(f\text{-number})^3$.

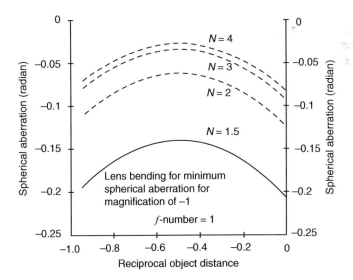

Figure 6.22 Variation of angular spherical aberration as a function of reciprocal object distance v_1 for various refractive indices when the lens is shaped for minimum spherical aberration for a magnification of -1. Spherical aberration for a specific *f*-number is determined by dividing the aberration value shown by $(f\text{-number})^3$.

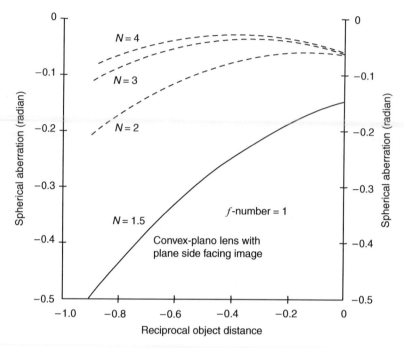

Figure 6.23 Variation of angular spherical aberration as a function of reciprocal object distance v_1 for various refractive indices when the lens has a convex-plano shape with the plano side facing the object. Spherical aberration for a specific *f*-number is determined by dividing the aberration value shown by $(f\text{-number})^3$.

6.4 THE IMAGE DISPLACEMENT CAUSED BY A PLANOPARALLEL PLATE

From Figure 6.24 it is clear that the longitudinal image displacement caused by the insertion of a thick planoparallel plate into the path of a ray having a convergence angle U is $S = BB'$, given by

$$S = \frac{Y}{\tan U'} - \frac{Y}{\tan U}$$
$$= \frac{Y}{\tan U'}\left(1 - \frac{\tan U'}{\tan U}\right)$$

But $Y/\tan U'$ is equal to t, the thickness of the plate. Therefore,

$$S = t\left(1 - \frac{\tan U'}{\tan U}\right) = \frac{t}{N}\left(N - \frac{\cos U}{\cos U'}\right)$$

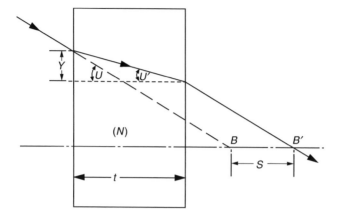

Figure 6.24 Image displacement caused by the insertion of a parallel plate.

where N is the refractive index of the plate. For a paraxial ray this reduces to

$$s = \frac{t}{N}(N-1)$$

Since $\sin U = n \sin U'$ and $\cos^2 U' + \sin^2 U' = 1$, it follows that

$$\frac{\cos U}{\cos U'} = \frac{n \cos U}{\sqrt{n^2 - \sin^2 U}}.$$

The exact spherical aberration is

$$S - s = \frac{t}{n}\left(1 - \frac{n \cos U}{\sqrt{n^2 - \sin^2 U}}\right).$$

The plate of glass occupies more space than its "air equivalent," which is defined as that thickness of air in which a paraxial ray drops or rises by the same amount as in the glass plate. Thus, the useful relationship,

air equivalent = glass thickness/refractive index

DESIGNER NOTE

The inclusion of a flat glass plate in an optical system can impact the ultimate image quality. Microscope cover glasses and dewar windows for infrared detectors are examples where the aberrations induced by a flat glass plate should be accounted for. Consider a flat glass plate placed between the plano-hyperboloid lens in Section 6.1.8 "Plane Surface in Front" and the image it forms. In this case, the added spherical aberration can be effectively mitigated by slightly weakening the conic constant of the lens.

DESIGNER NOTE

To the lens designer a reflecting prism in a system behaves as though it were a very thick parallel plate. In a converging beam a prism has the effect of overcorrecting the three astigmatic aberrations (spherical, chromatic, and astigmatism) while it undercorrects the comatic aberrations (coma and distortion), and lateral color.

6.5 SPHERICAL ABERRATION TOLERANCES

6.5.1 Primary Aberration

Conrady has shown[10] that if a lens suffers from a small amount of pure primary spherical aberration, the best-fitting reference sphere will touch the emerging wavefront at the center and edge, and the plane of best focus will lie midway between the marginal and paraxial image points. If the aberration is large compared with the Rayleigh limit, geometrical considerations dominate, and the geometrical circle of least confusion becomes the "best" focus. However, there is also a secondary best focus close to the paraxial focus; this has been amply verified by experiment.[11]

In the case of pure primary aberration, the magnitude of the maximum residual *OPD* at this best focus is equal to the Rayleigh quarter-wave limit when

$$LA' = 4\lambda/\sin^2 U'_m = 16\lambda(f\text{-number})^2 \qquad (6\text{-}17)$$

where f-number = focal length/diameter of aperture.

This aberration tolerance is surprisingly large, being four times the extent of the focal range. Some typical values for $\lambda = 0.0005$ mm are given in the following tabulation:

f-number	4.5	6	8	11	16	22
Primary aberration tolerance (mm)	0.2	0.3	0.5	1.0	2.0	3.9

6.5.2 Zonal Aberration

Conrady has also shown[12] that if a lens is spherically corrected for the marginal ray, the residual zonal aberration will reach the Rayleigh limit if its magnitude is

$$LZA = 6\lambda/\sin^2 U'_m \qquad (6\text{-}18)$$

or 1.5 times the tolerance for pure primary spherical aberration.

DESIGNER NOTE

For telescopes, microscopes, projection lenses, and other visual systems, it is best not to allow any overcorrection of the marginal spherical aberration, even though this would reduce the zonal residual. This is because overcorrection leads to an unpleasant haziness of the image, and the zonal tolerance is so large that it is not likely to be exceeded. Indeed, many projection lenses are deliberately undercorrected, even for the marginal ray, to give the cleanest possible image with maximum contrast. Photographic objectives, on the other hand, are generally given an amount of spherical overcorrection equal to two or three times the zonal undercorrection. The overcorrected haze is often too faint to be recorded on film, especially if the exposure is on the short side, and in any case the lens will generally be stopped down somewhat, which cuts off the marginal overcorrection, leaving a small and often quite insignificant zonal residual.

In this connection, it may well be pointed out that focusing a camera by unscrewing the front element has the effect of rapidly undercorrecting the spherical aberration. This leads to a loss of definition and some degree of focus shift at small apertures, but its convenience to the camera designer outweighs these objections. If the lens is known to be intended for this type of focusing, then it should be designed with a large amount of spherical overcorrection. If possible, the aberration should be well-corrected at a focus distance of about 15 to 20 feet.

6.5.3 Conrady's OPD'_m Formula

Probably the best way to ascertain if a lens is adequately corrected for zonal aberration is to calculate the optical path difference between the emerging wavefront and a reference sphere centered about the marginal image point. Conrady[13] has given a formula by which the contribution of each lens surface to this OPD can be found:

$$OPD'_m = \frac{Yn \sin I \sin \frac{1}{2}(U - U') \sin \frac{1}{2}(I - U')}{2 \cos \frac{1}{2}(U + I) \cos \frac{1}{2} U \cos \frac{1}{2} I \cos \frac{1}{2} U' \cos \frac{1}{2} I'} \tag{6-19a}$$

Referring back to Eq. (6-5), we see that by using the Q method of ray tracing, Conrady's expression can be greatly simplified to

$$OPD'_m = \frac{(Q - Q')n \sin I}{4 \cos \frac{1}{2} U \cos \frac{1}{2} I \cos \frac{1}{2} U' \cos \frac{1}{2} I'} \tag{6-19b}$$

This OPD term has the same sign as the spherical aberration contribution at any surface. If the lens is spherically corrected for the marginal ray, the magnitude of this sum is a measure of the zonal aberration, the sum being positive for a negative zone. The advantage of using the OPD formula is that the tolerance of the sum is known to be two wavelengths. Hence we have an immediate assessment of the significance of the zonal residual; this is much more accurate than the simple zonal tolerance given in Section 6.5.2, which is valid only for a mixture of primary and secondary aberrations.

If the spherical aberration is zero at both margin and 0.7 zone, as in the diagram of Figure 6-17b, then we can determine the seriousness of the two remaining small zones by calculating the *OPD* sum along the marginal ray (which should be zero) and also along the 0.7 zonal ray.

It is important for lens designers to know whether the interferometer software they are using produces Zernike coefficients[14] with a +1 or −1 multiplier with respect to the ray trace software being used. This issue arises from different *OPD* sign conventions being in use and one result in that programs may not agree on the sign of Zernike coefficients even though the absolute value of each like coefficient generated by such programs are the same. Conrady defined *OPD* as optical path length along the reference ray minus the optical path length along the ray under consideration.

ENDNOTES

[1] Transverse field-independent astigmatism is a function of odd orders of ρ while longitudinal field-independent astigmatism is a function of even orders of ρ. Consequently, longitudinal field-independent astigmatism has no defocus component and is purely spherical aberration of the form $\rho^2 + \rho^4 + \ldots$.

[2] E. Delano, "A general contribution formula for tangential rays," *J. Opt. Soc. Am.*, 42:631 (1952).

[3] G. S. Fulcher, "Telescope objective without spherical aberration for large apertures, consisting of four crown glass lenses," *J. Opt. Soc. Am.*, 37:47 (1947).

[4] R. K. Luneburg, *Mathematical Theory of Optics*, pp. 129–133, University of California Press, Berkeley, [reproduced by permission from mimeographed notes issued by Brown University in 1944] (1966).

[5] Robert C. Fisher and Allen D. Ziebur, *Calculus and Analytical Geometry*, pp. 167–201, Prentice-Hall, Englewood Cliffs (1963).

[6] This should not be confused with the so-called cat-eye sign reflector that utilizes spherical glass beads; however, the equation for r that precedes this endnote can be used letting $d = 2r$, which yields that $n = 2$ for retroreflection. However, if a material having $n = 2$ was available, spherical aberration would spoil the return beam. Since it is desirable to have a sign appear illuminated over some reasonably significant viewing angle, using a glass having $n = 1.75$, for example, allows the beam to diverge.

[7] R. Barry Johnson and Gary A. Jacobsen, "Advances in lenticular lens arrays for visual display (Invited Paper)," *Proc. SPIE*, 5874:06-1–11 (2005).

[8] A. E. Conrady, p. 95.

[9] Figures 6.20 to 6.23 after R. Barry Johnson, "Lenses," Section 1.10 in *Handbook of Optics, Second Edition*, Chapter 1, Vol. II, McGraw-Hill, New York (1995).

[10] A. E. Conrady, p. 628.

[11] H. G. Conrady, "An experimental study of the effects of varying amounts of primary spherical aberration on the location and quality of optical images," *Phot. J.*, 66:9 (1926).

[12] A. E. Conrady, p. 631.

[13] A. E. Conrady, p. 616.

[14] C-J Kim and Robert R. Shannon, "Catalog of Zernike Polynomials," Chap. 4 in *Applied Optics and Optical Engineering*, Vol. X, R. Shannon and J. Wyant, Eds., Academic Press (1987).

Chapter 7

Design of a Spherically Corrected Achromat

Since the chromatic aberration of a lens depends only on its power, whereas the spherical aberration varies with bending, it is obviously possible to select that bending of an achromat that will give us any desired spherical aberration (within limits). There are two possible approaches to this design. The first is the four-ray method, requiring no optical knowledge, and the second makes use of a thin-lens study based on primary aberration theory to guide us directly to the desired solution. The latter method is by far the most desirable since it also indicates how many possible solutions there are to any given problem.

7.1 THE FOUR-RAY METHOD

In this procedure we set up a likely first form, which can actually be rather far from the final solution, and determine the spherical aberration by tracing a marginal ray and a paraxial ray in D light, and we calculate the chromatic aberration by tracing 0.7 zonal rays in F and C light. We then make trial changes in c_2 and c_3, keeping c_1 fixed, using a *double graph* to indicate what changes should be made to reach the desired solution. This simple but effective procedure is sometimes called the brute force method; it is especially convenient if a computer is available for ray tracing.[1]

As an example we will use this procedure to design an achromatic doublet with a focal length of 10 and an aperture of 2.0 ($f/5$) using the glasses shown in Table 7.1. The thin-lens (c_a, c_b) formulas in Section 5.4 for an achromat give

$$c_a = 0.5090, \quad c_b = -0.2695$$

and if we assume that the crown element is equiconvex, our starting system will be

$$c_1 = 0.2545, \quad c_2 = -0.2545, \quad \text{and} \quad c_3 = 0.0150.$$

Table 7.1

Glasses for Achromatic Doublet

	n_C	n_D	n_F	Δ_n	V
(a) Crown	1.52036	1.523	1.52929	0.00893	58.6
(b) Flint	1.61218	1.617	1.62904	0.01686	36.6
					$V_a - V_b = 22.0$

Table 7.2

Aberrations for Setup A

$Y = 1$	$Y = 0.7$
$L'_D = 9.429133$	$L'_F = 9.426103$
$l'_D = 9.429716$	$L'_C = 9.430645$
Spherical aberration $= -0.000583$	Chromatic aberration $= -0.004542$

By means of a scale drawing of this lens (Setup A) we assign suitable thicknesses of 0.4 for the crown element and 0.16 for the flint. The results of ray tracing at the margin and zone are shown in Table 7.2.

We next make a trial change in c_3 by 0.002 (Setup B). This gives spherical aberration $= +0.001304$ and chromatic aberration $= -0.001533$. In addition, a further trial change in c_2 by 0.002 (Setup C) gives spherical aberration $= -0.002365$ and chromatic aberration $= -0.003027$. The initial setup and these two changes are plotted on a graph connecting chromatic aberration as ordinate with spherical aberration as abscissa (Figure 7.1). Next, line AB is drawn to show the change for Δc_3 and line BC to show the change for Δc_2.

Now drawing a line through the aim point (0, 0) parallel to the line AB, intersecting line BC at D, suggests that we should try the following changes from Setup B. Scale the initial Δc_2 by BD/BC, which yields that $\Delta c_2 = 0.00164$. But since c_2 was -0.2545, we therefore try $c_2 = -0.25286$. Denoting the aim point as E, the second step is to scale Δc_3 by DE/AB. We find that $\Delta c_3 = 0.00181$, and since c_3 was 0.0170, we consequently try $c_3 = 0.01881$. Ray tracing this system gives the following for the final setup:

c	d	n_D	V
0.2545			
	0.4	1.523	58.6
−0.25286			
	0.16	1.617	36.6
0.01881			

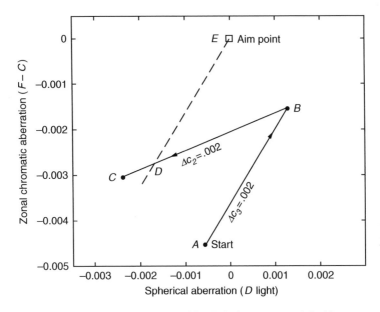

Figure 7.1 The four-ray method for designing a cemented doublet.

We have then $f' = 10.0916$, $l' = 9.6288$, $LA'(f/5) = -0.00005$, and $L'_{ch} = +0.00004$. Evidently the aberration changes are highly linear in this particular type of lens. We shall find many applications of this *double-graphing* technique whenever we are trying to correct two aberrations by making two simultaneous changes in the lens parameters.[2]

7.2 A THIN-LENS PREDESIGN

For the predesign of an ordinary cemented doublet, we start by determining the c_a and c_b values for thin-lens chromatic correction as described Section 5.4. We then set up the G-sum expressions for the primary spherical aberration of a thin system as described in Section 6.3.2. Since we shall be using c_1 as a bending parameter, we express everything in terms of c_1. For the crown element, c is c_a, c_1 and remains as c_1, and v_1 is the reciprocal of the object distance. For the flint element, $c_3 = c_1 - c_a$, since the two elements are to be cemented together, c is c_b, and $v_3 = v_1 + (n_a - 1)c_a$. The sum of the two G sums is now a quadratic in c_1, which can be solved either mathematically or graphically to give the two values of c_1 that meet the requirements of the problem. It can be seen that there are actually two and only two solutions; the four-ray method gave only the solution closest to the arbitrary starting setup and totally ignored the possibility of a second solution.

As an example we will use glasses similar to those used for the four-ray method, giving $c_a = 0.5085$ and $c_b = -0.2679$. For the G sums, with crown lens in front, we have $f^2 y^2 = 100$, $v_1 = 0$, $v_3 = 0.2659$, and $c_3 = c_1 - 0.5085$. Using these values, the spherical G sums give

$$SC_a = -30.759c_1^2 + 27.357c_1 - 7.9756$$
$$SC_b = 18.543c_1^2 - 23.698c_1 + 7.8392 \tag{7-1}$$
$$\text{total} = -12.216c_1^2 + 3.659c_1 - 0.1364$$

Evaluating this expression for a series of values of c_1 enables us to plot a graph of spherical aberration against c_1 (Figure 7.2) from which our two possible solutions can be picked off. It should be reiterated that this graph is incorrect for three reasons: It assumes thin lenses, it considers only paraxial chromatic aberration, and it considers only primary spherical aberration. Nevertheless, the two solutions come out to be surprisingly close to the final solutions.

7.2.1 Insertion of Thickness

Since we require zero spherical aberration, we read off the two solutions as

$$c_1 = 0.044 \text{ or } c_1 = 0.256.$$

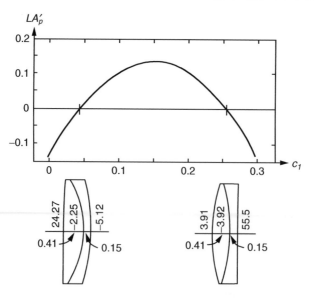

Figure 7.2   Thin-lens crown-in-front designs.

We now make a scale drawing of these systems and insert suitable thicknesses of 0.415 and 0.15, respectively. Next we trace a marginal ray in D light and calculate the last radius for perfect achromatism by the $D - d$ method, as explained in Section 5.9.1. We complete the trace of the marginal ray and add a paraxial ray so that the true spherical aberration can be found. Since this will not be quite the desired value, although it will generally be very close, we find dLA'/dc_1 by differentiating Eq. (7-1), and apply the coefficient to ascertain how much the c_1 must be changed to eliminate the spherical aberration residual. The results for the two solutions are shown in Table 7.3. The two final designs are shown in Table 7.4.

Scale drawings of the two systems are included in Figure 7.2. The decision as to which is the better design is based on the zonal aberration, which is nearly five times as large in the left-hand design as in the right-hand form. Furthermore, the surfaces in the right-hand design are weaker than in the left, resulting in economy in manufacture, and the fact that the crown element is almost equiconvex suggests that it should be made perfectly equiconvex to simplify the cementing operation. To do this requires a slight bending to the left, which would introduce a small spherical overcorrection, but it would probably be

Table 7.3

Spherical Aberration for Left-Hand and Right-Hand Crown-in-Front Configurations

c_1	0.044	0.256
Accurate LA'	0.0072	-0.0007
dLA'/dc_1	2.584	-2.596
Δc_1	-0.0028	-0.0003
New c_1	0.0412	0.2557
New LA'	-0.0001	0.0000
LZA'	-0.0171	-0.0045

Table 7.4

Solutions for Crown-in-Front Configurations

Left-hand solution			Right-hand solution		
c	d	n	c	d	n
0.0412			0.255755		
	0.412	1.523		0.415	1.523
-0.4442			-0.255037		
	0.15	1.617		0.15	1.617
-0.1953			0.018021		

$$f' = 9.9943$$
$$l' = 9.9545$$
$$(f/5) \begin{cases} LA' = -0.00007 \\ ZA = -0.01705 \end{cases}$$

$$f' = 9.99398$$
$$l' = 9.52719$$
$$(f/5) \begin{cases} LA' = -0.0000 \\ LZA' = -0.00450 \end{cases}$$

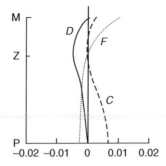

Figure 7.3   Spherochromatism of the right-hand *f*/5 crown-in-front solution.

better to hold the spherical correction by varying the last radius, and accept the slight chromatic residual would result. To complete the design, we calculate marginal, zonal, and paraxial rays in three wavelengths and plot the sphero-chromatism graph in Figure 7.3.

7.2.2 Flint-in-Front Solutions

There is no magic about having the crown element in front, and indeed for some applications a flint-in-front form is preferred. Repeating the predesign procedure with the flint glass as a and the crown glass as b gives

$$\text{spherical aberration} = -12.2162c_1^2 + 5.6493c_1 - 0.5399 \qquad (7\text{-}2)$$

This is plotted in Figure 7.4, from which we see that the two spherically cor-rected forms are shown in Table 7.5. The two final flint-in-front designs are shown in Table 7.6 (see page 216).

DESIGNER NOTE

Consideration of all four solutions indicates clearly that the right-hand crown-in-front form is in every way the best, although the zonal aberration of the left-hand flint-in-front form is not significantly greater. However, the weakness of the radii and the pos-sibility of making the crown element exactly equiconvex are sufficiently important to render the crown-in-front form generally preferable.

In recent years, significant effort has been given to developing a computer-based means of finding the lens configuration that yields the "best perfor-mance." Although this will be discussed in a later chapter, it is appropriate to mention a few pertinent points. One of the more difficult tasks of the lens designer is the construction of the merit function used by the lens design

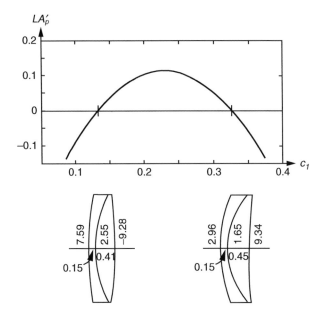

Figure 7.4   Flint-in-front solutions.

Table 7.5

Spherical Aberration for Left-Hand and Right-Hand Flint-in-Front Configurations

c_1	0.135	0.327
Accurate LA'	0.0078	0.0242
dLA'/dc_1	2.351	−2.340
Δc_1	−0.0033	0.0103
New c_1	0.1317	0.3373
Accurate LA'	−0.0002	0.0004
Accurate LZA'	−0.0052	−0.0194

program. Best performance was put in quotes above to indicate some uncertainty in what constitutes best performance. The same basic optical performance, such as f-number, resolution, spectral bandwidth, and field-of-view may be required by a two-lens system, yet the lens system may actually be rather different. The reason for this may be differences in operational environments, size and weight limitations, cost, fabrication tolerances, and so on.

The lens designer often needs to incorporate these other factors into the merit function and typically requires interaction with the mechanical engineer and optical and machine-shop personnel. The merit function can be viewed as a huge sheet in hyperspace that has a bizarre topology consisting of what can

Table 7.6

Solutions for Flint-in-Front Configurations

Left-hand solution			Right-hand solution		
c	d	n	c	d	n
0.1317			0.3373		
	0.15	1.617		0.15	1.617
0.3917			0.6052		
	0.414	1.523		0.454	1.523
−0.1079			0.108114		

<table>
<tr><td align="center">$f' = 9.9963$</td><td align="center">$f' = 10.0564$</td></tr>
<tr><td align="center">$l' = 9.7994$</td><td align="center">$l' = 9.4056$</td></tr>
<tr><td align="center">$(f/5) \quad LA' = -0.00015$</td><td align="center">$(f/5) \quad LA' = 0.00037$</td></tr>
<tr><td align="center">$LZA = -0.0052$</td><td align="center">$LZA = -0.0194$</td></tr>
</table>

be thought of as mountains, valleys, plains, and often pits. Possible solutions are found in the pits as they have the lower merit function values. With conventional optimization routines, the optical design program simply attempts to find the bottom of the pit local to the current location. However, the bottom of this pit may well not be the lowest and, consequently, not the optimum solution. (By optimum, we mean that the optical configuration solution having the smallest merit function value existing anywhere in the hyperspace; in other words, the global solution.)

Many of the optical design programs today include some form of what can generically be called global optimization. The objective of each searching approach these programs use is to locate the optimum solution or to give the designer a variety of potential solutions to consider. At times, "new" configurations have been found by allowing the number of elements and materials to vary. The achromat study just presented showed that there are exactly four perfect solutions for the merit function defined. A simple test that can be given to an optical design program is to find these four solutions. At least one optical design program is known to be able to automatically find these solutions.

7.3 CORRECTION OF ZONAL SPHERICAL ABERRATION

If the zonal aberration in a lens system is found to be excessive, it can often be reduced by splitting the system into two lenses, each having half the lens power, in a manner analogous to the reduction of the marginal aberration of a single lens (see Section 6.1.6).

Another method that is frequently employed in a cemented system is to separate the cemented interface by a narrow parallel airgap. For this procedure to be effective, there must be a large amount of spherical aberration in the airgap so that the marginal ray drops disproportionately rapidly as compared to the 0.7 zonal ray. *The airgap therefore undercorrects the marginal aberration more rapidly than the zonal aberration.* As the rear negative element is now not acting as strongly as before because of the reduction of incidence height, the last radius must be adjusted to restore the chromatic correction, ordinarily by use of the $D - d$ method. As the spherical aberration will now be strongly undercorrected, it must be restored by a bending of the whole lens. Using this procedure, it is often possible to correct both the marginal and the zonal aberrations simultaneously.

To determine the proper values of the airgap and the lens bending, we start with a cemented lens and introduce an arbitrary small parallel airgap, the last radius being found by the $D - d$ method. The whole lens is then bent by trial until the marginal aberration is correct and the zonal aberration is found. If it is still negative, a wider airgap is required. The desired values are quickly found by plotting suitable graphs.

As an example, we may consider the following three $f/3.3$ systems. They each have a focal length of 10.0, and they are made from K-3 and F-4 glasses, the last radius in each case being found by the usual $D - d$ procedure, as shown in Table 7.7.

System A is a well-corrected doublet of the ordinary type, but of unusually high aperture so as to illustrate the principle. The spherical aberration curve is shown as A in Figure 7.5. After introducing an airgap and suitably strengthening the last radius by the $D - d$ method, we have System B. The change in aberrations as a result of the introduction of this airgap is

$$\left. \begin{array}{l} \Delta LA_{marginal} = -0.116115 \\ \Delta LA_{zonal} = -0.034857 \end{array} \right\} \quad \text{ratio 3.33}$$

Table 7.7

Configurations of Three $f/3.3$ Achromatic Doublets

A			B			C		
c	d	n	c	d	n	c	d	n
0.259			0.259			0.236		
	0.75	1.51814		0.75	1.51814		0.75	1.51814
−0.2518			−0.2518			−0.2748		
	0.25	1.61644		0.0162	(air)		0.0162	(air)
0.018048			−0.2518			−0.2748		
				0.25	1.61644		0.25	1.61644
			0.022487			−0.005068		
$LA_{marginal} =$	0.001252		−0.114863			−0.000211		
$LA_{zonal} =$	−0.024094		−0.058951			0.000345		

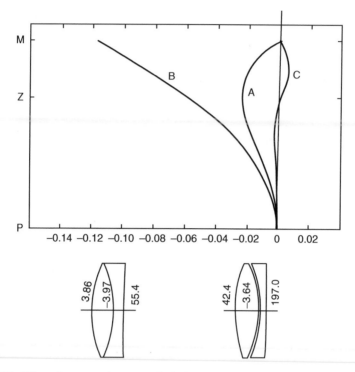

Figure 7.5 Effect of a narrow airgap on spherical aberration: (A) Cemented doublet; (B) effect of introducing a narrow airgap; (C) final solution.

We now bend the entire system to the left by $\Delta c = -0.023$ to restore the aberrations. The changes now are

$$\left.\begin{array}{l} \Delta LA_{\text{marginal}} = 0.114652 \\ \Delta LA_{\text{zonal}} = 0.059296 \end{array}\right\} \quad \text{ratio } 1.93$$

If everything were ideal and only primary and secondary aberration were present, the latter ratio would be 2.0, and so we see that the changes due to bending are fairly linear in this respect. Examination of curve C shows the presence of tertiary aberration.

Unfortunately, although the marginal and 0.7 zonal aberrations are virtually zero in System C, there are sizable intermediate zonal residuals remaining. By tracing a few additional zonal rays at various heights of incidence, we can plot the spherical aberration graph of this system as curve C in Figure 7.5. However, it is evident that these unavoidable residuals are much smaller than the 0.7 zonal aberration of the original cemented System A. The designer should be careful in adjusting the airgap to avoid the introduction of yet higher-order aberration terms. Somewhat improved performance can be achieved by shifting the zero zonal aberration point to a bit higher value of ρ. A problem likely to arise is that at least quintic aberration will now appear and have a rather significant value. The presence of the higher-order aberration makes the lens less tolerant to manufacturing and alignment errors.

When System A, after introducing a small airgap, was optimized by a typical lens design program using the same criteria as used in the preceding procedure, the resulting design was found to be quite similar, with the airgap being about one-third of System C. Figure 7.6 illustrates the longitudinal aberration and should be compared with curve C in Figure 7.5. It should be mentioned that there are many similar designs that have essentially the same performance as the airgap is varied and the curvatures are readjusted.

DESIGNER NOTE

An alternative procedure that can be applied to reduce zonal aberration is to thicken a lens element, provided there is a large amount of undercorrected aberration within the glass. This is done frequently in photographic objectives, such as in Double-Gauss lenses of high aperture. Of course, introducing an air space by breaking cemented surfaces can be done in concert with element thickening.

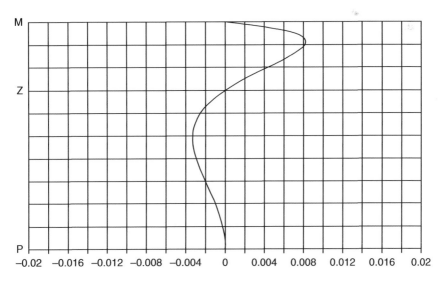

Figure 7.6 Longitudinal spherical aberration for an achromatic doublet having a small airgap designed using a computer optical design program.

7.4 DESIGN OF AN APOCHROMATIC OBJECTIVE

7.4.1 A Cemented Doublet

A simple cemented doublet can be made apochromatic if suitable glasses are chosen in which the partial dispersion ratios are equal. The combination of fluorite and dense barium crown mentioned in Section 5.5 is one possibility. Another is a doublet made from two Schott glasses such as in Table 7.8. The large V difference of 27.99 keeps the elements weak and reduces the zonal aberration.

Table 7.8

Glass Properties for Apochromatic Cemented Doublet

Glass	n_e	$\Delta n = (n_F - n_c)$	$V_e = \left(\dfrac{n_e - 1}{n_F - n_c} \right)$	P_{Fe}
FK-52	1.48747	0.00594	82.07	0.4562
KzFS-2	1.56028	0.01036	54.08	0.4562

7.4.2 A Triplet Apochromat

Historically the preferred form for an apochromatic telescope objective has been the apochromatic triplet or "photovisual" objective suggested by Taylor in 1892.[3] The preliminary thin-lens layout has already been described in Section 5.6, and we shall now proceed to insert thicknesses and find the bending of the lens that removes spherical aberration. The net curvatures and glass data of the thin system are also given in Section 5.6. The glass indices and other data are stated to seven decimal places by use of the interpolation formulas given in the Schott catalog; this extra precision is necessary if the computed tertiary spectrum figures are to be meaningful. Obviously, in any practical system such precision could never be attained.

A possible first thin-lens setup with a focal length of 10 is the following:

$$c_1 = 0.56 \text{ (say)} \qquad\qquad r_1 = 1.79 \text{ (approx.)}$$
$$c_a = 1.0090432$$
$$c_2 = c_1 - c_a = -0.4490432 \quad r_2 = -2.23$$
$$c_b = -0.7574313$$
$$c_3 = c_2 - c_b = 0.3083881 \qquad r_3 = 3.24$$
$$c_c = 0.1631915$$
$$c_4 = c_3 - c_c = 0.1451966 \qquad r_4 = 6.89$$

Tracing paraxial rays through this lens with all the thicknesses set at zero gives the image distances previously plotted in Figure 5.11.

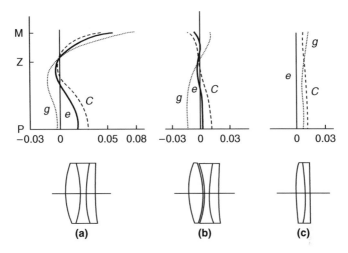

Figure 7.7 Apochromatic triplet objectives: (a) cemented triplet apochromat, (b) triplet apochromat with airgap, and (c) doublet achromat.

Since an aperture of $f/8$ is the absolute maximum for such a triplet apochromat, we draw a diagram of this setup at a diameter of 1.25, by means of which we assign suitable thicknesses, respectively 0.3, 0.13, and 0.18. This lens is shown in Figure 7.7a. Our next move is to trace a paraxial ray in e light through this thick system, and as we go along modify each surface curvature in such a way as to restore the paraxial chromatic aberration contribution to its thin-lens value. Since the chromatic contribution was shown (see Eq. (5-1b)) to be given by

$$L'_{ch}C = yni(\Delta n/n - \Delta n'/n')/u'^2_k$$

it is clear that all we have to do is to maintain the value of the product (yi) at each surface. The equations to be solved, therefore, are

$$i = \frac{thin - lens(yi)}{actual\ y}, \qquad c = \frac{u + i}{y}$$

When this is done, we have the following thick-lens paraxial setup:

c	d	n_e
0.40580124		
	0.4148	1.4879366
−0.36858873		
	0.17975	1.6166383
0.24679727		
	0.2489	1.7043823
0.11469327		
$f'_e = 10.000$	$l' = 9.0266$	

Tracing paraxial rays in other wavelengths reveals only very small departures from the thin-lens system. These are caused by the small assumptions that were made in deriving Eq. (5-1b).

We must next achromatize for the zonal rays by use of the $D - d$ method. For the Δn values, we use $(n_g - n_C)$ because we are endeavoring to unite C, e, and g at a common focus. When this is done, the fourth curvature becomes 0.14697738, and the focal length drops to 9.7209. However, the spherical aberration is found to be +0.35096, and we must bend the lens to the right to remove it. Repeating the design with $c_1 = 0.6$, and adding the marginal, zonal, and paraxial rays in all three wavelengths gives the spherochromatism curves shown in Figure 7.7a. Both the zonal aberration and the spherochromatism are clearly excessive, and so we adopt the device of introducing a narrow air space after the front element.

As this quickly undercorrects the spherical aberration, we return to the preceding setup, with the addition of an air space, and once more determine the last radius by the $D - d$ method:

c	d	n_e
0.39011389		
	0.4307	1.4879366
−0.35496974		
	0.0373	(air)
−0.35496974		
	0.1866	1.6166383
0.23767836		
	0.2584	1.7043823
0.11045547		
$f_e' = 10.000$	$l_e' = 8.8871$	

The spherochromatism curves are shown in Figure 7.7b, and the whole situation is greatly improved. This is about as far as we can go. Increasing the air space still further would lead to a considerable overcorrection of the zonal residual, and the result would be worse instead of better; however, if the air space is greatly increased, a different solution may be found as discussed later.

But first, it is of interest to compare this apochromatic system with a simple doublet made from ordinary glasses. An $f/8$ doublet was therefore designed using the regular procedure, the glasses being

	n_C	n_e	n_g
(a) Crown	1.52036	1.52520	1.53415
(b) Flint	1.61218	1.62115	1.63887

The final doublet system is shown in Table 7.9. The spherochromatism curves are shown in Figure 7.7c.

Table 7.9

Prescription of f/8 Doublet Shown in Figure 7.7c

c	d	
0.2549982		
	0.2	(crown)
−0.2557933		
	0.1	(flint)
0.00964734		

It is clear that the zonal aberration is negligible, the only real defect being the secondary spectrum. However, the effort to correct this in the three-lens apochromat has increased the zonal aberration and spherochromatism so much that it is doubtful if the final image would be actually improved thereby. An apochromat is useful only if some means can be found to eliminate the large spherochromatism that is characteristic of such systems.

7.4.3 Apochromatic Objective with an Air Lens

If the airgap is significantly increased and c_2 and c_3 are allowed to differ somewhat, an *air lens* is formed between these surfaces. By using a computer optimization program to achromatize the lens for g and C spectral lines, correct secondary spectrum using g and e spectral lines, correct marginal and zonal spherical aberration in the e spectral lines, and correct marginal spherochromatism for g and C spectral lines, diffraction-limited performance can be obtained. A representative lens is shown in Figure 7.8 that operates at f/8 and has the following prescription:

c	d	n_e
0.49149130		
	0.4286	1.4879367
−0.30739277		
	0.3593	(air)
−0.45082004		
	0.1857	1.6166386
0.29139083		
	0.2571	1.7043829
0.14851018		
$f'_e = 10.0086$	$l'_e = 7.4947$	

Figure 7.8 Layout of an f/8 apochromatic triplet objective lens having axial diffraction-limited performance and showing ray paths for axial, 1°, 2°, and 3° extraaxial object points.

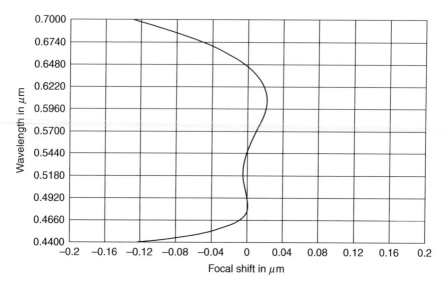

Figure 7.9 Chromatic focal shift.

The glasses used in this example are Schott N-FK51, N-KZFS4, and N-SF15, respectively. Figure 7.9 illustrates the achievable wide spectral bandwidth for this apochromatic triplet objective. Notice the characteristic shape of the central portion of the plot and the rapid chromatic undercorrection at each end of the spectral bandwidth.

The longitudinal meridional ray errors for light from 440 nm to 700 nm in steps of 20 nm is shown in Figure 7.10. The optimization criteria mentioned above yielded a highly corrected lens system. As can be seen, the marginal and axial chromatic error is negligible while some zonal aberration remains, although it is quite small. The spherochromatism comprises primary, secondary, and tertiary components having signs of minus, plus, and minus, respectively. Also, notice that the intercepts of the plots are wavelength dependent, which means that an amount of positive and negative zonal aberrations for each plot are wavelength dependent. The amount of positive and negative zonal aberrations for the *e* spectral line is essentially balanced (see arrow in Figure 7.10).

Does this apochromatic objective have excellent performance just on axis or does it have a useful field-of-view? Figure 7.11 presents the transverse ray fans for axial, 1°, 2°, and 3° extraaxial object points. The off-axis behavior will be discussed in later chapters, but recalling the discussions in Chapter 4, it is evident that (1) the lateral chromatic aberration grows as the field angle increases, (2) negative coma is dominant at 1° with very slight negative linear astigmatism, and (3) linear astigmatism is beginning to become dominant by 3°. The

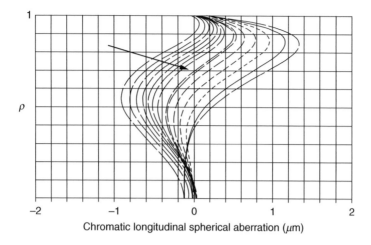

Figure 7.10 Longitudinal meridional ray errors for light from 440 nm to 700 nm in steps of 20 nm.

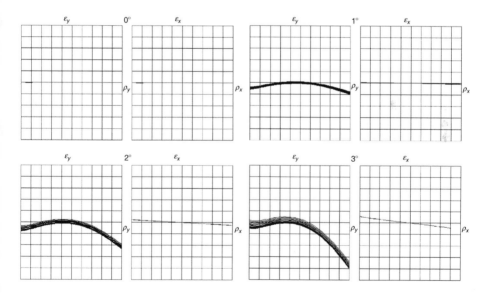

Figure 7.11 Transverse ray fans for axial, 1°, 2°, and 3° extraaxial object points. Scale is ±20 μm.

acceptability of the extraaxial image quality is, of course, dependent on the application.

The technique of incorporating air lenses in an optical system has been utilized for a long time. In fact, one could view the air space between lens elements

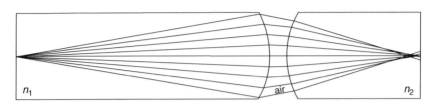

Figure 7.12   Air lens.

as air lenses. An air lens has no chromatic aberration, which is one reason the $D - d$ method of achromatizing works. Figure 7.12 illustrates a possible air lens. The source is located in a material having a refractive index of n_1 and the image is formed in a material having a refractive index of n_2. The space between the materials forms the air lens as is seen in the figure. If one or both of the bounding surfaces of the air lens are made to be conic surfaces, there is the possibility to dramatically control the marginal and zonal spherical aberrations magnitudes and sign.

Interestingly, in 2004, U.S. Patent No. 6,785,061 B2 entitled "Converging Air Lens Structures" was issued. The basic lens appears similar to the lens shown in Figure 7.12 except that an aperture stop was placed in the air space. The air lens concept has been used in various lens such as an Angenieux zoom lens compensated for temperature, vibration, and pressure.[4] Should the air space be replaced with another optical material, the resultant optical system that forms a complete imaging system is often referred to as a *solid optic*. Such systems have a variety of specialized applications.

It should be noted that the above solutions are far from being the only possible triplet apochromat that can be designed. We could assemble the three elements of our thin-lens solution in any order; we could introduce an airgap in the other interface; and of course we could use quite a different set of glasses. Anyone seriously engaged in designing such a system is well-advised to try out some of these other possibilities.

ENDNOTES

[1] It should be understood that one can use an optical design program to automatically optimize this lens by configuring the merit function appropriately; however, following the presented procedure provides insight into the lens' parametric behavior.

[2] This procedure was suggested to Dr. Kingslake by his colleague, Mr. H. F. Bennett.

[3] H. D. Taylor, Br. Patent 17994 (1892).

[4] Allen Mann, *Infrared Optics and Zoom Lenses, Second edition*, pp. 84–85, SPIE Press, Bellingham (2009).

Chapter 8

Oblique Beams

An oblique beam (also called a pencil) of rays from an extraaxial (or off-axis or nonaxial) object point contains meridional rays that can be traced by the ordinary computing procedures already described, and also a large number of skew rays that do not lie in the meridional plane. Each skew ray intersects the meridional plane at the object point and again at a "diapoint" in the image space, and nowhere else. Skew rays require special ray-tracing procedures, which will be discussed in Section 8.3. These are much more complex than for a meridional ray, and it is observed that skew rays were seldom used before the advent of electronic computers; now they are routinely traced by all lens designers since the available computing power of even the most common personal computer is extraordinarily great.

In Chapter 4, we discussed both axial and off-axis/nonaxial aberrations in an analytical sense rather than a causal sense. Axial aberrations have been investigated in some detail in the last several chapters. In this chapter, we will begin a more detailed study of field-dependent astigmatic and comatic aberrations. We begin by looking at the origin of coma and astigmatism, and then the role various types of stops have in lens systems. The remainder of the chapter discusses general ray tracing and graphical representation of skew ray aberrations.

8.1 PASSAGE OF AN OBLIQUE BEAM THROUGH A SPHERICAL SURFACE

8.1.1 Coma and Astigmatism

When a light beam is refracted obliquely through a spherical surface, several new aberrations arise that do not appear on the lens axis. To understand why this is so, we may consider the diagram in Figure 8.1 showing a single refracting surface and an aperture stop that admits a circular cone of rays from an off-axis object point B. We label the rays through the rim of the aperture by their position angles taken counterclockwise from the top as viewed from the image

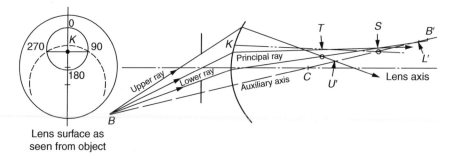

Lens surface as
seen from object

Figure 8.1 Origin of coma and astigmatism.

space, so that the upper ray is called 0° and the lower ray 180°, while the front
and rear sagittal rays become 90° and 270°, respectively. The line joining the
object point *B* to the center of curvature of the surface, *C*, is called the *auxiliary
axis*, and obviously there is complete rotational symmetry around this axis just
as there is rotational symmetry around the lens axis for an axial object point.

Moreover, because of this symmetry, every ray from the object point *B* passing through the aperture stop must cross the auxiliary axis somewhere in the
image space. If we could trace a paraxial ray from *B* along the auxiliary axis,
it would form an image of *B* at, say, *B'*. However, because of the spherical aberration arising at the surface, the intersection point for all other rays will move
along the auxiliary axis toward the surface by an amount proportional to the
square (approximately) of the height of incidence of the ray above the auxiliary
axis. Thus the upper limiting ray might cross the auxiliary axis at, say, *U'*, and
the lower limiting ray at *L'*. It is at once evident that the upper and lower rays
do not intersect each other on the principal ray but in general above or below it;
the height of the intersection point above or below the principal ray is called the
tangential coma (a relic of the old custom of calling meridional rays tangential
because they form a tangential focal line).

To find the point at which the two sagittal rays at 90° and 270° intersect the
auxiliary axis, we note that these rays are members of a hollow cone of rays centered about the auxiliary axis, all coming to the same focus on that axis. The
upper ray of this hollow cone strikes the refracting surface at *K*, slightly higher
than the principal ray, so that the spherical aberration of this ray will be a little
greater than that of the principal ray, forming an image at *S* on the auxiliary
axis (shown by the small circle). *S* lies below the principal ray on our diagram,
which indicates the presence of some negative sagittal coma, but not as much as
the tangential coma that we found previously. Indeed, it can be shown[1] that
for a very small aperture and obliquity, the tangential coma is three times the
sagittal coma; the exaggerations in our diagram do not make this relation obvious, but at least both comas do have the same sign.

We thus see that the extreme upper and lower rays of the marginal zone come to a focus at T, while the extreme front and rear rays come to a different focus at S. The longitudinal separation between S and T is called the astigmatism of the image, and evidently both coma and astigmatism arise whenever a light beam is refracted obliquely at a surface. It is essential to note that each surface in a lens has a different auxiliary axis, and that the proportion of coma and astigmatism therefore varies from surface to surface. It is thus possible to correct coma and astigmatism independently in a lens system provided there are sufficient degrees of freedom available.

In Sections 4.3.3 and 4.3.4, additional information about computing astigmatism and coma using exact ray tracing and the relationship to aberration coefficients was presented. In the next chapter, we will discuss coma, the Abbe sine condition, and offense against the sine condition. Astigmatism, Coddington equations, the Petzval theorem, distortion, and lateral color are explored in more depth in Chapter 11. Also in that chapter, the important symmetry principle will be introduced.

8.1.2 Principal Ray, Stops, and Pupils

At this point, it is necessary to define several important terms. The *aperture stop* or *stop* of a lens is the limiting aperture associated with the lens that determines how large an axial beam may pass through the lens. The stop can be an element within the lens system or a mechanical element such as a hole in a disk. A mechanical stop that can vary its opening size is also called an *iris*.

The *marginal ray*, also called the *rim ray*, is the extreme ray from the axial point of the object through the edge of the stop. As discussed in Section 4.2, the *entrance pupil* is the image of the stop formed by all lenses preceding it when viewed from object space. It also is the reference surface used to define ray coordinates, that is, (ρ, θ, H_y). By convention, the entrance pupil is aberration free. In a similar manner, the *exit pupil* is the image of the stop formed by all lenses following it when viewed from image space. The exit pupil is used as a reference surface for exiting wavefronts from the lens. Very often the rays incident at the exit pupil are not rectilinearly mapped onto the exit pupil due to pupil aberrations.

Consideration of the mapping error is necessary to properly compute image energy distribution, *MTF*, and diffraction. These two pupils and the stop are all geometric images of one another. The entrance and exit pupils can each be real or virtual images of the aperture stop located at finite distances or at infinity dependent on the optical configuration before and after the stop. For example, if an aperture stop is placed between the object and a singlet lens, and closer to the lens than the focal length, then the entrance pupil is clearly real and the exit pupil is virtual. In general, the spatial location of the pupils with respect to the stop can be in order (for example [exit, entrance, stop], [entrance, exit, stop], and [entrance, stop, exit]).

The *principal ray* is defined as the ray emanating from an off-axis object point that passes through the center of the stop. In the absence of pupil aberrations, the principal ray also passes through the center of the entrance and exit pupils. As the obliquity angle of the principal ray increases, the defining apertures of the components comprising the lens may limit the passage of some of the rays in the entering beam, thereby causing the stop not to be filled with rays. The failure of an off-axis beam to fill the aperture stop is called *vignetting*.[2] The ray centered between the upper and lower rays defining the oblique beam is called the *chief ray*. When the object moves to large off-axis locations, the entrance pupil often has a highly distorted shape, may be tilted, and/or displaced longitudinally and transversely, and no longer perpendicular to the lens axis.

Indeed, without this tilting of the entrance pupil a fisheye lens covering a full $\pm 90°$ in the object space would not transmit any light at the edge of the field. Due to the vignetting and pupil aberrations, the chief and principal rays may become displaced from one another. In some cases, the principal ray is vignetted while the chief ray is never vignetted as long as light passes through the lens at the considered obliquity angle. The terms principal ray and chief ray are frequently used interchangeably; however, once vignetting occurs, the distinction must be made.

DESIGNER NOTE

It is important that the lens designer understands how the optical design program being used handles the aiming of the chief ray. Typically, the chief ray is aimed toward the center of the (vignetted) entrance pupil, which is generally acceptable in the early stages of design. In the final stages, the chief ray should be aimed at the center of the (vignetted) stop. The reason for this is that additional computational time is required to aim at the (vignetted) stop. Since the stop is a real surface, the entrance pupil may well suffer aberrations. If the entrance pupil is considered unaberrated, then the stop is likely aberrated in theory at least. A design that may appear quite satisfactory using an unaberrated entrance pupil can perform in practice differently since the actual stop is unaberrated, thereby changing what rays actually pass through the lens system! Remember that the vignetted stop is made up of portions of the actual stop and boundaries of various lens elements (see Section 8.1.3).

The *field stop* is an aperture that limits the passage of principal rays beyond a certain field angle. The image of the field stop when viewed from object space is called the *entrance window* and is called the *exit window* when viewed from image space. The field stop effectively controls the field of view of the lens system. Should the field stop be coincident with an image formed within or by the lens system, the entrance and exit windows will be located at the object and/or image(s).

A *telecentric stop* is an aperture located such that the entrance and/or exit pupils are located at infinity (see Section 12.5.3). This is accomplished by placing the aperture in the focal plane. Consider a stop placed at the front focal plane of a lens. The stop image is located at infinity and the principal ray exits the lens parallel to the optical axis. This feature is often used in metrology since the measurement error is reduced when compared to conventional lens systems because the centroid of the blur remains at the same height from the optical axis even as the focus is varied.

8.1.3 Vignetting

In many lenses, and particularly those having a considerable axial length, an oblique pencil may be unable to traverse the lens without part of the beam being obstructed by the end lens apertures. For instance, in the triplet lens shown in Figure 8.2 the upper rays of the 20° oblique beam are cut off by the rear lens aperture, and the lower rays by the front aperture, so that the beam fails to fill the iris. This process is known as *vignetting*, the oblique beam is projected onto the plane perpendicular to the axis in the object space having the shape shown in the figure. Vignetting is one of the reasons why the illumination on the film/detector-array in a camera falls off at increasing transverse distances from the lens axis. Other reasons are (a) the cos[3] law, (b) distortion of the entrance pupil at high obliquities, and (c) image distortion. The effects of these various factors have been discussed elsewhere.[4]

To plot the vignetting diagram of a lens, the locations of the upper and lower "rim" rays are readily found by trial, but it is then necessary to determine the radii of the upper and lower limiting circular arcs. The lower arc obviously has the same radius as the front lens aperture, but the upper arc is the image of the rear aperture as seen through the lens. Its radius bears the same ratio to the radius of the entering axial beam as the diameter of the rear aperture itself bears to the diameter of the axial beam as it emerges from the rear of the lens.

In addition to the circles corresponding to the front and rear lens apertures, an oblique beam is limited also by the iris, and the image of the iris must therefore be projected into the object space along with the image of the rear aperture. To locate this iris image, we add a ray parallel to the upper and lower rim rays and passing through the center of the iris. This middle ray is projected into the vignetting diagram in Figure 8.2, and we draw a circle about it having the diameter of the entering axial beam because the axial beam necessarily fills the iris completely. The vignetted area of the oblique beam is shown shaded. The

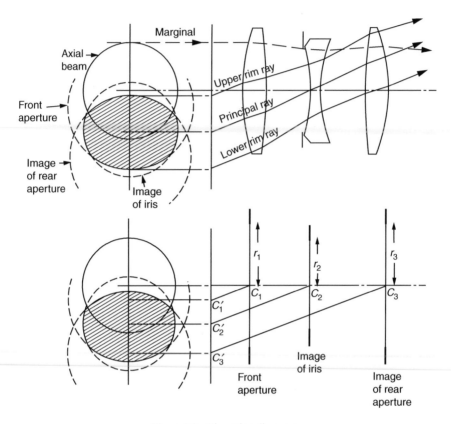

Figure 8.2   Vignetting diagrams.

"vignetting factor" is the ratio of the area of the oblique beam to the area of the axial beam, both measured in a plane perpendicular to the lens axis. It is, of course, an assumption that the images of the iris and of the rear lens aperture are circles; indeed, they are much more likely to be arcs of ellipses, but we make very little error by plotting them as circles.

An alternative method of plotting the vignetting diagram is shown in the lower diagram of Figure 8.2. We begin by determining the location and size of the images of the rear aperture and of the iris, projected into the object space, by use of paraxial rays traced right-to-left from the centers of those two apertures. The front aperture and the two images are shown at C_1, C_2, and C_3, their computed radii being, respectively, r_1, r_2, and r_3. We can now replace the lens by these three circles, and project their centers at any required obliquity onto a vertical reference plane as shown. Knowing the centers of the circles and their radii, it is a simple matter to draw the vignetting diagram directly. Of course, this procedure cannot be as accurate as the first method, but it is much simpler and generally sufficiently accurate for most purposes. This simple procedure cannot be used for wide-angle or fisheye lenses where the pupil is seriously distorted or tilted.

Another method that can be used relies on the linear nature of paraxial ray tracing. It is easy to show that if two paraxial rays have been traced through a lens system and at each surface y and $\bar{y}$ are known, then the intercept height $\bar{\bar{y}}$ for any other ray at any surface can be computed without ray tracing.[4] The general equation is

$$\bar{\bar{y}}_j - \bar{y}_j = (\bar{\bar{y}}_i - \bar{y}_i)\frac{y_j}{y_i}. \tag{8-1}$$

Calculation of the coordinate of any ray on the entrance pupil ($j = 1$) having the coordinate $\bar{\bar{y}}_i$ on the ith surface, with $\bar{y}_1 = 0$, then Eq. (8-1) becomes

$$\bar{\bar{y}}_1 = (\bar{\bar{y}}_i - \bar{y}_i)\frac{y_1}{y_i}. \tag{8-2}$$

Table 8.1 contains ray trace data for a simple two-lens system having an internal stop. Lens A diameter is 3.0, the diameter of Lens B is 2.0, and the diameter of the stop is 2.0. From Table 8.1, the entrance pupil is located a distance of 2.5 behind Lens A (or 0.5 to the right of the stop). The marginal ray and two principal rays ($u = 0.1$ and 0.2) were traced. The size of the entrance pupil can be determined in a couple of ways from the data in the table. First, the marginal ray has a height of 1.25 for a stop radius of 1.0. Remember that the linear nature of paraxial ray tracing implies that a stop diameter of 0.80, or the magnification of the stop to form the entrance pupil, is 1.25. A second way is to observe that the principal angle at the entrance pupil is 0.1 and is 0.125 at the stop. Hence, the magnification of the stop to form the entrance

Table 8.1

Ray Trace Data for Lens Systems Demonstrating Vignetting

	Entrance pupil		Lens A		Stop		Lens B
Surf #	1		2		3		4
$-\phi$	0		-0.1		0		-0.1
t		-2.5		2		1	
y marginal	1.25		1.25		1		0.875
u		0		-0.125		-0.125		-0.2125
$\bar{y}$ principal	0		-0.2500		0		0.125
$\bar{u}$		0.1		0.125		0.125		0.1125
$\bar{y}$ principal	0		-0.5000		0		0.2500
$\bar{u}$		0.2		0.25		0.25		0.2250

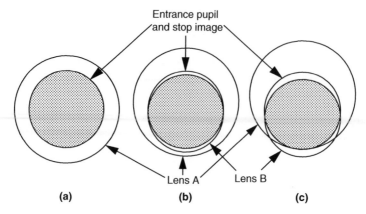

Figure 8.3 Vignetting diagram for the lens system shown in Table 8.1 where (a) is for an axial object, (b) is for a distant object having a field angle of 0.1, and (c) is for a distant object having a field angle of 0.2.

pupil is again found to be 1.25 (u_3/u_1). To compute the shift, often called shear, of the ith element's projected center onto the entrance pupil, set $\bar{\bar{y}}_i = 0$ and use Eq. (8-2) with the data in the table. Determination of the projected size of the lens onto the entrance pupil once again uses Eq. (8-2).

Figure 8.3 depicts the circular apertures for the stop and both lenses contained in Table 8.1. where Figure 8.3a is for an axial object, Figure 8.3b is for a distant object having field angle of 0.1, and Figure 8.3c is for a distant object having a field angle of 0.2. The shaded area illustrates the portion of the lens that can pass light at these three angles. Notice in Figure 8.3b that Lens B is starting to vignette while Lens A is relatively far from vignetting. In Figure 8.3c, Lens B is vignetting more and Lens A is just starting to vignette. The vignetted entrance pupil appears to shift down and become elliptically shaped.

8.2 TRACING OBLIQUE MERIDIONAL RAYS

For any given object point, or for any given obliquity angle if the object is at infinity, a specific meridional ray must be defined by some convenient ray parameter. This may be the height A at which the ray intersects the tangent plane at the first lens vertex, or it may be the intersection length L of the ray relative to the front lens surface. For a ray proceedings uphill from left to right and entering the lens above the axis, A will be positive and L negative.

Whatever ray parameter is chosen, it is necessary to use appropriate "opening equations" to convert the given ray data into the familiar (Q, U) values to trace the ray.

1. *A Finite Object*

   If the object point is defined by its H and d_0 (Figure 8.4), then

   $$\tan U = -(A - H)/d_0 \quad \text{and} \quad Q = A \cos U$$

   If the ray is defined by its L value, then

   $$\tan U = -H/(L - d_0) \quad \text{and} \quad Q = -L \sin U$$

2. *A Very Distant Object*

   The slope angle of all entering rays is now the same, being equal to the principal-ray slope U_{pr}; we use only the second of the opening equations to find Q.

3. *Closing Equations*

   Having traced an oblique ray through a lens, we generally wish to know the height at which it crosses the paraxial image plane. This is given by (Figure 8.5)

   $$H' = (Q' + l' \sin U')/\cos U'$$

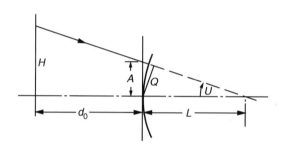

Figure 8.4 Opening equations.

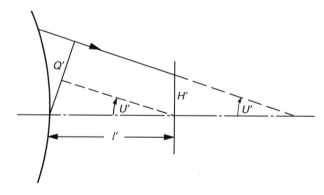

Figure 8.5 Closing equations.

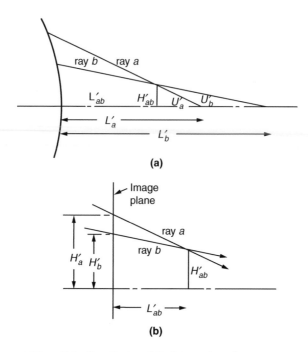

Figure 8.6 Coordinates of the intersection of two rays.

Sometimes we want to know the coordinates of the intersection point of two traced rays, knowing their L' or their Q' value and also their slope angles U'. The formulas to be used are (Figure 8.6a)

$$L'_{ab} = \frac{L'_a \tan U'_a - L'_b \tan U'_b}{\tan U'_a - \tan U'_b}, \quad \text{where} \quad L' = -Q'/\sin U'$$

$$H'_{ab} = -(L'_a - L'_{ab}) \tan U'_a = -(L'_b - L'_{ab}) \tan U'_b \tag{8-3a}$$

8.2.1 The Meridional Ray Plot

Having traced a number of oblique rays through a lens from a given object point, we need some way to plot the results and interpret the mixture of aberrations that exists in the image. This mixture will contain spherical aberration, of course, and also the oblique aberrations coma and meridional field curvature. Astigmatism as such will not appear because it involves sagittal rays, which are not traced in a meridional beam. The two chromatic aberrations will not appear unless colored oblique rays are being traced.

The usual procedure is to plot the intercept height H' of the ray at the paraxial image plane as the ordinate, with some reasonable ray parameter as the abscissa.

For the latter we may use the Q value of the ray at the front lens surface, or the incidence height A of the ray at the tangent plane to the front vertex, or the intersection length L of the ray at the first lens surface. Sometimes we use the height of the ray at the paraxial entrance pupil plane, or its height in the stop. However, there is good reason to use as abscissa the tangent of the ray slope angle U' in the image space. When this is done a perfect image point plots as a straight line, whose slope is a measure of the distance from the paraxial image plane to the oblique image point. The reason for this can be seen in Figure 8.6b, which shows two rays in an oblique pencil having heights H'_a and H'_b at the image plane and emerging slope angles U'_a and U'_b, respectively. The longitudinal distance L'_{ab} from the image plane to the intersection of these rays with one another is given by

$$H'_{ab} = H'_a + L'_{ab} \tan U'_a \text{ and } H'_{ab} = H'_b + L'_{ab} \tan U'_b$$

Eliminating H'_{ab} gives

$$L'_{ab} = \frac{H'_a - H'_b}{\tan U'_b - \tan U'_a} \tag{8-3b}$$

If the data of the two rays are plotted on a graph connecting H' with $\tan U'$, the slope of the line joining the two ray points will be a direct measure of L'_{ab}. Consequently if all the rays in the beam have the same L'_{ab}, their ray points will all lie on a straight line, with the lower rim ray at the left and the upper at the right. The principal ray will fall about midway between the two rim rays. A perfect lens with a flat field will plot as a horizontal straight line (Figure 8.7a). A perfect lens with an inward-curving field plots as a straight line sloping down from left to right (Figure 8.7b). Primary coma is represented by a parabolic graph, the ends being up in the case of positive coma (Figure 8.7c), and down for negative coma (Figure 8.7d). Primary spherical aberration is represented by a cubic

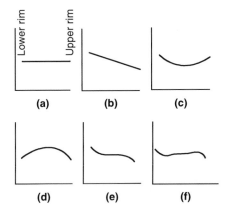

Figure 8.7 Some typical $H - \tan U$ curves: (a) a perfect lens, (b) inward-curving field, (c) positive coma, (d) negative coma, (e) spherical undercorrection, and (f) zonal spherical aberration.

curve, and if the image along the principal ray lies in the paraxial image plane, the middle of the cubic curve will be horizontal (Figure 8.7e). Zonal spherical aberration is revealed by a curve with a double bend, which is a combination of a cubic curve for the primary aberration component and a fifth-order curve for the secondary aberration (Figure 8.7f). Of course, any imaginable mixture of these aberrations can occur, and the experienced designer soon gets to recognize the presence of the different aberrations by the shape of the curve.

8.3 TRACING A SKEW RAY

A skew ray[5,6] is one that starts out from an extraaxial object point and enters a lens in front of or behind the meridional plane. It should be noted that for every skew ray there is another skew ray that is an image of the first, formed as if the meridional plane were a plane mirror. Thus, having traced one skew ray we have really traced two, the ray in front of the meridional plane and the corresponding ray behind it. These two skew rays intersect each other at the same diapoint (see Figure 2.1).

In tracing a skew ray, we denote a known point on the ray as X_0, Y_0, Z_0, and the direction cosines of the ray as K, L, M. Of course, in the object space the point X_0, Y_0, Z_0 can be the original object point, and we must somehow specify the direction cosines of the particular entering ray that we wish to trace. This is often done by specifying the point at which the entering ray pierces the tangent plane at the first lens vertex. Then, knowing X_0, Y_0, Z_0, and K, L, M, we can determine the point X, Y, Z at which the ray strikes the following lens surface, and after refraction it will have a new set of direction cosines K', L', M' and proceed on its way. The ray-tracing problem thus reduces to two steps: the transfer of the ray from some known point to the next surface, and the refraction of the ray at the next surface.

8.3.1 Transfer Formulas

Since the direction cosines of a line are defined as the differences between the X, Y, Z coordinates of two points lying on the line divided by the distance between these points, it is clear from Figure 8.8 that

$$K = \frac{X - X_0}{D}, \qquad L = \frac{Y - Y_0}{D}, \qquad M = \frac{Z - Z_0 + d}{D}$$

where D is the distance along the ray from the point X_0 Y_0 Z_0 to the point of incidence X, Y, Z, and d is the axial separation of the surfaces. By means of these relationships we see that

$$X = KD + X_0, Y = LD + Y_0, Z = MD + (Z_0 - d) \qquad (8\text{-}4a)$$

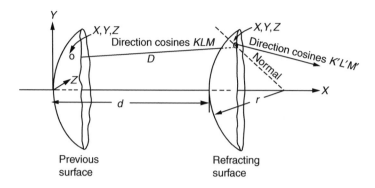

Figure 8.8 Transfer of a skew ray from one surface to the next.

The equation of the next refracting surface is, of course, known. For a sphere of radius r it is

$$X^2 + Y^2 + Z^2 - 2rX = 0 \qquad (8\text{-}4\text{b})$$

and substituting Eqs. (8-3a) in (8-3b) gives the equation to be solved for D as

$$D^2 - 2rF \cdot D + rG = 0$$

where

$$F = M - \frac{KX_0 + LY_0 + M(Z_0 - d)}{r}$$

$$G = \frac{X_0^2 + Y_0^2 + (Z_0 - d)^2}{r} - 2(Z_0 - d) \qquad (8\text{-}5)$$

The solution is, of course,

$$D = r\left[F \pm \left(F^2 - \frac{G}{r} \right)^{1/2} \right]$$

The ambiguous sign of the root indicates the two possible points of intersection of the ray with a complete sphere of radius r. Only one of these is useful, and the appropriate sign must be chosen. Remember that D must always be positive. Knowing D we return to Eq. (8-4a) and calculate X, Y, and Z, the coordinates of the point of incidence. For a plane surface,

$$D = G/2F = -(Z_0 - d)/M$$

8.3.2 The Angles of Incidence

It is a well-known property of direction cosines that the angle between two intersecting lines is given by

$$\cos I = Kk + Ll + Mm$$

Here K, L, M are the direction cosines of the ray, and k, l, m the direction cosines of the normal at the point of incidence. For a spherical surface,

$$k = -\frac{X}{r}, \; l = -\frac{Y}{r}, \; m = 1 - \frac{Z}{r} \tag{8-6}$$

Hence,

$$\cos I = F - \frac{D}{r}$$
$$\cos I' = [1 - (n/n')^2(1 - \cos^2 I)]^{1/2} \tag{8-7}$$

For a plane, $\cos I = K$.

8.3.3 Refraction Equations

To derive the refraction equations, we refer back to Figure 2.3, used in connection with the process of graphical ray tracing. It is reproduced and enhanced in Figure 8.9. In the vector triangle OAB, OA is a vector of magnitude n in the direction of the incident ray, OB is a vector of magnitude n' in the direction of the refracted ray, while AB is a vector of magnitude $n' \cos I' - n \cos I'$ in the direction of the normal. Hence, we may construct the vector equation

$$n'\mathbf{R}' = n\mathbf{R} + (n' \cos \mathbf{I'} - n \cos \mathbf{I})\mathbf{N}$$

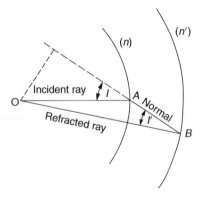

Figure 8.9   Refraction of a skew ray.

where $\mathbf{R'}$, $\mathbf{R}$, and $\mathbf{N}$ are unit vectors. Since the components of a unit vector are simply the direction cosines of the vector, we can resolve the vector equation into its three component equations:

$$n'K' = nK + (n'\cos I' - n\cos I)k$$
$$n'L' = nL + (n'\cos I' - n\cos n\cos I)l \tag{8-8}$$
$$n'M' = nM + (n'\cos I' - n\cos I)m$$

The direction cosines of the normal, k, l, and m, are given in Eq. (8-6). Hence Eq. (8-8) becomes

$$n'K' = nK - JX$$
$$n'L' = nL - JY \tag{8-9}$$
$$n'M' = nM - J(Z - r)$$

where $J = (n'\cos I' - n\cos I)/r$. As a check on our work, we can verify that $(K'^2 + L'^2 + M'^2) = 1$. For refraction at a plane surface these relations become

$$M = \cos I, \quad M' = \cos I'$$
$$n'K' = nK, \quad n'L' = nL, \quad J = 0$$

8.3.4 Transfer to the Next Surface

This has already been described. The direction cosines K', L', M' become the new K, L, M, and we calculate the new point of incidence by Eqs. (8-4), (8-5), (8-7), and (8-9), in order.

8.3.5 Opening Equations

1. *Distant Object*
   Here we have a parallel beam incident on the lens inclined at an angle U_{pr} to the lens axis. Then

$$K = 0, \quad L = \sin U_{pr}, \quad M = \cos U_{pr}$$

The point of incidence of the particular skew ray must be determined in some way so that the X, Y, Z can be found. It is common to define the ray by its point of incidence with the tangent plane at the vertex of the first surface. If this is done, it is convenient to regard this tangent plane as the first lens surface with air on both sides of it, and use the general transfer equations to go from the tangent plane to the first refracting surface in the ordinary way.

2. *Near Object*

Here again we assume a tangent plane at the first lens surface, and we specify the point Y, Z at which the skew ray is to pierce that plane. The X_0, Y_0, Z_0 of the object point are, of course, known and also the distance d between the object and the front vertex. Then

$$K = \frac{X - X_0}{D}, \quad L = \frac{Y - Y_0}{D}, \quad M = \frac{d}{D}$$

where

$$D^2 = d^2 + (X - X_0)^2 + (Y - Y_0)^2.$$

8.3.6 Closing Equations

The closing equations for a skew ray are trivial, since the ray can be transferred to the final image plane by the ordinary transfer equations. This process gives the X', Y' coordinates of the intersection of the ray with the image plane directly. The d is, of course, nothing but the back focal distance from the rear vertex of the lens to the image plane.

8.3.7 Diapoint Location

For some purposes we may desire to determine the diapoint location of the skew ray. As has been stated, this is the point where the ray pierces the meridional plane. The X coordinate of the diapoint is therefore zero, but the other coordinates must be found. By means of a diagram such as that in Figure 8.10 it is easy to show that

$$L'_d = -X'M'/K' \quad \text{and} \quad H'_d = Y' - (X'L'/K')$$

where K', L', M' are the direction cosines of the ray as it emerges from the lens, and Y', Z' are the coordinates of the point where the ray pierces the image plane. The L'_d, H'_d are the required coordinates of the diapoint relative to the midpoint of the image plane and the optical axis of the lens.

8.3.8 Example of a Skew-Ray Trace

To illustrate the kind of record required in the manual tracing of a skew ray by these formulas, we will trace a ray through our old familiar cemented doublet objective, entering at an upward slope of 3° through a point at unit distance

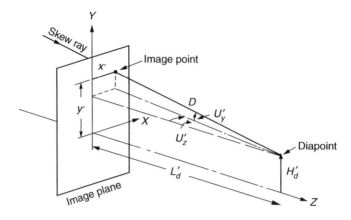

Figure 8.10 Diapoint calculation.

behind the meridional plane and on the same level as the principal ray. Regarding the tangent plane at the first vertex as a refracting surface, the starting data at that surface are

$$Z = 0, \quad M = \cos(-3°) = 0.9986295$$
$$Y = 0, \quad L = \sin(-3°) = -0.0523360$$
$$X = 1, \quad K = 0$$

We now transfer the ray from the tangent plane to the first spherical refracting surface in the usual way. The results of the trace are shown in Table 8.2.

8.4 GRAPHICAL REPRESENTATION OF SKEW-RAY ABERRATIONS

8.4.1 The Sagittal Ray Plot

The name *sagittal* is generally given to the 90° and 270° skew rays that lie in a plane perpendicular to the meridional plane, containing the principal ray. This is not one single plane throughout a lens but it changes its tilt after each surface refraction (Figure 8.1). The point of intersection of a sagittal ray with the paraxial image plane may have both a vertical error and a horizontal error relative to the point of intersection of the principal ray, and both these errors can be plotted separately against some suitable ray parameter. This parameter is often the entrance pupil coordinate for $\theta = 90°$ or the horizontal distance x from the meridional plane to the point where the entering ray pierces the tangent plane at

Table 8.2
Manual Tracing of a Skew Ray

	Tangent plane				*Image plane*
r	∞	7.3895	−5.1784	−16.2225	∞
d	0	1.05	0.4	11.28584	
n	1	1.517	1.649	1	
$(n/n')^2$		0.4345390	0.8463106	2.719201	
F	0.9986295	0.8000638	0.9673926		0.9952001
G	0.1353271	1.584704	0.9077069		22.627015
D	0.0680706	0.8939223	0.4623396		11.368061
X	1	1.0	0.9584830	0.9457557	−0.0033456
Y	0	0.0035625	0.0342546	0.0491091	0.6289086
Z	0	0.0679773	−0.0895928	−0.0276675	0
cos I		0.9894178	0.9726889	0.9958924	
cos I'		0.9954155	0.9769360	0.9887907	
J		0.0704550	−0.0261467	0.0402796	
K	0	−0.0046436	−0.0275280	−0.0834884	
L	0.0523360	0.0343342	0.0321289	0.0510025	
M	0.9986295	0.9983305	0.9991040	0.9952011	

the first lens vertex. Figure 8.11 shows the layout for an $f/2.8$ triplet photographic objective[7] having a focal length of 10, and Figure 8.12 shows the set of meridional and sagittal ray plots for this lens having the following prescription:

Radius	t	n	V
4.7350			
	0.6372	1.7440	44.9 (LaF2)
148.835 (stop)			
	1.0015	(air)	
−5.8459			
	0.2705	1.7400	28.2 (SF3)
5.1414			
	0.9253	(air)	
33.1041			
	0.6979	1.7440	44.9 (LaF2)
−4.4969			
	8.4894	(air)	

The meridional plot, of course, has no symmetry, but the two sagittal ray plots do have symmetry. The vertical errors are identical for rays entering at equal distances in front of and behind the meridional plane, these errors being forms of sagittal coma. The horizontal errors are antisymmetrical, so that the error of the 90° ray is equal and opposite to the horizontal error of the 270° ray; these errors represent sagittal field curvature and sagittal oblique spherical aberration, strictly analogous to the effects of tangential field curvature and tangential oblique spherical aberration in the ordinary meridional ray plot.

8.4.2 A Spot Diagram

The meridional and sagittal ray plots already discussed take account of only the rays passing through a cross-shaped aperture over the lens. To include every possible skew ray, it is necessary to divide the lens aperture into a checkerboard of squares, or a rectangular grid, and to trace a ray through every intersection of the lines. Assuming that each ray carries the same amount of light energy, the assembly of the intersection points of all such rays with the image plane will be a fair representation of the type of image that may be expected when the lens has been fabricated, assembled, and tested.

Actually it requires a large number of rays, say hundreds, to provide a fair approximation to the actual image. Of course, it is unnecessary to trace rays

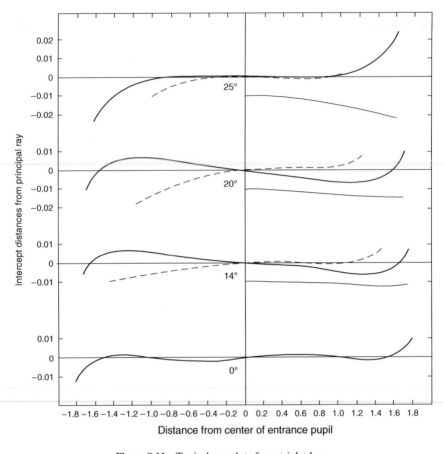

Figure 8.11 Typical ray plots for a triplet lens.

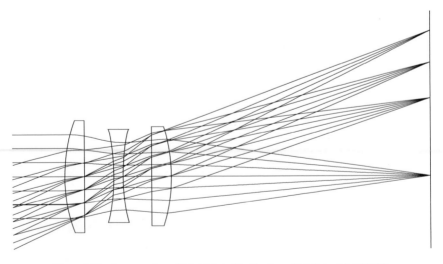

Figure 8.12 Layout of an f/2.8 triplet objective lens, U.S Patent 2,966,825.

both behind and in front of the meridional plane since they are identical, but it is necessary to plot both rays in the image plane.[8] Such dot patterns are called spot diagrams, and they were obviously never plotted before the advent of high-speed computers to do the ray tracing.[9]

A typical spot diagram for the aforementioned $f/2.8$ triplet photographic objective is shown in Figure 8.13a, where a rectangular pattern was used; for this pattern over a thousand skew rays were traced through each side of the lens aperture. Figure 8.13b shows the same spot diagram except that a hexapolar pattern was used, and a dithered pattern was used in Figure 8.13c. The dithered pattern traces a pseudorandom distribution of rays through the entrance pupil. The intent of using a dithered pattern is to mitigate the symmetry artifacts caused by using either the rectangular or the hexapolar pattern.

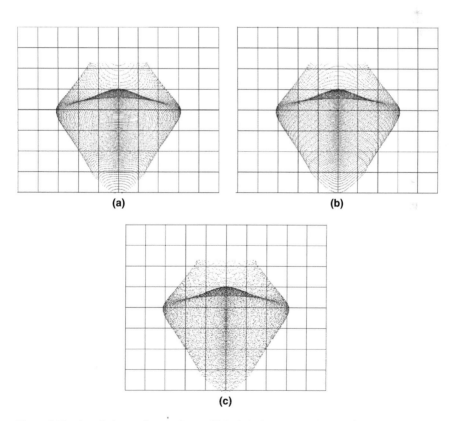

(a)

(b)

(c)

Figure 8.13 A typical spot diagram for an $f/2.8$ triplet lens at $14°$ off axis ($f' = 10$) for d light. The ray pattern used at the entrance pupil was (a) rectangular, (b) hexapolar, and (c) dithered.

The comparison of Figures 8.13a, b, and c clearly demonstrates the presence of such artifacts and their mitigation. It should be recognized that there is no perfect or best ray pattern. Also, the lens designer should be cautious about making inferences about aberrations in the lens as a consequence of observing such induced artifacts. And finally, spot diagrams are strictly geometric; however, an image of the point source accounting for both aberrations and diffraction can computed and displayed. Figure 8.14 presents the *point spread function* (PSF) for the same image shown in Figure 8.13. Notice the similarities and differences between them, but realize that the PSF is closer to what will actually be observed.

DESIGNER NOTE

A rough rule of thumb, often called the "three-to-one rule," can be used to decide if diffraction or geometric aberration dominates. If the rms geometric blur diameter is less than one-third of the diffraction blur diameter, then diffraction is dominant. The converse is also true. In the "in-between" region, both must be considered. The 3:1 ratio can be 5:1 or whatever the designer desires, but not less than 3:1.

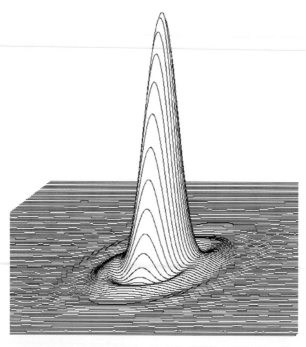

Figure 8.14 Point spread function for the same lens and object as in Figure 8.13.

8.4.3 Encircled Energy Plot

By counting the rays enclosed by a succession of circles of increasing size laid over the spot diagram, it is possible, by a suitable computer program, to plot a graph of the *encircled energy* of a given lens at several obliquities. This assumes that each ray carries the same amount of light energy, a justifiable assumption if the rays are incident in a checkerboard pattern at the entrance pupil of the lens. To include chromatic effects it is necessary to trace many rays in other colors, the size of the checkerboard squares for each wavelength being dependent on the spectral response of the detector intended for that particular lens.

The encircled energy graphs of the *f*/2.8 triplet used above are shown in Figure 8.15. As can be seen, this one plot shows the performance of the lens at several obliquities. Graphs of this type provide the designer with a great deal of useful information, particularly in comparing one design with another. It is noted that Figure 8.15 is based on only geometric ray trace data. Encircled energy plots can be made where the geometric data are multiplied by the diffraction-limited values to produce perhaps a better expectation of what may be observed. Diffraction-encircled energy plots can also be produced which properly account for both diffraction and aberrations; however, computation is more time-consuming than for the other versions.

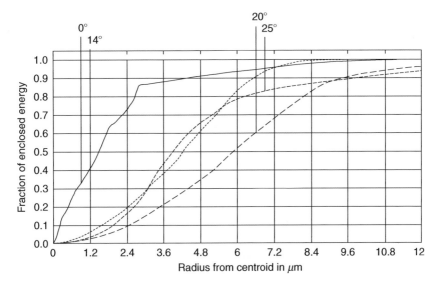

Figure 8.15 Encircled energy plot.

8.4.4 Modulation Transfer Function

An important addition to the tools of the lens designer was the development of the modulation transfer function (*MTF*) for optical systems; serious interest in *MTF* began in the 1950s although it was not actually accepted by most practitioners until the 1970s. Today, the *MTF* is arguably the dominant method for describing the performance of lenses. Willams and Becklund have presented a comprehensive history and study of the optical transfer function (*OTF*).[10] The *MTF* = |*OTF*| where the *OTF* is a complex value. Analogous to electronic communication systems, optical systems can also be considered as linear systems and utilize similar theory.[11] In an electrical system, the *MTF* is essentially the ratio of the output to the input of a linear system as a function of frequency (i.e., $MTF(f) = \text{output}(f)/\text{input}(f)$).

In a like manner, the *MTF* for a lens is $MTF(v) = \text{output}(v)/\text{input}(v)$, where *v* is spatial frequency and may be multidimensional, unlike temporal frequency used in electronics. In a simple sense, the *MTF* is a measure of the fidelity the lens-formed image has to the object. Although the measurement should be, and has been, accomplished using sinusoidal spatial targets, most often *MTF* is measured using alternating black and white bars. The resulting *MTF* is known as square-wave *MTF* and the plots are not the same as the sine-wave *MTF*. Figure 8.16 shows the geometrical *MTF* for the triplet lens multiplied by

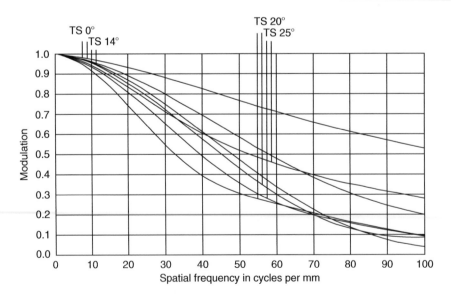

Figure 8.16 Geometrical *MTF* multiplied by the diffraction-limited *MTF*.

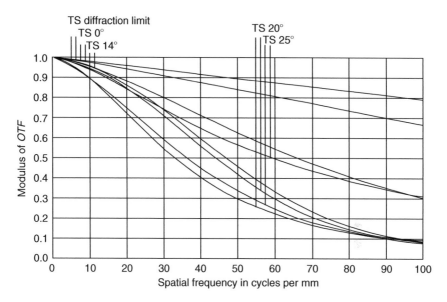

Figure 8.17 Diffraction-based *MTF* and curve showing the diffraction-limited *MTF* for comparison.

the *MTF* for a diffraction-limited lens. Compare this *MTF* to that in Figure 8.17, which shows the diffraction-based *MTF*. As can be observed, the axial geometric *MTF* is underestimated while being overestimated for 14° and is about the same for the greater off-axis object points. Multiplication of the geometric *MTF* by the *MTF* for a diffraction-limited lens should be used only when the geometric blur is not greater than about one-third that of the diffraction blur.

When the lens is to be used over some finite spectral bandwidth, a method has been determined for estimating the blur size and shape for a polychromatic source and an aberration-free lens system.[12] The perfect-image irradiance distribution of a polychromatic point source can be written as

$$E(r) = C_1 \int_0^\infty \tilde{\Re}(\lambda) \left[\frac{2J_1\left(kD_{ep}r/2\right)}{kD_{ep}r} \right]^2 d\lambda$$

where $\tilde{\Re}(\lambda)$ is the peak normalized spectral weighting factor and C_1 is a scaling factor. By invoking the central limit theorem to approximate this distribution by a Gaussian function, we obtain

$$E(r) \approx C_2 e^{-\left(r^2/2\sigma^2\right)}$$

where C_2 is a scaling constant and σ^2 is the estimated variance of the irradiance distribution. When $\tilde{\Re}(\lambda) = 1$ in the spectral interval λ_{short} to λ_{long} and zero otherwise with $\lambda_{short} < \lambda_{long}$, an estimate of sigma can be written as

$$\sigma = \frac{M\lambda_{long}}{\pi D_{ep}}$$

where $M = 1.335 - 0.625b - 0.25b^2 - 0.0465b^3$ with $b = (\lambda_{long}/\lambda_{short}) - 1$. Should $\tilde{\Re}(\lambda) = \lambda/\lambda_{long}$ in the spectral interval λ_{short} to λ_{long} and zero otherwise, which approximates the behavior of a quantum detector, $M = 1.335 - 0.65b + 0.385b^2 - 0.099b^3$.

The Gaussian estimate of residual error is less than a few percent for $b = 0.5$ and remains useful even as $b \to 0$. A useful estimation of the modulation transfer function for this diffraction-limited *polychromatic* lens system is given by

$$MTF(v) \approx e^{-2(\pi\sigma v)^2}$$

where v is the spatial frequency. This approximation can provide a useful insight into expected performance limits.

8.5 RAY DISTRIBUTION FROM A SINGLE ZONE OF A LENS

The nature of the various oblique aberrations of a lens may be better understood if we trace a family of rays passing through a single zone of a lens, both on and off axis. We take the cemented telescope doublet used many times before and isolate a single zone of radius one unit. On axis, all the rays from this zone will, of course, intersect at a single point, forming a perfect focus. At an obliquity of only one degree, however, the rays from the zone form a succession of complicated loop figures as shown in Figure 8.18. As before, the upper and lower rim rays are labeled 0° and 180°, while the sagittal rays are 90° and 270°.

Referring to this diagram, the tangential focus is at the intersection of rays 0° and 180°, giving 0.000084 for the amount of tangential coma. The sagittal focus is where the 90° and 270° rays intersect, forming a sagittal coma of magnitude 0.000035, about one-third of the tangential coma. It can be proved that in the absence of higher-order aberrations this ratio should be exactly 3:1.[13] The presence of field curvature is indicated by the tangential and sagittal foci not being in the same plane as the axial image of the zone. This series of patterns arises at each zone of the lens, and it is clear that when all zones are open together, the resulting image is very complicated indeed.

It should be noted that the diapoint locus of the zone is also indicated in Figure 8.18. It is bisected by the sagittal image, and it ends at upper and lower tangential rays. Conrady refers to the diapoint locus as the *characteristic focal line* of a zone.

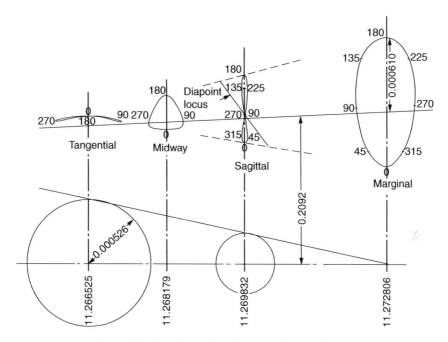

Figure 8.18 Ray distribution from a single zone of a lens.

ENDNOTES

[1] A. E. Conrady, pp. 284, 742.

[2] The word *vignetting* is pronounced as "vǐn yět' ting". It is used in the context that the image formed shades off gradually toward the edges as a consequence of more and more rays in the beam being vignetted as the obliquity angle increases.

[3] MIL-HDBK-141, *Optical Design*, Section 6.11.8, Defense Supply Agency, Washington, DC (1962).

[4] R. Kingslake, "Illumination in optical images," *in Applied Optics and Optical Engineering*, II:195, Academic Press, New York (1965).

[5] MIL-HDBK 141, *Optical Design*, Chapter 5, Defense Supply Agency, Washington, DC (1962).

[6] Daniel Malacara and Zacarias Malacara, *Handbook of Lens Design*, Chapter 2, Marcel Dekker, New York (1994).

[7] C. Baur and C. Otzen, U.S. Patent 2,966,825, filed in February (1957).

[8] The context of "behind and in front of the meridional plane" is that one is considered to be viewing the lens from the side with light propagation going from left to right. Front is the space between the observer and the meridional plane and contains entrance pupil coordinates containing θ values from 0° to 180°.

[9] As a historical note, once high-speed computers became available in the late 1950s and into the 1970s, plotters were often not readily available and were rather slow as well. A clever method was devised to use the text printer to generate a useful spot diagram. The technique was to consider each print character position as a ray bin and a character was assigned to each bin to represent the number of rays "hitting" that bin. Of course, a blank meant no rays, but then the count went as 1, 2, 3, ..., 9, A, B, C, ..., Z. Although such a spot diagram is inferior to the spot diagrams of today, a lens designer quickly learned to "read" such character-based spot diagrams. Printers even in the 1950s were quite fast because a primary utilization of computers was for business and government accounting, inventory, and payroll.

[10] Charles S. Williams and Orville A. Becklund, *Introduction to the Optical Transfer Function*, Wiley, New York (1989).

[11] R. Barry Johnson, "Radar analogies for optics," *Proc. SPIE*, 128:75–83 (1977).

[12] R. Barry Johnson, "Lenses," in *Handbook of Optics, Second Edition,* Vol. II, Chapter 1, pp. 1.39–1.41, Michael Bass (Ed.), McGraw-Hill, New York (1995).

[13] *Hint:* Consider the ray aberration expansion equations in Section 4.2, assume all aberrations are zero except for third-order coma (σ_2), and consider the ratio of the values of ε_y for $\theta = 0°$ and $90°$.

Chapter 9

Coma and the Sine Condition

9.1 THE OPTICAL SINE THEOREM

The Lagrange theorem applies only to paraxial rays, while the optical sine theorem is the equivalent for marginal rays. The optical sine theorem provides an expression for the image height formed by a pair of sagittal rays passing through a single zone of a lens. It is valid for a zone of any size but only at very small obliquity. This obliquity limitation effectively removes all aberrations except coma, which is represented by a difference between the image height for the selected zone and the paraxial image height given by the theorem of Lagrange. Recall that coma can be considered a variation in magnification from one zone to another zone as discussed in Section 4.3.4.

To derive the optical sine theorem we consider the perspective diagram in Figure 9.1a, which shows a pair of sagittal rays passing through a single refracting surface, and Figure 9.1b, which shows the path of the marginal ray through the same zone. The entering and emerging marginal ray slopes are U and U', respectively, in the usual way.

It was pointed out in Section 8.1.1 that a pair of sagittal rays intersect on the auxiliary axis drawn through the object point and the center of curvature of the surface. Hence, by the similar triangles shown in Figure 9.1,

$$\frac{h'_s}{h} = \frac{CS}{CB_0} = \frac{L' - r}{L - r} = \left(\frac{P'}{\sin U'}\right)\left(\frac{\sin U}{P}\right)$$

$$= \frac{n \sin U}{n' \sin U'}$$

Hence,

$$h'_s \, n' \sin U' = hn \sin U \tag{9-1}$$

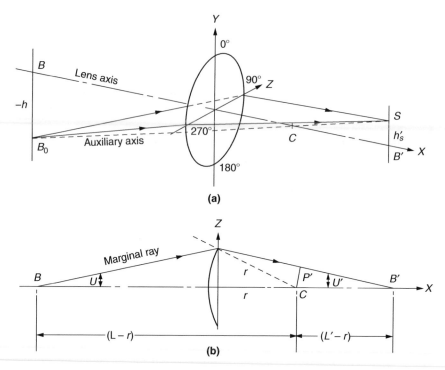

Figure 9.1   Derivation of the sine theorem: (a) oblique view and (b) plan view.

It is essential to remember that h'_s is the height of the sagittal image for the zone, namely, the intersection of the 90° and 270° rays, and has no relation whatsoever to the height of any other rays from the zone. It is, in particular, not related to the meridional rays in any way.

9.2 THE ABBE SINE CONDITION

Abbe regarded coma as a consequence of a difference in image height from one lens zone to another, and he thus realized that a spherically corrected lens (in his case a microscope objective) would be free from coma near the center of the field if the paraxial and marginal magnifications $m = nu/n'u'$ and $M = n \sin U/n' \sin U'$ were equal, that is,

$$u/u' = \sin U / \sin U' \qquad (9\text{-}2)$$

This is known as the *Abbe sine condition*.

For a very distant object, the sine condition takes a different form. As was shown in Section 3.3.4, the Lagrange equation for a distant object can be written as

$$h' = -(n/n')f' \tan U_{\text{pr}}$$

where f' is the distance from the principal plane to the focal point measured along the paraxial ray, or $f' = y_1/u'_k$. A similar expression can be written for the focal length of a marginal ray, namely,

$$F' = Y_1 / \sin U'_k \tag{9-3}$$

where F' is the distance measured along the marginal ray from the equivalent refracting locus to the point where the ray crosses the lens axis. Thus for a spherically corrected lens and a distant object, Abbe's sine condition reduces to

$$F' = f'$$

This relation tells us that in such a lens, called by Abbe an *aplanat*, the equivalent refracting locus is part of a hemisphere centered about the focal point. The maximum possible aperture of an aplanat is therefore $f/0.5$, although this aperture is never achieved in practice. The greatest practical aperture is about $f/0.65$ when the emerging ray slope is about $50°$.

There is no equivalent rule for a lens that is aplanatic for a near object, such as a microscope objective. We can, if we wish, assume that in such a case the two principal planes are really parts of spheres centered about the axial conjugate points, but we could just as easily make any other suitable assumption provided the marginal ray moves from one principal "plane" to the other along a line lying parallel to the lens axis, as indicated for paraxial rays in Figure 3.10.

If the refractive index of either the object space or the image space is other than 1.0, we must include the actual refractive index in the f-number:

$$f\text{-number} = \frac{\text{focal length } f'}{\text{entering aperture } 2y} \left(\frac{n}{n'}\right)$$

Thus if the image space were filled with a medium of refractive index 1.5, the highest possible relative aperture would be $f/0.33$. To realize the benefit of this high aperture, the receiver, film, CCD, or photocell must be actually immersed in the dense medium. Similarly, when a camera is used for underwater photography, the effective aperture of the lens is reduced by a factor of 1.33, which is the refractive index of water.

When the object is not located at infinity, the effective f-number is given by

$$f\text{-number}_{\text{effective}} = f\text{-number}_\infty (1 - m).$$

If, for example, a lens is being used at unity magnification ($m = -1$), then $f\text{-number}_{\text{effective}} = 2f\text{-number}_\infty$. The *numerical aperture* of a lens is $NA = n' \sin U'$. If the lens is aplanatic, $f\text{-number}_{\text{effective}} = \frac{1}{2NA}$.

9.2.1 Coma for the Three Cases of Zero Spherical Aberration

It was shown in Section 6.1.1 that there are three cases in which a spherical surface has zero spherical aberration: (a) when the object is at the surface itself, (b) when the object is at the center of curvature of the surface, and (c) when the object is at the aplanatic point. It so happens that each of these possible situations also satisfies the Abbe sine condition, thus justifying the name *aplanatic* for all of them. The reason for this is that in each case the ratio sin U/sin U' is a constant. Thus, we have the following:

- Case (a), object at surface: $U = I$, $U' = I'$; hence sin U/sin $U' = n'/n$
- Case (b), object at center: $U = U'$; hence sin U/sin $U' = 1$
- Case (c), object at aplanatic point: $I = U'$, $I = U$; hence sin U/sin $U' = n/n'$

The aplanatic single-lens elements discussed in Section 6.1.2 are corrected for both spherical aberration and coma, and hence fully justify the name *aplanatic*. It should be added that such a lens introduces both chromatic aberration and astigmatism in the sense that would be expected from a single positive element.

9.3 OFFENSE AGAINST THE SINE CONDITION

It is clear that we ought to be able to derive some useful information about the magnitude of the coma from a knowledge of the paraxial and marginal magnifications, even though the lens does have some spherical aberration. This situation is indicated in Figure 9.2. In this diagram B' represents an oblique image point in the paraxial image plane P of a lens at very small obliquity, its height h' being given by the Lagrange equation. The point S represents the sagittal image formed by a single zone of the lens, its height h'_s being computable by the sine theorem. The point S is assumed to lie in the same focal plane as the marginal image M. At the very small obliquity considered here, the principal ray must be traced by paraxial formulas; it emerges through the center of the exit pupil EP' as shown.

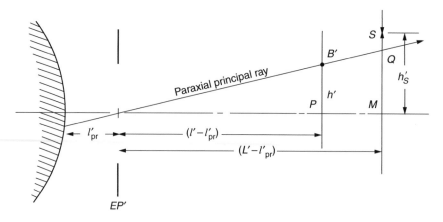

Figure 9.2 Offense against the sine condition (*OSC*).

We may express the magnitude of the sagittal coma by the dimensionless ratio QS/QM in the marginal image plane, and we call this ratio the "offense against the sine condition," or OSC (see also Section 4.3.4 and Eq. (10-3)). Thus

$$OSC = \frac{QS}{QM} = \frac{SM - QM}{QM} = \frac{SM}{QM} - 1$$

The length SM is the h'_s given by the sine theorem; the length QM is obtainable from the paraxial image height h' by

$$QM = h' \left(\frac{L' - l'_{pr}}{l' - l'_{pr}} \right)$$

Hence

$$OSC = \frac{h'_s}{h'} \left(\frac{l' - l'_{pr}}{L' - l'_{pr}} \right) - 1$$

For a near object we can insert the values of h' and h'_s by the Lagrange and sine theorems, respectively, giving

$$OSC = \frac{u' \sin U}{u \sin U'} \left(\frac{l' - l'_{pr}}{L' - l'_{pr}} \right) - 1$$

$$= \frac{M}{m} \left(\frac{l' - l'_{pr}}{L' - l'_{pr}} \right) - 1 \tag{9-4}$$

where M and m are, respectively, the image magnification for the finite and paraxial rays.

The bracketed quantity, which contains data relating both to the spherical aberration of the lens and the position of the exit pupil, can be readily modified to

$$\left(1 - \frac{LA'}{L' - l'_{pr}} \right)$$

and for a very distant object, M/m can be replaced by F'/f'. Hence for a distant object, Eq. (9-4) becomes

$$OSC = \frac{F'}{f'} \left(1 - \frac{LA'}{L' - l'_{pr}} \right) - 1 \tag{9-5}$$

Conrady[1] states that the maximum permissible tolerance for OSC is 0.0025 for telescope and microscope objectives. This large tolerance is because in those instruments the object of principal interest can always be moved into the center of the field for detailed study. A very much smaller tolerance applies to photographic objectives.

9.3.1 Solution for Stop Position for a Given *OSC*

Since the exit-pupil position (l'_{pr}) appears in the formulas for *OSC*, it is clear that as we shift the stop along the axis the *OSC* will change provided there is some spherical aberration in the lens. *If the spherical aberration has been corrected, then shifting the stop will have no effect on the coma.* We can thus solve for the value of l'_{pr} to give any desired *OSC* by inverting Eqs. (9-4) and (9-5).

For a near object,

$$l'_{pr} = L' - \frac{LA'}{(\Delta m/M) - (m\,OSC/M)}$$

For a distant object,

$$l'_{pr} = L' - \frac{LA'}{\Delta F/F' - (f'\,OSC/F')}$$

These formulas find use in the design of simple eyepieces and landscape lenses for low-cost cameras.

9.3.2 Surface Contribution to the *OSC*

By a process similar to that used for determining the surface contribution to spherical aberration (Section 6.1), we can develop a formula giving the surface contribution to the *OSC*. For this derivation, we trace a marginal ray and the paraxial principal ray. The development given in Section 6.1 indicates that in our present case we have

$$(Snu_{pr})'_k - (Snu_{pr})_1 = \sum (Q - Q')ni_{pr} \tag{9-6}$$

We can see from the diagram in Figure 9.3 that $S' = (L' - l'_{pr})\sin U'$, and similarly for the incident ray. Hence, dividing Eq. (9-6) by the Lagrange invariant and substituting for S and S' we get

$$\left[\frac{(L' - l'_{pr})\sin U' n' u'_{pr}}{h'n'u'}\right]_k - \left[\frac{(L - l_{pr})\sin U n u_{pr}}{hnu}\right]_1 = \sum \frac{(Q - Q')ni_{pr}}{(h'n'u')_k} \tag{9-7}$$

Now $h'/u'_{pr} = (l' - l'_{pr})$, and $h/u_{pr} = (l - l_{pr})$. Also, by the Lagrange and sine theorems we have

$$\left[\frac{\sin U'}{u'}\right]_k = \frac{hn\sin U}{h'_s n'}\left(\frac{h'n'}{hnu}\right) = \frac{\sin U_1}{u_1}\left(\frac{h'}{h'_s}\right)_k$$

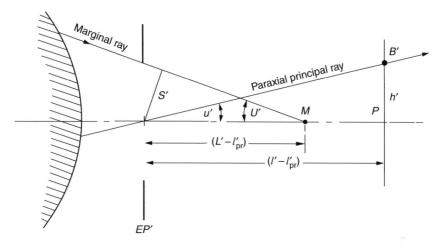

Figure 9.3 Surface contribution to *OSC*.

Substituting all this in Eq. (9-7) gives

$$-\left[\left(\frac{L'-l'_{\text{pr}}}{l'-l'_{\text{pr}}}\cdot\frac{h'}{h'_s}\right)_k\frac{\sin U_1}{u_1}\right]+\left[\frac{L-l_{\text{pr}}}{l-l_{\text{pr}}}\cdot\frac{\sin U}{u}\right]_1=\sum\frac{(Q-Q')ni_{\text{pr}}}{(h'n'u')_k} \qquad (9\text{-}8)$$

Now by Figure 9.2 we see that

$$\left(\frac{L'-l'_{\text{pr}}}{l'-l'_{\text{pr}}}\cdot\frac{h'}{h'_s}\right)_k=\frac{QM}{SM}=\frac{SM-QS}{SM}=1-\frac{\text{coma}'_s}{SM}=1-OSC \qquad \text{(approx.)}$$

Thus Eq. (9-8) becomes

$$(OSC-1)+\left(\frac{L-l_{\text{pr}}}{l-l_{\text{pr}}}\right)_1=\frac{u_1}{\sin U_1}\sum\frac{(Q-Q')ni_{\text{pr}}}{(h'n'u')_k}$$

and hence

$$OSC=\left[1-\frac{L-l_{\text{pr}}}{l-l_{\text{pr}}}\right]_1+\frac{u_1}{\sin U_1}\sum\frac{(Q-Q')ni_{\text{pr}}}{(h'n'u')_k}$$

$$=\frac{-LA_1}{(l-l_{\text{pr}})_1}+\frac{u_1}{\sin U_1}\sum\frac{(Q-Q')ni_{\text{pr}}}{(h'n'u')_k} \qquad (9\text{-}9)$$

It should be noted that any spherical aberration in the object leads to a contribution to the OSC. Also, the factor outside the summation, $u_1/\sin U_1$, becomes y_1/Q_1 for a distant object.

Example

As an example of the use of this contribution formula, we will take our old familiar telescope doublet (Section 2.5) and trace a paraxial principal ray through the front vertex at an entering angle of, say, $-5°$ $(\tan(-5°) = -0.0874887)$, with the results shown in Table 9.1. Hence,

$$l'_{pr} = 0.9580946$$

$$OSC\frac{u'}{\sin U'}\left(\frac{l' - l'_{pr}}{L' - l'_{pr}}\right) - 1 = -0.000171$$

For the OSC contribution formula, we pick up the data of the marginal ray from Table 6.1, giving the tabulation shown in Table 9.2.

Table 9.1

Trace of Paraxial Principal Ray

$y_{pr}(nu)_{pr}$	0		0.0605558		0.0821525	
$(nu)_{pr}$		-0.0874887		-0.0874887	-0.0890323	-0.0857457
u_{pr}		-0.0874887		-0.0576721	-0.0539916	
$i_{pr} = (y_{pr}\,c - u_{pr})$		0.0874887		0.0459782	0.0489275	

Table 9.2

OSC Tabluation for Example Doublet

$Q - Q'$	-0.017118	-0.022061	0.037258
n	1	1.517	1.649
i_{pr}	0.0874887	0.0459782	0.0489275
Constant	5.715023	5.715023	5.715023
OSC contribution	-0.008559	-0.008794	0.017179	$\sum = -0.000173$

For this formula, the Lagrange invariant has the value $(h'n'u') = 0.1749774$. The excellent agreement between the direct and contribution calculations is evident. Also see Section 4.3.4 for an alternative OSC formula using the y-coordinate ray intercept data from a sagittal ray and a principal ray:

$$\frac{Y(\rho, 90^0, H, \xi) - Y(0, 0^0, H, \xi)}{Y(0, 0^0, H, \xi)}.$$

9.3.3 Orders of Coma

The coma of a pencil of rays at finite aperture and field may be analyzed into orders (see Section 4.3.4) as follows:

$$\text{coma} = a_1 Y^2 H' + a_2 Y^4 H' + a_3 Y^6 H' + \ldots$$
$$+ b_1 Y^2 H'^3 + b_2 Y^4 H'^3 + b_3 Y^6 H'^3 + \ldots$$
$$+ c_1 Y^2 H'^5 + c_2 Y^4 H'^5 + c_3 Y^6 H'^5 + \ldots$$
$$+ \cdots$$

The first term, $a_1 Y^2 H'$, is the primary term, and it evidently varies as aperture squared and obliquity to the first power. The whole top row of terms included in the *OSC* is applicable to any aperture but only to a small field. The higher-order terms represent forms of coma that appear in photographic lenses of high aperture at angles of considerable obliquity.

9.3.4 The Coma *G* Sum

There is a G-sum expression for the primary coma of a thin lens analogous to that for primary spherical aberration.[2] It varies with aperture squared and image height to the first power. The coma of the object, if any, is transferred to the final image by the ordinary transverse magnification, whereas primary spherical aberration, being a longitudinal quantity, is transferred by the longitudinal magnification rule.

It should be noted that this coma G-sum expression is valid only if the stop is located at the thin lens. The formula is

$$\text{coma}'_s = \text{coma}_s(h'/h) + h'y^2(-\tfrac{1}{4}G_5 cc_1 + G_7 cv_1 + G_8 c^2) \tag{9-10}$$

where

$$G_5 = 2(n^2 - 1)/n \quad G_7 = (2n + 1)(n - 1)/2n = G_2/n$$
$$G_8 = n(n - 1)/2 = G_1/n$$

As before, with a thin doublet we assume an infinitely thin air layer between the elements, and then the G sums may be directly added. Hence

$$OSC = \text{coma}'_s/h' = y^2[(G \text{ sum})_a + (G \text{ sum})_b]$$

9.3.5 Spherical Aberration and *OSC*

It should be clear by now that the spherical aberration of a lens is determined by the location of the intersection point of a ray with the lens axis, whereas the coma is determined by the slope angle of the ray at the image. If the shape of a

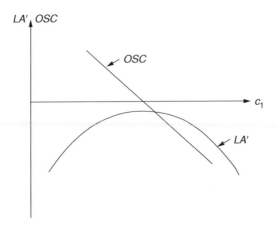

Figure 9.4   Typical effect of bending a single thin lens.

lens is such that the equivalent refracting locus is too flat, the marginal focal length will be too long and the OSC will be positive. A thin lens bent to the left meets this condition. Similarly, if the rim of the lens is bent to the right the OSC will be negative. Plotting spherical aberration and OSC against bending for such a lens gives a graph like the one in Figure 9.4.

It should be noted that in any reasonably thin lens, the lens bending for which the spherical aberration reaches an algebraic maximum is almost exactly the same bending as that which makes the OSC zero. For the primary aberrations of a single thin lens, this is easily verified by comparing the value of c_1 that makes $\partial LA'_p / \partial c_1 = 0$ (Section 6.3.2) with the value of c_1 that makes the coma$_p$ $= 0$ [Eq. (9-10)]. It will be found that for a variety of refractive indices and a variety of object distances, the c_1 for zero coma is always slightly greater than the c_1 for maximum spherical aberration.

DESIGNER NOTE

There is, of course, no aperture limit for a nonaplanatic system. A parabolic mirror, for example, has zero spherical aberration for a distant axial object point, but the focal length of each ray is the distance from the mirror surface to the image point, measured along the ray. The focal length continuously increases with increasing incidence height, which means the magnification is changing, as explained in Section 4.3.4. Consequently, the image is afflicted with enormous positive coma. Consider an $f/0.25$ parabolic mirror as illustrated in Figure 9.5. Notice that the marginal ray heads toward the image orthogonal to the optical axis just as it does for an aplanatic lens (Section 9.2); however, the focal length of the aplanatic lens remains constant as a function of incidence height while the marginal focal length of the parabola is twice that of its axial focal length.

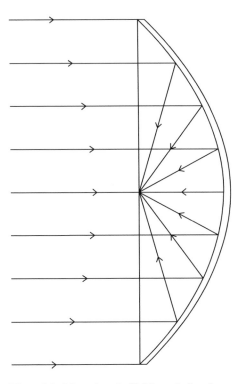

Figure 9.5 Nonaplanatic *f*/0.25 parabolic mirror.

Figure 9.6 shows the geometric spot diagram for an *f*/0.26 parabolic mirror with the paraxial image located only one-third of an Airy disk radius from the optical axis. The small circle in this figure represents the Airy disk. As can be observed, the coma is huge (contains many higher-order terms) compared to the diffraction blur, assuming no aberrations, although the shift in the object is just a fraction of the diffraction disk. This means that any calculations relying on this type of optical system behaving as a linear system will be seriously flawed. In contrast, a well-behaved system will have aberrations that are relatively slow to change as the field angle changes, thereby having regions in the image plane that are spatially stable, such that the shape (aberrations or wavefront) of the point-source image remains constant over an area of at least several Airy disk diameters. Such an image region is called an isoplanatic region or patch. Beware that some optical design programs may blindly compute diffraction-based *MTF*, point spread functions, etc. and produce erroneous results as a consequence of not using a correct modeling construct! Use common sense and test your program with an appropriate example to assure yourself of valid results.[3]

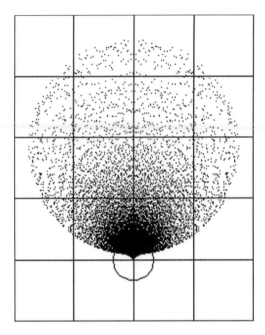

Figure 9.6 Image of point source located 1/3 of Airy radius from axis. The circle represents the Airy disk.

9.4 ILLUSTRATION OF COMATIC ERROR

As seen in the preceding Designer Note, a parabolic mirror is free of spherical aberration but suffers from coma near the axis. Figure 9.7 shows the ray fan plot at 2.25° off-axis for an $f/1.7$ parabolic mirror having a focal length of 12. The corresponding spot diagram for this mirror is illustrated in Figure 9.8. Examination of both figures indicates that the coma is essentially primary or third-order linear coma (σ_2). The secondary coma is over a factor of 20 times less as can be observed from the aberration coefficients.

Computing transverse coma using real rays was discussed in Section 4.3.4. The tangential component is given by

$$TCMA(\rho, H) = \left[\frac{Y(\rho, 0^0, H) + Y(\rho, 180^0, H)}{2}\right] - Y(0, 0^0, H) = 0.031504$$

and the sagittal component by

$$SCMA(\rho, H) = Y(\rho, 90^0, H) - Y(0, 0^0, H) = 0.010164$$

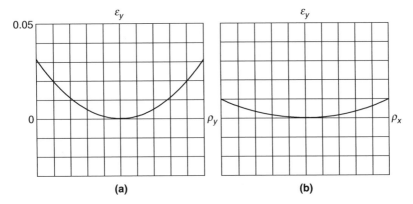

Figure 9.7 Ray fan plots at 2.25° off-axis for $f/1.7$ parabolic mirror having a focal length of 12: (a) tangential coma component; (b) sagittal coma component.

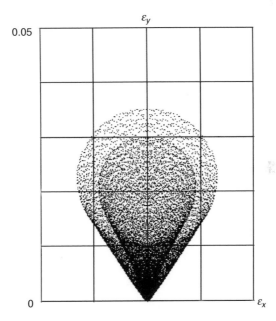

Figure 9.8 Spot diagram at 2.25° off-axis for $f/1.7$ parabolic mirror having a focal length of 12.

which compare appropriately to Figure 9.7. The ratio $TCMA/SCMA = 3.099$ which is about the 3.00 ratio expected for linear coma (see Section 4.3.4). Removal of the very small amount of astigmatic aberration was achieved by a slight defocus of 0.02 towards the mirror. Otherwise, for example, the ends of the curve in Figure 9.7a would be at different values (see Figure 4.4).

Figure 9.9 (a) Original photograph.[4] (b) Image showing coma formed by $f/1.7$ parabolic mirror.

To observe the degradation in image formation by coma caused by the parabola, we can generate a simulated image of a photograph using an analysis feature available in some lens design programs. Figure 9.9a shows the original photograph and Figure 9.9b the resultant image formed by the parabolic mirror. The linear growth of the comatic blur as a function of field angle is illustrated. Notice the fine detail reproduction in the center of Figure 9.9b since spherical aberration is absent; however, details such as scratches and specks rapidly blur away from the center of the image due to coma. Compare this image with the quadratic blur growth due to astigmatism shown in Figure 11.15.

ENDNOTES

[1] A. E. Conrady, p. 395.

[2] A. E. Conrady, p. 324.

[3] R. B. Johnson and W. Swantner, "MTF computational uncertainities," *OE Reports*, 104, August (1992).

[4] Circa 1910.

Chapter 10

Design of Aplanatic Objectives

It has already been mentioned that Abbe used the term *aplanatic* to refer to a lens system corrected for both spherical aberration and *OSC*. We shall use the term *aplanat* for a relatively thin spherically corrected achromat that is also corrected for *OSC* and thus satisfies the Abbe sine condition. As we have seen, a cemented doublet has three degrees of freedom, which are typically used to maintain the focal length and control the spherical and chromatic aberrations. To include the *OSC* correction requires an additional degree of freedom, which can be obtained in various ways. The principal types of aplanat will now be considered.

10.1 BROKEN-CONTACT TYPE

In this type of aplanat the powers of the two lens elements are determined for chromatic correction by the ordinary (c_a, c_b) formulas given by Eq. (5-4), and then each element is separately bent to correct the spherical aberration and *OSC*. Obviously such a lens cannot be cemented, and this type is used mainly in large sizes. It is possible to perform a thin-lens predesign by the use of Seidel aberration contributions, but the subsequent insertion of finite thicknesses causes such an upset that the preliminary study turns out to be useless.

Since bending a lens affects the spherical aberration most, the *OSC* much less, and the chromatic aberration scarcely at all, we select the bending of an achromatic doublet that corresponds to the peak of the spherical-aberration curve, since this is known to be close to the zero-*OSC* form. We then make small trial bendings of each element separately, and plot a double graph by which we can correct LA' and the *OSC* using the bending parameters c_1 and c_3.

Following this procedure, we see that the graph in Figure 7.2 for a crown-in-front achromat reaches its maximum at $c_1 = 0.15$. Then, at $c_a = 0.5090$, we find $c_2 = c_1 - c_a = -0.3590$. We start with $c_3 = c_2$ and a narrow air space such as 0.01. Suitable lens thicknesses are 0.42 and 0.15 for an aperture of 2.0 and a trim diameter of 2.2 (*f*/5). We shall, of course, achromatize every trial system by solving for the last radius by the $D - d$ method (see Section 5.9.2). Our starting System A is found to have

$$LA' = 0.1057, \quad OSC = 0.00062$$

the l'_{pr} for the *OSC* formula being taken as zero; that is, the stop is assumed to be in contact with the rear surface and consequently the exit pupil is the stop.

To build up our double graph, we next apply trial bendings of 0.01 to each of the two lens elements separately. Bending the crown element gives System B, with

$$LA' = 0.1057, \quad OSC = -0.00270$$

Restoring c_1 to its initial value and bending the flint element gives System C, with

$$LA' = -0.1245, \quad OSC = 0.00304$$

These values are plotted in Figure 10.1, with LA' as abscissa and OSC as ordinate. Inspection of the graph suggests that we ought to reach the aim point (0, 0), assuming that all aberrations are linear. The process requires that a line be

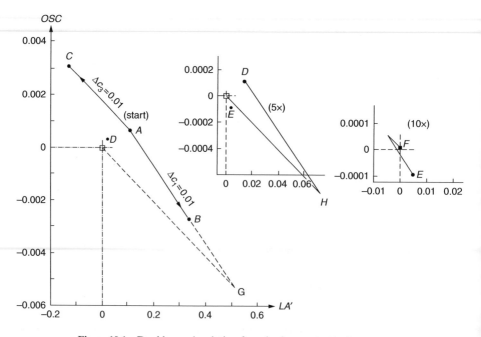

Figure 10.1   Double-graph solution for a broken-contact aplanat.

drawn from the aim point parallel to the line AC. Line AB is extended to intersect the line from the aim point at point G. We now apply to the original System A new changes for Δc_1 and Δc_3:

$$\Delta c_1 = \frac{AG}{AB} \cdot 0.01 = 0.0172 \quad c_3 = \frac{G(\text{aim point})}{AC} \cdot 0.01 = 0.0212.$$

These changes, with $c_a = 0.5090$ and the usual solution of the last radius, give System D, with

$$LA' = 0.01522, \quad OSC = 0.00010$$

The coma is satisfactory but the spherical aberration is still much too large.

Since we are too close to the aim point for the first graph to be useful, we enlarge both scales by a factor of 5, and drawing lines parallel to the original lines ($HE \parallel AC$ and $AB \parallel DH$) suggests that we try

$$\Delta c_1 = 0.0025, \quad \Delta c_3 = 0.0031$$

Applying these changes gives System E, with

$$LA' = 0.0052, \quad OSC = -0.00010$$

To remove these residuals resulting from slight nonlinearity of the adjustments, we draw an even larger-scale graph (10×), giving

$$\Delta c_1 = -0.00045, \quad \Delta c_3 = -0.00020$$

The final System F has $LA' = -0.00027$ and $OSC = 0$. The zonal aberration is $+0.0040$, of the unusual overcorrected type. As was pointed out in Section 7.3, this is to be expected in view of the narrow air space in this lens. Notice that the zonal aberration was undercorrected in that case, as is more commonly expected.

The final system (Figure 10.2) is given in the following tabulation:

c	d	n_D	V
0.16925			
	0.42	1.523	58.6
−0.33975			
	0.01	(air)	
−0.33490			
	0.15	1.617	36.6
−0.06682			

where $f' = 9.9302$, $l' = 9.6058$, $Y = 1.0$, trim diameter $= 2.2$, $LA' = -0.00027$, $LZA = +0.0040$, and the $OSC = 0$.

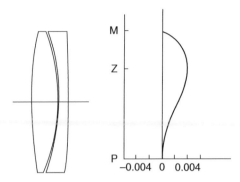

Figure 10.2   Broken-contact aplanat.

DESIGNER NOTE

It is worth noting that the air space in this lens has the form of a negative element, the equivalent of a positive glass lens that undercorrects the spherical aberration. Increasing the airgap will increase the zonal aberration noticeably, while decreasing it at this separation will reduce the zonal aberration only slightly. A broken-contact lens of this type requires the utmost care in mounting, and particularly in centering one element relative to the other.

In a large lens it is best to mount each element into a separate metal ring, using push–pull screws to secure and adjust the separation to give the best possible definition. For a small lens, the air space is too narrow for a loose spacer to be used, and it is best to mount the two elements on opposite sides of a fixed metal flange with separate clamping rings to hold them in place.

10.2 PARALLEL AIR-SPACE TYPE

As an alternative to the broken-contact type just discussed, we may prefer to keep the two inner radii equal to save the cost of a pair of test plates, and vary the air space to correct the spherical aberration. Then if the coma is excessive, we can correct it by bending the whole lens.

As before, we start at the maximum of the bending curve, with $c_1 = 0.15$ and $c_a = 0.5090$, giving $c_2 = c_3 = -0.3590$. In Section 10.1 our starting setup had an air space of 0.01, giving $LA' = 0.10566$ and $OSC = 0.00062$ (Setup A in Figure 10.1). If we increase the air space to 0.04, with the usual $D - d$ solution for the last radius (Section 5.9.2), we obtain Setup B:

$$LA' = -0.01466, \quad OSC = 0.00305$$

We next apply a trial bending of 0.01 to the entire lens, with the 0.04 air space, and we get $LA' = -0.00646$ and the $OSC = 0.00201$ (Setup C). These values are plotted on a double graph with spherical aberration as abscissa and OSC as ordinate, as before (Figure 10.3). Evidently a further bending by 0.0198 should bring us close to the aim point. Actually this bending gave $LA' = 0.00014$ and $OSC = 0$ (Setup D).

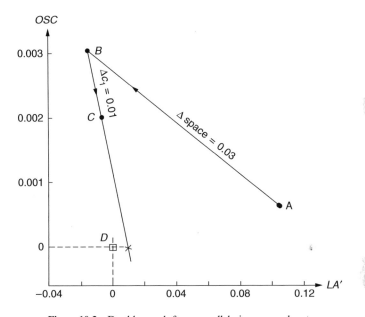

Figure 10.3 Double graph for a parallel air-space aplanat.

As the zonal aberration of this air-spaced lens is liable to be strongly overcorrected, we prefer to have a small negative value for the marginal aberration. Since our trial change in air space gave $\partial LA'/\partial(\text{space}) = 4.0$, we try increasing the air space by 0.0001. This gives the final setup as follows for trim diameter $= 2.2$, $f' = 10.1324$, $l' = 9.7012$, $LA' = -0.00017$, $LZA = +0.00666$, $OSC = 0$. It should be noted that the overcorrected zonal aberration is now 1.6 times as great as in the broken-contact design, and this is the principal reason why the previous type is generally to be preferred (Figure 10.4). However, the air space is now wider, which may be of help in designing the lens mount. Nevertheless, the LA' is very sensitive to changes in the airgap. Figure 10.4 shows the excellent state of zonal chromatic correction. The prescription for the final design is as follows.

c	d	n	V
0.1798			
	0.42	1.523	58.6
−0.3292			
	0.0401		
−0.3292			
	0.15	1.617	36.6
−0.0553			

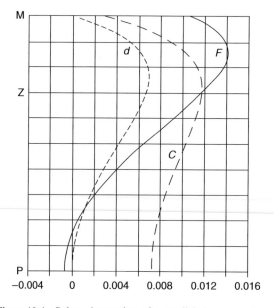

Figure 10.4   Spherochromatism of a parallel air-space aplanat.

The spherical aberration for the parallel air space aplanat shown in Figure 10.4 almost entirely comprises primary and secondary (third- and fifth-order) contributions. The spherochromatism for each the C and F also has the same general shapes. If now the first surface of this aplanat is made aspheric to reduce the spherical aberration in d light, a dramatic reduction can be obtained as illustrated in Figure 10.5. Notice that the spherical aberration contributions now contain a tertiary (seventh-order) term. This is a nice example to illustrate that the inclusion of an aspheric surface can cause remarkable variation in spherochromatism. Overall, the image quality of the lens with the aspheric first surface is superior. Also, as you should expect, the zonal chromatic correction is unchanged as is the zonal secondary color.

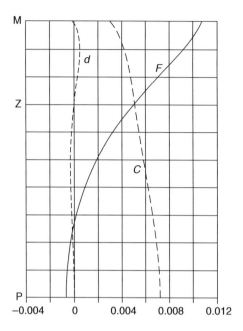

Figure 10.5 Spherochromatism of a parallel air-space aplanat with aspheric first surface.

10.3 AN APLANATIC CEMENTED DOUBLET

In Section 9.3.5 it was pointed out that the bending of a cemented doublet that yields zero *OSC* almost coincides with the bending for maximum spherical aberration. Consequently, if we can find two types of glass for which the spherical aberration curve just reaches zero at the top of the bending parabola, this peak bending will also be very nearly aplanatic.

Some guidance as to likely types of glass can be obtained by calculating the spherical *G* sums, and plotting the thin-lens bending curve as in Figure 7.2, relying on the fact that the true thick-lens curve coincides closely with the approximate thin-lens graph shown there. A few trials along these lines indicate that the spherical aberration curve will be bodily lowered if we increase the *V* difference between the glasses or if we reduce the *n* difference between them.

Since we have only three degrees of freedom in a cemented doublet, which must be used for focal length, spherical aberration, and *OSC*, it is clear that we must leave the final choice of glass until the end in order to secure achromatism by the *D – d* method. Since there are more crowns than flints in the Schott

catalog, we will adopt some specific flint and try several crowns to see how
the chromatic condition is operating. Taking as our flint Schott's SF-9 with
$n_D = 1.66662$ and $V_D = 33.08$, we select three possible crowns, and with each
we adopt an approximate value of $c_a = 0.3755$ for $f' = 10$.

Thus we find by a series of trials the value of c_3 that corrects the spherical
aberration at $f/5$. The whole lens is then bent, again by a series of trials, to eliminate the *OSC*. Then the $D - d$ sum of the aplanat is found for the marginal ray;
finally we find, also by the $D - d$ method, what value the crown V_D should have
to produce a perfect achromat. Repeating the process with each of the three
crowns enables us to plot a locus of possible crowns on the glass chart
(Figure 5.5), and if this locus happens to pass through an actual glass, that glass
will be used to complete the design. Figure 10.6 is a magnified portion of the
glass chart containing this locus.

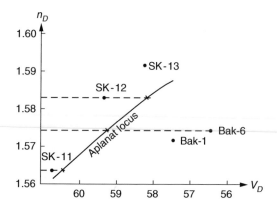

Figure 10.6   Locus of crown glasses for a cemented doublet aplanat, using SF-19 as a flint.

Our three trials give the results shown in Table 10.1. Even without plotting a
curved locus on the blowup of the glass chart as shown in Figure 10.6, we can
see that the third selection, SK-11, gives a close achromat with our chosen flint.
The final design after solving the last radius to give a zero $D - d$ sum is as follow:

c	d	Glass	n_D	V_D
0.1509				
	0.32	SK-11	1.56376	60.75
−0.2246				
	0.15	SF-19	1.66662	33.08
−0.052351				

Table 10.1

Crown Glass Selection

Crown glass type	n_D	$n_F - n_C$	V_D	$\sum (D - d) \Delta n$	Desired crown V for perfect achromatism
SK-12	1.58305	0.00983	59.31	0.0000365	58.16
BaK-6	1.57436	0.01018	56.42	−0.0000906	59.23
SK-11	1.56376	0.00928	60.75	0.0000083	60.46

for trim diameter $= 2.2$, $f' = 10.3663$, $l' = 10.1227$, $LA' = -0.00010$, $LZA = -0.00176$, and $OSC = 0.00060$. This would make an excellent objective. The undercorrected zonal residual is very small, being only 40% of that found for the common achromat in Section 7.2.1, Table 7.3.

DESIGNER NOTE

The comparative smallness of this zone is due to the use of higher-index glasses. It is found that with a given flint, the zonal residual is large for low-index crowns, drops to a minimum for some medium-index crowns, and then rises again for high-index crowns. There are clearly two opposing tendencies. Raising the crown index weakens the front radius, but it also lowers the index difference across the cemented interface thereby requiring a stronger curvature at the interface.

Somewhat in defiance of this well-known behavior, Ditteon and Feng developed an analytical method for designing a cemented aplanatic doublet where they discovered a pair of glasses, namely FK-54 and BaSF-52, that corrected both coma and secondary spectrum at the same time.[1] The FK-54 (437907) is a very-low index crown while the BaSF-52 (702410) is a medium index flint. The approach was to lower the spherical aberration parabola (see Figure 7.2) by having a large difference in Abbe values to have a single zero spherical aberration solution rather than two. Coma is essentially zero as we learned in Chapter 7. The secondary spectrum is corrected by having the partial dispersions of the glasses be essentially equal.

10.4 A TRIPLE CEMENTED APLANAT

Another way to obtain the additional degree of freedom necessary for OSC correction is to divide the flint component of a cemented doublet into two and place one part in front of and the other part behind the crown component to make a cemented triplet. Of course, alternatively the crown component could be divided in this way, but it is generally better to divide the flint.

Table 10.2

Thin Lens Formulas for Triple Cemented Aplanat

	Lens a	Lens b	Lens c
Net curvature (c)	x	Cr	Fl $- x$
Front curvature (c_1)	$y + x$	y	$y - $ Cr
Reciprocal object distance (v)	$v_1 = 1/l_1$	$v_1 + (n_a - 1)x$	$v_1 + (n_a - 1)x + (n_b - 1)$Cr

Conrady[2] has given a very complete study of this system on the basis of the spherical and coma G sums. To apply such an analysis, for each element we need net curvature c, bending parameter c_1, and reciprocal object distance v. In this triplet lens we have two absolute degrees of freedom, which we may call x and y: the amount of flint power in the front element and a bending of the whole lens. We therefore define $x = c_1 - c_2 = c_a$, and $y = c_2$.

The total powers of crown and flint are found by the ordinary (c_a, c_b) formulas (Eq. 5-4); they will be referred to here as Cr and Fl. Hence for the three thin-lens elements we have what is shown in Table 10.2. Here $n_a = n_c$ is the flint index and n_b the crown index. To draw a section of the lens to determine suitable thicknesses, we note that the thin-lens value of c_4 is $x + y - ($Cr $+ $ Fl$)$.

In performing the G sum analysis, we find that the spherical aberration expression is quadratic, while the coma expression is linear. Hence there will be two solutions to the problem. To reduce the zonal residual and to have as many lenses as possible on a block, we choose that solution in which the strongest surface has the longer radius. (A "block" refers to the tool to which the lens blanks are affixed for grinding and polishing. The working diameter of a short radius tool is less than that for a longer radius tool, which means that, for a given lens diameter, more elements can be mounted on the longer radius block.[3])

As an example, we will design a low-power cemented triplet microscope objective, with magnification 5× and tube length 160 mm. This represents a focal length of 26.67 mm. The numerical aperture, $\sin U'_4$, is to be 0.125; therefore the entering ray slope is $\sin U_1 = 0.025$. We will use the following common glass types:

(a) Flint: F–3, $n_e = 1.61685$, $\Delta n = 0.01659$, $V_e = 37.18$
(b) Crown: BaK–2, $n_e = 1.54211$, $\Delta n = 0.00905$, $V_e = 59.90$

with $V_b - V_a = 22.72$. The (c_a, c_b) formulas give for the total crown and flint powers

$$\text{Cr} = 0.1824, \quad \text{Fl} = -0.0995$$

Conrady's G-sum analysis gives the following approximate solutions:

x	−0.088	−0.019
y	+0.158	+0.072

Hence		
$c_1 = x + y$	0.070	0.053
$c_2 = y$	0.158	0.072
$c_3 = y - \text{Cr}$	−0.0244	−0.1104
$c_4 = x + y - (\text{Cr} + \text{Fl})$	−0.0129	−0.0299 (or −0.03035 by $D - d$)

The strongest curve in the first solution is $c_2 = 0.158$, whereas the strongest surface in the second solution is $c_3 = -0.1104$. We therefore continue work on the second solution. Since the radii are approximately 18.9, 13.9, −9.1, and −33.4, we can draw a diagram of the lens. The semiaperture is to be 5.0 since the Y of the marginal ray is $160 \times 0.025 = 4.0$. Suitable thicknesses are found to be 1.0, 3.5, and 1.0, respectively (Figure 10.7), all dimensions in millimeters.

We begin by tracing a marginal ray with $L_1 = -160$ and $\sin U_1 = 0.025$, solving the last radius by the $D - d$ method as usual. We calculate the LA' and OSC of this ray (Setup A), assuming that the aperture stop is located at the rear lens surface, where $OSC = (Ml'/mL') - 1$. We then make small experimental changes in x and y, and plot the usual double graph with OSC as ordinate and LA' as abscissa (Figure 10.7).

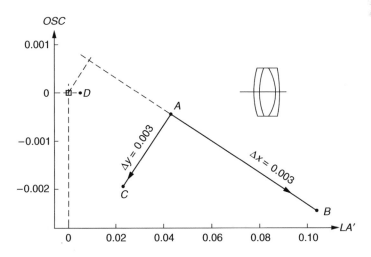

Figure 10.7 Double graph for a triple cemented aplanat.

This graph indicates the required very small changes in x and y from the starting setup to make both aberrations zero, namely, $\Delta x = -0.0017$ and $\Delta y = +0.0014$. These changes give the following solution to the problem:

	c	d	n_e
	0.0527		
		1.0	1.61685
	0.0734		
		3.5	1.54211
	−0.1090		
		1.0	1.61685
$(D-d)$	−0.030667		

with $L_1 = -160$ mm, $\sin U_1 = 0.025$, $LA' = 0.00042$ mm, $LZA = -0.04688$ mm, $OSC = -0.00002$, $l' = 30.145$ mm, $1/m = -4.927$, and $NA = 0.123$.

Since the numerical aperture is slightly below our desired value of 0.125, we may shorten the focal length in the ratio $0.123/0.125 = 0.984$ by strengthening all the radii in this proportion. The small zonal residual is far less than the Rayleigh tolerance of 0.21 mm and is negligible.

10.5 AN APLANAT WITH A BURIED ACHROMATIZING SURFACE

The idea of a "buried" achromatizing surface was suggested by Paul Rudolph in the late 1890s.[4] Such a surface has glass of the same refractive index on both sides, but because the dispersive powers are different, it can be used to control the chromatic aberration of the lens. Thus achromatism can be left to the end and the available degrees of freedom can be used for the correction of other aberrations.

Some possible matched pairs of glasses from the 2009 Schott catalog are shown in Table 10.3.

Table 10.3

Matched Glass Pairs Suitable for Buried Achromatizing Surface

		n_D	$n_F - n_c$	V_D	V difference
(1)	N-SK16	1.62032	0.01029	60.28	
	F2	1.61989	0.01705	33.37	23.91
(2)	N-SK14	1.60302	0.00933	60.60	
	F5	1.60328	0.01587	38.03	22.57
(3)	N-SSK2	1.62229	0.01168	53.27	
	F2	1.61989	0.01705	36.43	16.84
(4)	SK-7	1.65103	0.01165	55.89	
	N-BaF51	1.65211	0.01451	44.96	10.93

DESIGNER NOTE

An exact index match is unnecessary; particularly since the actual index of standard delivery fine annealed glass may depart from the catalog values by as much as ± 0.0005 and the Abbe number by $\pm 0.8\%$. Within a given lot of fine annealed glass, the refractive index variation is ± 0.0001 and about twice that within a lot of pressings. The lens designer should be attentive to the impact such variations may have on the lens performance. Also, for critical applications it is wise to use the actual melt data for the glass purchased and make final adjustments to the curvatures and thicknesses of the lens before making the lens elements. (Glass is typically delivered with a test report according to ISO 10474.)

The measurements are performed with an accuracy of $\pm 3 \times 10^{-5}$ for refractive index and $\pm 2 \times 10^{-5}$ for dispersion. Data are provided to five decimal places. The reported values are the median value of the samples taken from the lot. Consequently, the actual value of a part made from the lot may vary by the aforementioned refractive index tolerance. Most lens design programs include tolerancing analysis and some include tolerance sensitivity mitigation during lens optimization.

As an example of the use of a buried surface, we will design a triple aplanat in which the third surface will be buried. The remaining three radii will be used for spherical aberration and coma, the last radius being in all cases solved for the required focal length. We will maintain a focal length of 10.0 and an aperture of $Y = 1.0$ ($f/5$), allowing sufficient thicknesses for a trim diameter of 2.2 for the crown and for the insertion of the buried surface in the flint. For the crown lens we will use K-5 glass ($n_D = 1.5224$, $\Delta n = 0.00876$, and $V_D = 59.63$). For the flint we will use the glasses in selection (3) in Table 10.3, performing the ray tracing with the average index of 1.6222.

A convenient starting system is

c	d	n_D
0.16		
	0.42	1.5224
−0.26		
	0.35	1.6222
(solve for f')		

The last curvature comes out to be -0.069605, giving $LA' = 0.01013$ and $OSC = 0$. We now make a trial change in c_1 by 0.01, giving $LA' = 0.02059$ and $OSC = -0.00096$. Returning to the original setup and changing c_2 by 0.01 gives $LA' = -0.00241$ and $OSC = 0.00016$. These values are plotted on the double graph of Figure 10.8, and we conclude that we should make a further

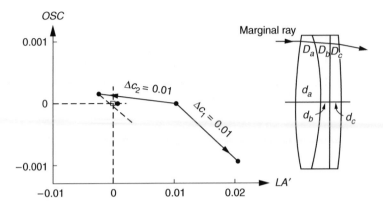

Figure 10.8 Double graph for a buried-surface triple aplanat.

change in c_1 by 0.0022. This completes the design so far as spherical aberration and coma are concerned.

We must now introduce the buried surface for achromatism. To do this we calculate the D values of the two elements along the marginal ray and divide the axial thickness of the second lens suitably, say at 0.15 and 0.20, to form the new second and third elements. We tabulate the four surfaces with as much information as we have. Knowing the value of $(D - d)\,\Delta n$ for the front element, we see that the sum of the remaining two elements must be equal and opposite to it. We also know the sum of the two D values for the last two lenses. Solving the two simultaneous equations tells us that D_b for lens b must be 0.2779594 and D_c for lens c must be 0.1650274. Knowing the Y values at the various surfaces, we finally ascertain that c_3 must be 0.0080508. This completes the design, which is as follows:

c	d	n_D	V_D
0.1622			
	0.42	1.5224	59.63
−0.25			
	0.15	1.62222	36.07
0.008051			
	0.20	1.62218	53.13
−0.066105			

for $f' = 10.0$, $l' = 9.6360$, $LA' = -0.00008$, $LZA = -0.00232$, and $OSC = -0.00005$. Compare the performance of this design with the aplanatic cemented doublet in Section 10.3.

10.6 THE MATCHING PRINCIPLE

If we wish to design an aplanatic lens of such a high aperture that a single doublet is impossible, we resort to the use of two achromats in succession. We now have four degrees of freedom. The subdivision of power between the two components and the air space between them is arbitrary, while the two bendings can be used for the correction of spherical aberration and *OSC*. Any reasonable types of glass can be used, and by achromatizing each component separately we automatically correct both the chromatic aberration and the lateral color.

The design of lenses of this type has been described in detail by Conrady,[5] in particular when used as a microscope objective of medium power. Having decided on suitable values for the two arbitrary quantities, we trace a marginal ray through the system from front to back, solving r_3 and r_6 by the $D - d$ method, and we then add two paraxial rays, one through the front component from left to right using $l_1 = L_1$ and $u_1 = \sin U_1'$, and the other through the rear component from right to left, taking $u_6' = \sin U_6'$ and $l_6' = L_6'$ as starting data. *If we can now find such a pair of bendings that the two paraxial rays match in the air space between the lenses, the system will be corrected for both spherical aberration and OSC. This is what is meant by the matching principle.*

To make the required trial bendings, we have no problem with the front component, but we must adopt standard entry data for the rear component. We can easily adopt a fixed value for L_4 by always choosing a suitable air space between the lenses, but any standard value of U_4 that we may adopt will never agree exactly with the emerging slope U_3' from the front component. Consequently it becomes necessary to match actual aberrations in the air space rather than trying to match lengths and angles.[6] So far as lengths are concerned, we have always $L_4 = L_3' - d$, and we require that $l_4 = l_3' - d$. Subtracting these tells us that we must select bendings such that

$$LA_4 = LA_3' \tag{10-1}$$

To match the slope angles of the paraxial rays in the air space, we have approximately $\sin U_4 = \sin U_3'$, and we require that u_4 should also be equal to u_3'. Dividing these gives

$$OSC_4 = OSC_3' \tag{10-2}$$

where the *OSC* is defined as

$$OSC = (u/\sin U) - 1 \tag{10-3}$$

Since this kind of *OSC* does not contain the usual correcting factor for spherical aberration and exit-pupil position (see Sections 4.3.4 and 9.3), we refer to it as *uncorrected OSC* in the present context. Recall that we are trying to match the ray slope angles in the space between the front and rear components; hence the uncorrected *OSC* is just a convenient gauge of the relation linking the paraxial and marginal rays. To reiterate, should the matching principle concept of requiring Eqs. (10-1) and (10-2) to be satisfied not be fulfilled, it is certain that the lens system suffers imperfect correction of its aberrations.

As an example to illustrate the matching principle, we will design a 10× microscope objective of numerical aperture 0.25, so that the entering ray slope at the long conjugate end is 0.025. Assuming an object distance of −170 mm, we can trace any desired rays into the front component of the system. It should be noted that, as always, we calculate a microscope objective from the long conjugate to the short, because the long conjugate distance is fixed while the short is not, so that the long-conjugate end becomes the "front" of our system. This conflicts with ordinary microscope parlance, which regards the front of a microscope objective as the short conjugate end; this is a unique exception and we shall ignore it here.

Our first problem is to deal with the two arbitrary degrees of freedom, namely, the subdivision of refracting power between the two components, and the air space between them. For this, it is common to require that the paraxial ray suffers equal deviation at each component, and to place the rear component approximately midway between the front component and its image. This makes the object distance for all rear-element bendings about 20 mm, and we shall adopt that value here.[7]

As the overall paraxial deviation is $0.25 + 0.025 = 0.275$, we must allow each component to deviate the paraxial ray by 0.1375, which makes the ray slope between the components equal to 0.1125. We shall therefore adopt this value of $\sin U_4$ in making all trial bendings of the rear component. For both lenses we use the following common types of glass:

(a) Crown: $n_e = 1.52520$, $n_F - n_C = 0.00893$, $V_e = 58.81$
(b) Flint: $n_e = 1.62115$, $n_F - n_C = 0.01686$, $V_e = 36.84$

with $V_a - V_b = 21.97$. The thin-lens data of the two components (Figure 10.9) are as shown in Table 10.4.

After determining the last radius by the $D - d$ method in every case, the results of several bendings of each component are found to be as shown in Table 10.5. These results are plotted side by side on one graph in Figure 10.10.

Figure 10.9   A Lister-type microscope design.

Table 10.4

Thin-Lens Data for Lister-type Microscope Shown in Figure 10.9

Object distance (mm)	Image distance (mm)	Focal length (mm)	Clear aperture (mm)	c_a	c_b	Suitable thicknesses (mm)
−170	37.77	30.90	8.5	0.1649	−0.0874	3.2, 1.0
20.0	9.00	16.36	4.5	0.3116	−0.1650	2.0, 0.8

Table 10.5

Aberrations Versus Bendings for Lister-type Microscope

Front component
$L_1 = l_1 = -170.0$ $\sin U_1 = u_1 = 0.025$

c_1	0	0.02	0.04	0.06	0.08
c_3 by $D - d$	−0.08273	−0.06254	−0.04130	−0.01870	+0.00558
L_3'	33.149	34.666	35.465	35.567	35.005
LA_3'	−0.1474	0.9529	1.3282	0.8596	−0.6049
uncorrected OSC_3'	0.01164	0.03753	0.03727	0.01360	−0.03567

Rear component
$L_4 = 20.00$ $\sin U_4 = -0.1125$

c_4	0.05	0.10	0.15	0.20	0.25
c_6 by $D - d$	−0.11360	−0.05405	0.01450	0.09511	0.19249
$L_6' = l_6'$	7.3552	7.3700	7.2695	7.0888	6.8570
$\sin U_6' = u_6'$	−0.25862	−0.24760	−0.23939	−0.23259	−0.22588
l_4	18.8706	20.2829	20.7971	20.4545	19.3870
LA_4	1.1294	−0.2829	−0.7971	−0.4545	0.6130
uncorrected OSC_4	0.03222	−0.03055	−0.04323	−0.01363	0.05653

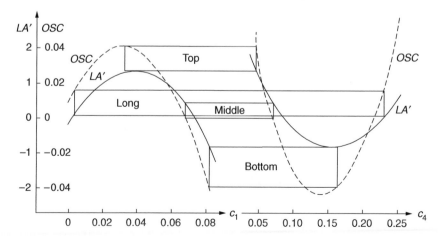

Figure 10.10   The matching principle.

Table 10.6

Matching Solutions

Rectangle	c_1	c_2	c_3	c_4	c_5	c_6
A (top)	0.032	−0.133	−0.050	0.045	−0.266	−0.119
B (middle)	0.067	−0.098	−0.010	0.070	−0.242	−0.091
C (bottom)	0.081	−0.084	0.007	0.165	−0.147	0.037
D (long)	0.003	−0.162	−0.080	0.228	−0.084	0.147

It is our aim to select such values of c_1 and c_4 that $LA'_3 = LA_4$ and simultaneously $OSC'_3 = OSC_4$. This is done by searching for rectangles that just fit into the four curves, with spherical aberration and coma points, respectively, each being on the same level. In this case, there are four such rectangles to be found, indicating that the curves represent quadratic expressions. The four solutions are shown in Table 10.6.

For many reasons we shall continue the design using solution C. All the other solutions contain stronger surfaces, and moreover both components of solution C contain almost equiconvex crown elements. This starting setup is as follows:

c	d	n_e
0.081		
	3.2	1.52520
−0.08394		
	1.0	1.62115
0.00685		
	14.9603	(air)
0.165		
	2.0	1.52520
−0.14654		
	0.8	1.62115
0.03730		

with $l'_6 = 7.2095$, $LA'_6 = 0.01383$, $u'_6 = 0.2361$, and $OSC'_6 = -0.00297$. For the final OSC'_6 calculation we assumed that the exit pupil is in such a position that $(l' - l'_{pr})$ is about 17.0. This puts the exit pupil about 10 mm inside the rear vertex of the objective.

Although this solution is close, we must improve both aberrations by means of a double graph. Changing c_1 by 0.001 and maintaining the $D - d$ solutions, and $L_4 = 20.0$, we find that the aberrations become

$$LA'_6 = -0.000829, \quad OSC'_6 = -0.002404$$

Restoring the original c_1 and changing c_4 by 0.001 gives

$$LA'_6 = 0.001306, \quad OSC'_6 = -0.003279$$

Inspection of the graph suggests that we change the original c_1 by 0.001 and c_4 by –0.01. These changes give

$$LA'_6 = -0.000403, \quad OSC'_6 = +0.000474$$

Unfortunately the numerical aperture of the system is now 0.2381, whereas it should be 0.25. We therefore scale all radii down by 4%, which gives

$$LA'_6 = 0.001114, \quad OSC'_6 = 0.000221$$

Further reference to the double graph suggests that we try $\Delta c_1 = 0.0005$ and $\Delta c_4 = 0.002$. This change gives the almost perfect solution drawn to scale in Figure 10.9, namely,

c	d	n_e
0.08578		
	3.2	1.52520
0.08576		
	1.0	1.62115
0.009152		
	13.8043	(air)
0.16320		
	2.0	1.52520
−0.16080		
	0.8	1.62115
0.02602		

with $l'_6 = 6.8925$, $u'_6 = 0.2500$, $LA'_6 = 0.000004$, $OSC'_6 = -0.000095$, and $LZA'_6 = -0.00289$. In practice, of course, we should apply trifling further bendings to both components to render the crown elements exactly equiconvex. These changes are so slight that they have no significant effect on any of the aberrations.

The zonal aberration tolerance is $6\lambda/\sin^2 U'_m = 0.053$, so that the zonal residual of our objective is about half the Rayleigh limit. To improve it, we would have to go to a flint of somewhat higher index, but the present design would be acceptable as it stands.

DESIGNER NOTE

In searching for matching rectangles, there is no *a priori* certainty of how many rectangles may be found. In the example given, four solutions were found; however, there may be three, two, one, or even no useful solutions, particularly if the chosen glasses are very abnormal. The lens designer should observe from the design of this Lister-type microscope objective that multiple solutions exist and that there are reasons why one solution should be preferred over another. When attempting to design this or other lens systems using an automatic optical design program, the lens designer should be attentive to exploring alternative solutions that the optical design program may not find. This is not dissimilar to the difficulty most optical design programs would have in finding multiple solutions of the spherically corrected achromat discussed in Section 7.2.[8]

ENDNOTES

[1] Richard Ditteon and Feng Guan, "Cemented aplanatic doublet corrected for secondary spectrum," *OSA Proc. of the International Optical Design Conference*, Vol. 22, G. W. Forbes (Ed.), pp. 107–111 (1994).

[2] A. E. Conrady, p. 557.

[3] See Douglas F. Horne, *Optical Production Technology, Second Edition*, Adam Higler, Bristol (1972); R. M. Scott, "Optical manufacturing," in *Applied Optics and Optical Engineering*, Vol. III, Rudolf Kingslake (Ed.), Academic Press, New York (1965). Of particular interest are the writings of more than three decades by Frank Cooke: *Optics Cooke Book, Second Edition*, Stephen D. Fantone (Ed.), Optical Society of America, Washington, DC (1991).

[4] P. Rudolph, U.S. Patent 576, 896, filed July (1896).

[5] A. E. Conrady, p. 662.

[6] A related approach, called XSYS, examines portions of a variety of lenses designed by experts and then attempts to combine certain of the "modules" to form a new lens. Just as the matching principle attempts to match aberrations of the two parts of the lens system, XSYS attempts to match the wavefronts of the modules using scaling. The objective is to determine a viable starting point for the design process. Donald C. Dilworth, "Expert systems in lens design," International Lens Design Conference, George. N. Lawrence (Ed.), *Proc. SPIE*, 1354:359–370 (1990).

[7] There is, as Conrady taught, certain theoretical justification for having slightly more deviation in the front component (see Figure 10.9) to perhaps reduce the zonal aberration by a modest weakening of the power of the rear lens being that it is in more convergent light. Rather than a 50:50 division of the ray deviation, one could try allocating 60% to the front and 40% to the rear component.

[8] At least two optical design programs that have demonstrated the ability to locate the multiple solutions for the spherically corrected achromat problem are CODE V (Optical Research Associates) and SYNOPSYS (Optical Systems Design, Inc.). Each of these programs uses proprietary global exploration algorithms.

Chapter 11

The Oblique Aberrations

In Chapter 4 we introduced the subject of the oblique aberrations of a lens, and in Chapter 8 discussed in detail the origin and computation of coma. In this chapter we continue the discussion, giving computing procedures for the remaining oblique aberrations, namely, astigmatism, field curvature, distortion, and lateral color.

11.1 ASTIGMATISM AND THE CODDINGTON EQUATIONS

When a narrow beam of light is obliquely incident on a refracting surface, astigmatism is introduced, and the image of a point source formed by a small lens aperture becomes a pair of focal lines, a series of beam sections being indicated in Figure 11.1. One focal line (sagittal) is radial to the field and points toward the lens axis, while the other focal line (tangential) is tangential to the field. Both focal lines are perpendicular to the principal ray, and their locations can be calculated once the principal ray has been traced. The astigmatic images formed by the first surface become the objects for the second, and so on through the system. The locations of the focal lines are found by the two Coddington[1,2] equations, which will now be derived.

11.1.1 The Tangential Image

In Figure 11.2, BP is an entering principal ray, B being the tangential object point distance t from the point of incidence P; the length t being measured along the principal ray, negative if the object point lies to the left of the surface as usual. The line BG represents a neighbor ray close to the principal ray, lying in the meridian plane, so close in fact that the short arc $PG = r\,d\theta$ can be regarded as tangent to the refracting surface itself.

The central angle $\theta = I - U$, and hence

$$d\theta = dI - dU \tag{11-1}$$

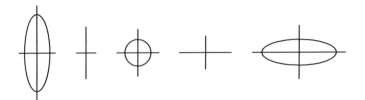

Figure 11.1 The astigmatic focal lines.

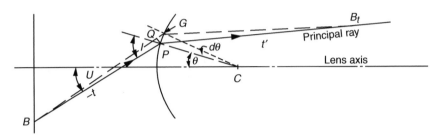

Figure 11.2 The tangential focus.

The short line PQ, perpendicular to the incident ray, is given by

$$PQ = -t\, dU = PG\cos I = r\cos I\, d\theta$$

But by Eq. (11-1) we have $dU = dI - d\theta$. Therefore, $PQ = -t(dI - d\theta) = r\cos I\, d\theta$, hence

$$dI = \left\{1 - \frac{r\cos I}{t}\right\} d\theta \qquad (11\text{-}1a)$$

Similarly for the refracted ray we have

$$dI' = \left\{1 - \frac{r\cos I'}{t'}\right\} d\theta \qquad (11\text{-}1b)$$

By differentiating the law of refraction we obtain

$$n\cos I\, dI = n'\cos I' dI' \qquad (11\text{-}1c)$$

and inserting (11-1a) and (11-1b) into (11-1c) we get

$$\frac{n'\cos^2 I'_{\text{pr}}}{t'} - \frac{n\cos^2 I_{\text{pr}}}{t} = \frac{n'\cos I'_{\text{pr}} - n\cos I_{\text{pr}}}{r} \qquad (11\text{-}2)$$

The term on the right degenerates to the surface power $(n' - n)/r$ when the object point lies on the lens axis so that $I'_{\text{pr}} = I_{\text{pr}} = 0$. It may be regarded as the *oblique power* of the refracting surface for the principal ray. The oblique power is always slightly greater than the axial power, which provides a convenient check on the calculation.

11.1.2 The Sagittal Image

The other focal line is located at the sagittal image point B_s. This is a paraxial-type image formed by a pair of sagittal (skew) rays lying close to the principal ray. As explained in Section 8.1.1, the image of a point formed by a pair of sagittal rays always lies on the auxiliary axis joining the object point to the center of curvature of the surface. This property of sagittal rays enables us to derive the second Coddington equation locating the sagittal focal line.

In Figure 11.3 we show the principal ray, the sagittal object point B, and the sagittal image B_s, with the auxiliary axis joining B, C, and B_s. Now, the area of any triangle ABC is given by $\frac{1}{2}ab \sin C$; and since triangle BPB_s = triangle BPC plus triangle PCB_s, we have

$$-\frac{1}{2}ss'\sin(180° - I + I') = -\frac{1}{2}sr\sin(180° - I) + \frac{1}{2}s'r\sin I'$$

where

$$-ss'\sin(I - I') = -sr\sin I + s'r\sin I'$$

Expanding $\sin(I - I')$ and multiplying by $(n'/ss'r)$ gives

$$-\frac{n'\sin I\cos I' - n'\cos I\sin I'}{r} = -\frac{n'\sin I}{s'} + \frac{n'\sin I'}{s}$$

But by the law of refraction, $n'\sin I'$ can be everywhere replaced by $n\sin I$. When this is done, the $\sin I$ cancels out, giving

$$\frac{n'}{s'} - \frac{n}{s} = \frac{n'\cos I'_{pr} - n\cos I_{pr}}{r} \tag{11-3}$$

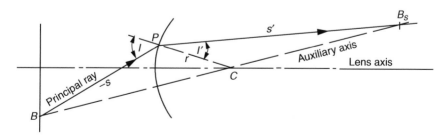

Figure 11.3 The sagittal focus.

The term on the right is the same oblique power of the surface that we found for the tangential image in Eq. (11-2). Thus the only difference between the formulas for tangential and sagittal foci is the presence of the $\cos^2$ terms in the tangential formula.

Conrady[3] has also given a very direct derivation of these formulas by a method depending on the equality of optical paths at a focus. However, the purely geometrical derivations given here are easier to follow and are quite valid.

11.1.3 Astigmatic Calculation

To use these formulas to calculate the astigmatism of a lens, we begin by tracing a principal ray at the required obliquity. We calculate the starting values of s and t from the object to the point of incidence measured along the principal ray (see Opening Equations section).

Oblique Power

We next determine the oblique power of each surface by

$$\phi = c(n' \cos I' - n \cos I)$$

for a spherical surface of curvature c. If the surface is aspheric, then it is necessary to calculate the separate sagittal and tangential surface curvatures at the point of incidence by

$$c_s = \sin(I - U)/Y, c_t = (d^2Z/dY^2) \cos^3(I - U)$$

The second derivative d^2Z/dY^2 is found from the equation of the aspheric surface. The rest of the data refers to the principal ray itself at the surface. It is common to find a great difference between the sagittal and tangential surface curvatures; indeed, they may even have opposite sign.

Oblique Separations

The third step is to calculate the oblique separation between successive pairs of surfaces, measured along the principal ray, by

$$D = (d + Z_2 - Z_1)/\cos U_1'$$

where the Z values of the principal ray at the various surfaces are found by

$$Z = \frac{1 - \cos(I - U)}{c}$$

or better by

$$Z = G \sin(I - U)$$

Sagittal Ray

We then trace the sagittal neighbor ray by applying at each surface

$$s' = \frac{n'}{(n/s + \phi)}$$

Transfer is $s_2 = s'_1 - D$.

Tangential Ray

The formula for tracing the tangential neighbor ray is

$$t' = \frac{n' \cos^2 I'}{[(n \cos^2 I)/t] + \phi}$$

Transfer is $t_2 = t'_1 - D$. The process for tracing the tangential ray can be made similar to that for the sagittal ray by listing across the page the values of $n \cos^2 I$ and $n' \cos^2 I$, and then treating these products as if they were the actual refractive indices of the glasses.

Opening Equations

If the object is at infinity, the opening values of both s and t are infinity. If the object is at a distance B from the front lens vertex (negative if to the left), then we must calculate (see Figure 11.4a)

$$s = t = (B - Z_{pr})/\cos U_{pr} = (H_0 - Y_{pr})/\sin U_{pr}.$$

Closing Equations

Having traced the sagittal and tangential neighbor rays, we generally wish to know the axial distances of the sagittal and tangential focal lines from the paraxial image plane. These are given by (see Figure 11.4b):

$$Z'_s = s' \cos U'_{pr} + Z - l'$$

$$Z'_t = t' \cos U'_{pr} + Z - l' \tag{11-4}$$

where Z is the sag of the rear lens surface computed for the principal ray.

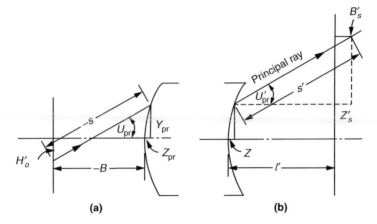

Figure 11.4 (a) Opening equations. (b) Closing equations.

Example

As an example of the use of these formulas, we will trace a principal ray through the Section 2.5 cemented doublet used several times before. The principal ray will enter the front vertex at an angle of 3°. The tabular layout of the computation, as perhaps performed on a small pocket calculator, is given in Table 11.1. The closing equations give $Z'_s = -0.02674$ and $Z'_t = -0.05641$ recalling that $l' = 11.28586$ (Section 6.1). The tangential focal line is thus about twice as far from the paraxial focal plane as the radial focal line, and both are inside the focal plane.

11.1.4 Graphical Determination of the Astigmatic Images

The location of the sagittal focus along a traced principal ray is easily found because the image lies on an auxiliary axis drawn from the object point through the center of curvature of the surface.

T. Smith[4] credits Thomas Young with the discovery of a similar procedure for locating the tangential image point. Young's method for the construction of the refracted ray itself involves drawing two auxiliary circles about the center of curvature of the refracting surface, one with radius rn/n' and the other with radius rn'/n (Figure 11.5). The incident ray is extended to cross the second of these auxiliary circles at E, and then E is joined to the center of curvature C. This line crosses the first auxiliary circle at E'; then the refracted ray is drawn from P through E'.

Table 11.1

Calculation of Astigmatism Along a Principal Ray

c	0.1353271		−0.1931098		−0.0616427
d		1.05		0.4	
n		1.517		1.649	
Tracing of 3° principal ray					
Q	0		0.0362246		0.0491463
Q'	0		0.0362270		0.0491086
I	3.00000		1.57608		1.67722
I'	1.97708		1.44989		2.76642
U	3.0	1.97708		1.85089	2.94009
Tabulation of cosines					
$\cos I$	0.9986295		0.9996217		0.9995716
$\cos I'$	0.9994047		0.9996798		0.9988346
$\cos U$		0.9994047		0.9994783	0.9986837
Oblique powers of surfaces					
ϕ	0.0700274		−0.0254994		0.0400344
Oblique separations					
Z	0		−0.0001268		−0.0000745
D		1.050499		0.4002611	
Sagittal ray					
s	∞		20.612450		33.884695
s'	21.662949		34.284956		11.274028
Tangential ray					
$n \cos^2 I$	0.9972609		1.5158524		1.6475874
$n' \cos^2 I'$	1.5151944		1.6479443		0.9976705
t	∞		20.586666		33.836812
t'	20.637165		34.237073		11.244328

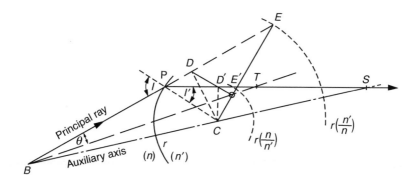

Figure 11.5 Young's construction for the sagittal and tangential foci ($n = 1$, $n' = 1.7$).

To locate the tangential image of any point B situated on the incident ray, we drop perpendiculars from C onto the two sections of the ray, striking them at D and D', respectively. Then the point of intersection of DD' and EE' is the point O. The line DD' is found to be perpendicular to EE'. This point O replaces C when graphically locating the tangential image of B, so that the line from B to O crosses the refracted ray at the tangential image, while the line from B through the center of curvature C locates the sagittal image.

The proof of this is difficult. It is best to assume that the angle θ in triangle BDO is equal to the angle θ in triangle BPT; then the geometry of the two triangles leads to the regular Coddington equation for the tangential image.

11.1.5 Astigmatism for the Three Cases of Zero Spherical Aberration

In Section 6.1.1 it was pointed out that a single spherical surface contributes no spherical aberration when the object is at (a) the surface itself, (b) the center of curvature of the surface, and (c) the aplanatic point. In Section 9.2.1 it was shown that the OSC also is zero for these three object points.

By means of the Coddington equations it is easy to show that at small obliquity the astigmatism contribution will be zero in cases (a) and (c), but when the object is at the center of curvature the astigmatism contribution is large and in the unexpected sense—that is, the convex front surface of a positive lens, for instance, contributes positive astigmatism when we would ordinarily have expected it to lead to an inward-curving field. This result is often of great significance, and it explains many anomalies, such as the flat tangential field of a Huygenian eyepiece.

11.1.6 Astigmatism at a Tilted Surface

If a lens surface is tilted through a given angle, the procedure outlined in Section 2.6 can be used to trace the principal ray, and the ordinary Coddington equations can be used to locate the astigmatic images along the principal ray. However, because of the asymmetry, the astigmatism at some angle, say $15°$, above the axis will not be the same as the astigmatism at $15°$ below the axis, and to plot the fields it is now necessary to trace several principal rays with both positive and negative entering obliquity angles.

As an example, we will refer ahead to the design of a Protar lens (Section 14.4), and pick up the principal-ray data at several obliquities. We will next suppose that the rear lens surface has been tilted clockwise through an angle of $0.10°$ (6 arcmin), so that $\alpha = 0.1$. By comparing the field curves given in Figure 11.6

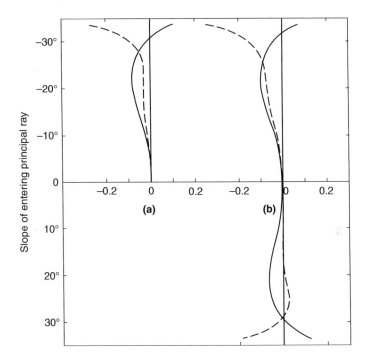

Figure 11.6 Fields of a Protar lens, (a) centered and (b) rear surface tilted clockwise by 6 arcmin.

before and after the last surface was tilted, the effect of the tilt can be readily seen. Briefly, it causes the field to tilt in a counterclockwise direction, the tangential field being tilted and distorted much more than the sagittal field. Limiting ourselves to one field angle, say 17.2°, we find that the tangential field has been tilted by 35.2 arcmin while the sagittal field has been tilted through 13.3 arcmin, both considerably more than the surface tilt that caused the problem. Actually, the effects of a tilt as small as 5 arcmin can generally be detected, and it is customary to try to limit accidental surface tilts in any good lens to about one arcmin. *Surface tilt does more damage to an image than any other manufacturing error, and in assembling a lens it is essential to avoid tilted surfaces at any cost.*

11.2 THE PETZVAL THEOREM

From very simple considerations, it is clear that a positive lens ought to have an inward-curving field. The extraaxial or off-axis points on a flat object are further from the lens than the axial point, and consequently their images should be closer to the lens than the axial image, leading at once to an inward-curving field.

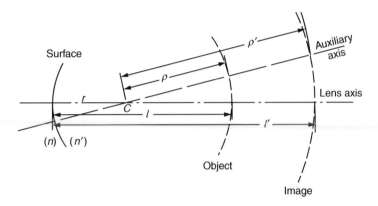

Figure 11.7   The Petzval theorem.

The exact amount of this natural field curvature can be calculated by the following argument. Suppose we place a small stop at the center of curvature C of a single spherical refracting surface (Figure 11.7). This will automatically eliminate coma and astigmatism by forcing the oblique light to be refracted along an auxiliary axis as if it were an axial beam. If the stop is small enough to eliminate spherical aberration also, we shall be left with nothing but the basic field curvature that we are trying to evaluate.

It is, of course, obvious that under these conditions an object having the form of a sphere centered about C must be imaged as a sphere also centered about C. If the radii of curvature of object and image are represented by ρ and ρ', then[5]

$$\rho = l - r, \quad \rho' = l' - r$$

and since for a single surface $n'/l' - n/l = (n' - n)/r$, we can readily show that, for one surface,

$$\frac{1}{n'\rho'} - \frac{1}{n\rho} = \frac{n' - n}{nn'r}$$

We can now write this expression for every surface in the lens and add them up, but this procedure will be valid only if we can assume that all traces of astigmatism have somehow been eliminated. Nevertheless, for such a lens having k surfaces, we find that

$$\frac{1}{n'_k\rho'_k} - \frac{1}{n_1\rho_1} = \sum \frac{n' - n}{nn'r}$$

This expression relates the radius of curvature of the image with the radius of curvature of the object, provided there is no astigmatism present. It is clear, then, that the radius of curvature of the image of a plane object with $\rho_1 = \infty$, is given by

$$\frac{1}{\rho_k'} = n_k' \sum \frac{n' - n}{nn'r} \tag{11-5}$$

It should be noted that a positive value of ρ corresponds to a negative sag, or an inward-curving image. Hence the sag of the curved image of a plane object, in the absence of astigmatism, will be given by

$$Z_{ptz}' = -\frac{1}{2} h_k' 2 n_k' \sum \frac{n' - n}{nn'r} \tag{11-6}$$

This is the famous *Petzval theorem*, and we shall have many occasions to refer to it since *it is only possible to design a flat-field lens free from astigmatism by reducing the Petzval sum; thus the Petzval theorem dominates the entire design processes for flat-field photographic lenses.*

The quantity under the summation in these different expressions is called the *Petzval sum*, and the radius of curvature of the image is evidently the reciprocal of the Petzval sum. Another useful term is the Petzval ratio, which is the ratio of the Petzval radius to the focal length of the lens. It is given by

$$\rho'/f' = 1/f'\Sigma$$

where Σ is the Petzval sum. Note the reciprocal relationship here. A long focal length lens tends to have a small Petzval sum, while the sum is large in a strong lens of short focal length.

11.2.1 Relation Between the Petzval Sum and Astigmatism

It can be shown[6] that at very small obliquity angles the tangential astigmatism—that is, the longitudinal distance from the Petzval surface to the tangential focal line—is three times as great as the corresponding sagittal astigmatism. Thus, if the astigmatism in any lens can be made zero, the two focal lines will coalesce on the Petzval surface. In all other cases the locus of the tangential foci at various obliquities is called the tangential field of a lens, and similarly for the sagittal field. As the Petzval surface in most simple lenses is inward-curving, it is often possible to flatten the tangential field by the deliberate introduction of overcorrected astigmatism, leaving the sagittal image to fall between the Petzval surface and the tangential image. However, when designing an "anastigmat" having a flat field free from astigmatism, it is necessary to reduce the Petzval sum drastically.

If it is necessary to design a lens having an inward-curving field to meet some customer requirement, the astigmatism can easily be removed and the Petzval sum adjusted to give the desired field curvature. On the other hand, if the field must be backward-curving, it is difficult to avoid an excessive amount of over-corrected astigmatism. It is worth noting that in some types of lens, if the Petzval sum is made too small the separation between the astigmatic fields becomes excessively large at intermediate field angles.

Many decades ago, lens designers taught that the tangential astigmatic field should be flattened to obtain the smallest spot size.[7,8,9] This can be easily understood by considering Eqs. (4-6) and (4-7) and assuming that all aberration coefficients are zero other than primary astigmatism (σ_3) and Petzval (σ_4). The sagittal and tangential astigmatic ray errors in the paraxial image plane are $(\sigma_3 + \sigma_4)\rho\bar{H}^2 \sin\theta$ and $(3\sigma_3 + \sigma_4)\rho\bar{H}^2 \cos\theta$, respectively. Three basic cases to contemplate are a flat sagittal field, a flat tangential field, and equally balanced fields about the paraxial image plane. For a flat sagittal field, $\sigma_3 + \sigma_4 = 0$ or $\sigma_3 = -\sigma_4$, which means that the residual tangential astigmatism in the paraxial image plane is $3\sigma_3 + \sigma_4 = -2\sigma_4$. When the tangential field is flat, $3\sigma_3 + \sigma_4 = 0$, which implies that $\sigma_3 = -\sigma_4/3$ and the residual sagittal astigmatism is $2\sigma_4/3$. When the errors are balanced, the tangential astigmatism is equal to the negative of the sagittal astigmatism, or $\sigma_3 = -\sigma_4/2$.

In the balance-fields case, the values of the residual sagittal and tangential astigmatism are observed each to be smaller than the residual values of the prior two cases. This might lead one to select this condition as the optimal minimum spot size[10]; however, such a conclusion is erroneous.[11] It is a general practice by lens designers to adjust the astigmatic surfaces such that the tangential field is flat and then adjust the position of the image plane to the location of the smallest blur at the edge of the field. The definition of the imagery is relatively uniform over the whole image area. In a balanced-field case, the image definition is quite superior in the central region of the image to that of the flat tangential field case, and inferior in the outer portions of the imagery.[12] As B. K. Johnson stated, "It therefore depends much on the requirements for which the lens is to be used, as to which criterion is to be adopted."

11.2.2 Methods for Reducing the Petzval Sum

There are several methods by which the Petzval sum can be reduced, and one or more of these appear in every type of photographic objective. These methods can also be applied to a wide variety of optical systems.

A Thick Meniscus

If we have a single lens in which both radii of curvature are equal and of the same sign, the Petzval sum will be zero, while the lens power is proportional to the thickness. Cemented interfaces in such a lens have very little effect on the

Petzval sum. This property has been used in many symmetrical lenses such as the Dagor and Orthostigmat.

Separated Thin Elements

In a system containing several widely separated thin elements, the Petzval sum is given by

$$\text{Ptz} = \sum \phi/n \qquad (11\text{-}7)$$

where ϕ is the power of an element. If there is about as much negative as positive power in such a system, the Petzval sum can be made as small as desired. This property has been used in many lenses of the dialyte type (see Section 13.2).

Negative Lens Field Flattener

An interesting special case is that in which a negative lens element is placed at or near an image plane, as illustrated in Figure 11.8. This element has little or no effect on the focal length or the aberrations, but it contributes its full power to the Petzval sum. (See Section 11.7.4.)

Conversely, if it is necessary to insert a positive lens in an image plane to act as a field lens, then this lens has a large adverse effect on the Petzval sum. For this reason it is almost impossible to reduce the Petzval sum in a long periscope having several internal images and field lenses. However, by using photographic-type lenses as field lenses it is sometimes possible to reduce the sum appreciably.

It should be noted that in a lens having a long central air space, the Petzval sum is increased if both components are positive (as in the Petzval portrait lens) because the rear component acts partly as a positive field lens. On the other hand, if the rear component is negative (as in a telephoto), then the Petzval

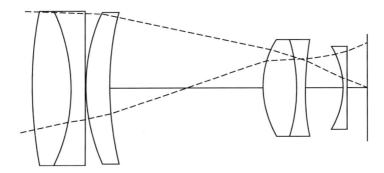

Figure 11.8 Negative lens element is placed at or near an image plane.

sum is reduced, and in an extreme telephoto it may actually become negative, requiring some degree of undercorrected astigmatism to offset it.

A Concentric Lens Field Flattener

The preceding field flattener has several inherent problems that may make it difficult or impossible to use. There are situations where a field flattener needs to be remote from the image plane—for example, an infrared detector array located inside a vacuum dewar. Rosin[13] described the use of a concentric lens centered about the focus of diverging or converging axial rays as illustrated in Figure 11.9. Since any of these axial rays are incident normal to the lens surfaces, the position of the image does not change with the introduction of this lens nor does it change the image size.

It can be shown that aberration contributions of this lens have the following characteristics.

- Zero spherical aberration
- Zero tangential and sagittal coma
- Zero axial color
- Zero lateral color
- Distortion is unchanged
- Sagittal field curvature is unaffected
- Tangential field curvature can be independently controlled

The tangential field contribution is proportional to $(R_2 - R_1)\left(\frac{N-1}{N}\right)$, which means that the effect cause is based on the thickness of the lens and the distance to its center of curvature. Higher-order aberrations will generally remain corrected for spherical aberration, axial color, distortion, sagittal coma, sagittal field curvature, spherochromatism, zonal spherical aberration, and sagittal

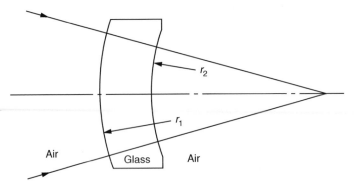

Figure 11.9 Concentric lens used to flatten tangential field.

oblique spherical aberration. In cases where the stop is significantly distant from the lens, certain higher-order aberrations can become bothersome, namely, tangential coma, lateral color, tangential field curvature, and tangential oblique spherical aberration. In addition to these possible limitations, the lens curvatures toward the image plane may restrict the size of the field due to the geometric size of the field flattener lens. It should be noted that since these lens aberrations are all about zero, spatial positioning of this lens does not need to be nearly as precise as a lens contributing large amounts of aberrations.

The concentric field flattener lens was independently discovered[14] and successfully employed, beginning in 1968, to enhance the performance of a variety of thermal infrared optical systems having low f-numbers and moderate fields-of-view. However, the exact concentricity and positioning of the lenses were deviated from the above lens specifications to mitigate potential ghost images that could be formed at the image plane due to reflections from surfaces R_1 and R_2 when centered on the axial image point. In some cases, this nearly concentric field flattener was used as the dewar window. The amount of aberrations induced by breaking exact concentricity and positioning can be reasonably small while still providing predominate control of the tangential field curvature.

Mann used a concentric field flattener in a 3:1 infrared zoom lens and significantly reduced the field curvature over the zoom range, and generally achieved balanced astigmatic fields (see Section 11.2.1).[15] His design technique for the field flattener was to first design the zoom lens and then to place a flat plate where the dewar window was to be located. He then allowed the computer program to vary the curvatures, and somewhat the flattener's thickness, which naturally became near concentric about the axial image. The final system was near diffraction-limited.

Figure 11.10 illustrates a typical configuration shown by Rosin where the concentric field flattener is located in the image space of a Petzval-type lens. In this case, he followed a design procedure of reducing the Petzval and sagittal field curvatures while rather ignoring the tangential field curvature as shown in

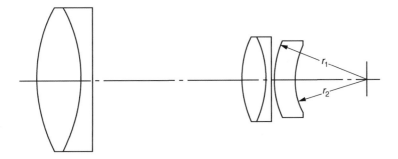

Figure 11.10   Concentric lens behind Petzval lens.

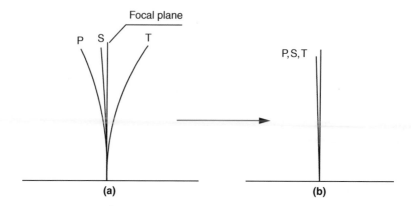

Figure 11.11 Flattening of tangential field using concentric lens where lens was designed to have flat sagittal field. (a) Initial design with flat sagittal field; (b) Tangential field bought into coincidence with the sagittal and Petzval curvatures using the concentric field flattener.

Figure 11.11a. With the introduction of the concentric field flattener lens, the tangential (T) field curvature was brought into coincidence with the Petzval (P) and sagittal (S) field curvatures as depicted in Figure 11.11b. A substantial improvement in resolution uniformity and contrast over the field of view was obtained. An alternative location for the concentric lens is to place it between the front and rear elements. In this case, the radii of the concentric lens are centered at the focus location of the front element.

A concentric field flattener lens may be introduced in other than the image space of a lens; however, this can shift the spatial location of the image plane. Another form of the concentric field flattener can be realized by considering a solid glass plate placed in image space, as illustrated in Figure 11.12a, which has been shown in Sections 3.4.4 and 6.4 to shift the image location and introduce aberrations.[16] A concentric air lens[17] is now formed by removal of the middle section of the glass plate—that is, a plano-concave lens followed by a convex-plano lens as illustrated in Figure 11.12b. The internal surface curvatures are centered on the image location that would occur should the glass plate have contained the image. The design of a lens being combined with this concentric field flattener element should include the aberrations resulting from a glass plate having a thickness equal to the distance between the plano surfaces of the concentric field flattener lens. The air concentric field flattener lens has a manufacturing advantage over the form shown in Figure 11.9 since it is more difficult to colocate the centers of the surfaces of a concentric lens element.

Another field flattener approach[18] uses a concentric shell centered about the stop (convex toward the image). The beam passing through this lens has the chief ray always perpendicular to the surfaces of the shell so its induced

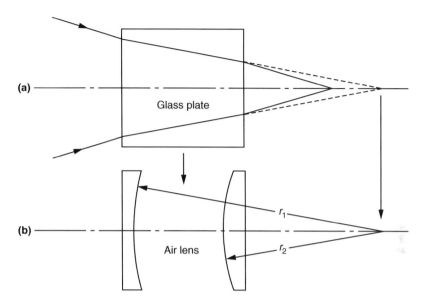

Figure 11.12 (a) Image shift caused by glass plate. (b) Creation of concentric air lens from the glass plate.

aberrations are constant as a function of field angle. As it does have some negative power, the shell affects the Petzval sum. It also shifts the image somewhat.

> **PROBLEM:** For a concentric field flattener lens, show that spherical aberration, tangential and sagittal coma, axial color, and lateral color are zero, and that distortion and sagittal field curvature are unaffected.

> **PROBLEM:** Consider the field diagrams shown in Figure 11.11 and explain the astigmatic aberrations depicted in both using Seidel aberration coefficients (σ_3 and σ_4).

A New-Achromat Combination

By 1886, Abbe and Schott in Jena, Germany, had developed barium crown glasses having just the required property to reduce the Petzval sum, and these glasses were immediately adopted by Schroder in 1888 in his Ross Concentric lens.[19] These glasses provided the sought after method for controlling the Petzval sum by using a crown glass of low dispersion and high refractive index in combination with a flint glass of higher dispersion and a low refractive index. This is precisely opposite to the choice of glasses used in telescope doublets and other ordinary achromats. Lenses of this type are therefore known as "new achromats." They have been used in the Protar (Section 14.4) and many other types of photographic objectives.

11.3 ILLUSTRATION OF ASTIGMATIC ERROR

As has been observed, our dutiful cemented doublet (Section 2.5) suffers from astigmatism, as is evident by examination of the ray fans plots in Figure 11.13, the focusing ray bundles in Figure 11.14a, and the field curves shown in Figure 11.14b. Also, both the tangential and sagittal fields are inward curving, and the maximum zonal spherical aberration is less than 10% of the peak astigmatic error at 5°. By comparing the 3.5° plot with the 5° plot, we can see that the aberration plots are linear with ρ and the ratio of the errors between these plots is about 2:1.

From our study, we recognize that the aberration is primary linear astigmatism since $(H'_{5°}/H'_{3.5°})^2 \approx 2$ and the linear behavior with ρ. Consequently, only σ_3 and σ_4 of the field-dependent aberration coefficients have significant values. Recalling Eqs. (4-6) and (4-7), it is easy to compute that $\sigma_3 \approx 0.79\sigma_4$, which means the Petzval curve is also inward curving but lies between the sagittal curve and the image plane. Computing the transverse astigmatism using real rays was discussed in Section 4.3.3. The tangential component is given by

$$TAST(\rho, H) = Y(\rho, 0°, H) - Y(\rho, 180°, H) - 2Y(\rho, 0°, 0) = -0.053159$$

and the sagittal component by

$$SAST(\rho, H) = 2[X(\rho, 90°, H) - Y(\rho, 0°, 0)] = -0.025650$$

which compare favorably to Figure 11.14. Coma is insignificant.

To observe what degradation in image formation the astigmatism in our lens will cause, we can generate a simulated image of a photograph using an analysis feature available in some lens design programs. Figure 11.15a shows the original and Figure 11.15b the resultant image (see page 309). The quadratic growth of the blur as a function of field angle is demonstratively illustrated. Compare this image with the linear blur growth due to coma shown in Figure 9.9b. Notice that the fine detail is observable over a larger central area than coma as a consequence of the quadratic growth of the blur. However, the image degradation at the top/bottom center and left/right sides is similar since the blur sizes are roughly the same although the shapes differ. The blurring in the corners is worse for Figure 11.15b than for Figure 9.9b.

11.4 DISTORTION

Distortion is a peculiar aberration in that it does not cause any loss of definition but merely a radial displacement of an image point toward or away from the lens axis. Distortion is calculated by determining the height H'_{pr} at which the

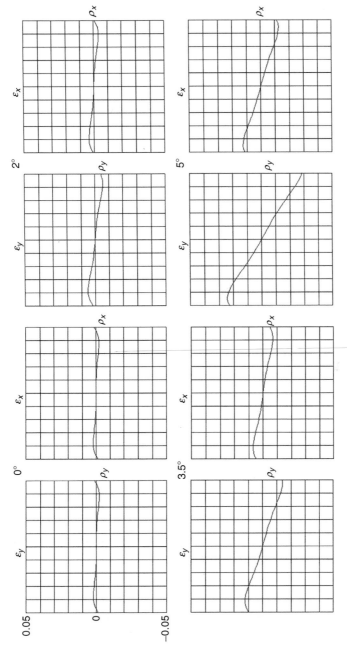

Figure 11.13 Monochromatic ray fans for Section 2.5 cemented doublet.

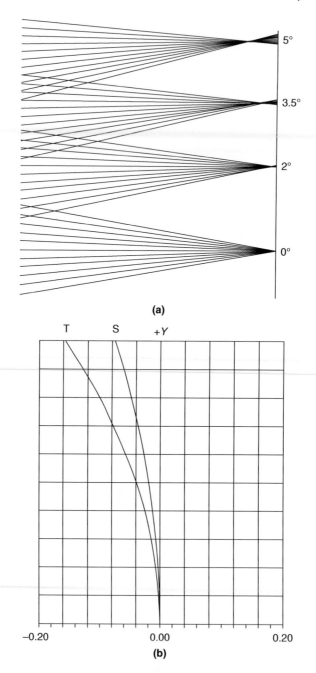

(a)

(b)

Figure 11.14 (a) Monochromatic focusing behavior for Section 2.5 cemented doublet. (b) Field curves for the tangential and sagittal foci.

(a)   (b)

Figure 11.15 (a) Original photograph. (b) Image formed by Section 2.5 cemented doublet showing the effect of astigmatism.

principal ray intersects the image plane, and comparing that height with the ideal Lagrangian or Gaussian image height calculated by paraxial formulas. Thus

$$\text{distortion} = H'_{\text{pr}} - h'$$

where h' for a distant object is given by $(f \tan U_{\text{pr}})$, or for a near object by (Hm), where m is the image magnification.

As discussed in Section 4.3.5, distortion is aperture-independent coma and can be resolved into a series of powers of H', namely,

$$\text{distortion} = \sigma_5 H'^3 + \mu_{12} H'^5 + \tau_{20} H'^7 + \ldots \tag{11-8}$$

However, very few lenses exhibit much distortion beyond the first cubic term. Because of the cubic law, distortion increases rapidly once it begins to appear, and this makes the corners of the image of a square, for example, stretch out for positive (pincushion) distortion, or pull in with negative (barrel) distortion.

The magnitude of distortion is generally expressed as a percentage of the image height, at the corners of a picture. Figure 11.16 shows two typical cases of moderate amounts of pincushion distortion, namely, 4% and 10%, respectively. The diagrams represent images that should be 50 mm squares, the quantity d beneath each figure being the lateral displacement of the midpoints of the

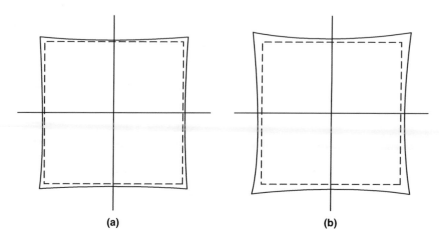

Figure 11.16   Pincushion distortion. (a) 4%, $d = 0.50$ mm, $r = 676$ mm; (b) 10%, $d = 1.25$ mm, $r = 302$ mm.

sides of the square due to distortion. The quantity r is the radius of curvature of the sides of the images, which should, of course, be straight. As can be seen, 4% distortion is just noticeable, whereas 10% is definitely objectionable. Consequently, *we generally set the distortion tolerance at about 1% since few observers can detect such a small amount.* For specialized applications such as aerial surveying and map copying, the slightest trace of distortion is objectionable, and the greatest care must be taken in the design and manufacture of lenses for these purposes to eliminate distortion completely.

11.4.1 Measuring Distortion

Since distortion varies across the field of a lens, it is difficult to determine the ideal Gaussian image height with which the observed image height is to be compared. One method is to photograph the images of a row of distant objects located at known angles from the lens axis and measure the image heights on the film. Since focal length is equal to the ratio of the image height to the tangent of the subtense angle, we can plot focal length against object position and extrapolate to zero object subtense to determine the axial focal length with which all the other focal lengths are to be compared. If the lens is to be used with a near object, we substitute object size for angular subtense and magnification for focal length. The determination can be performed at several object field positions and the coefficients σ_5, μ_{12}, and τ_{20} for Eq. (11-8) can be found.

11.4.2 Distortion Contribution Formulas

To develop an expression for the contribution of each lens surface to the distortion, we repeat the spherical-aberration contribution development from Section 6.1 but using the principal ray instead of the marginal ray. Thus Eq. (6-3) becomes

$$(S'n'u')_k - (Snu)_1 = \sum ni(Q' - Q)$$

where capital letters now refer to the data of the traced principal ray. Figure 11.17a shows that at the final image $S'_{pr} = H'_{pr} \cos U'_{pr}$, and similarly for the object. Hence if there are k surfaces in the lens,

$$H' = H \left\{ \frac{nu \cos U}{n'_k u'_k \cos U'_k} \right\} + \sum \frac{ni(Q' - Q)}{n'_k u'_k \cos U'_k}$$

For a distant object, the first term in this expression reduces to

$$f'(\sin U / \cos U'_k)_{pr}$$

To relate this formula to the distortion, we note that $Dist = H' - h'$, where h', the Lagrangian image height, is equal to $f' \tan U_1$. Hence

$$\text{distortion} = h'_{\text{Lagrangian}} \left(\frac{\cos U_1}{\cos U'_k} - 1 \right)_{pr} + \sum \frac{ni(Q' - Q)_{pr}}{(n'_k u'_k \cos U'_k)_{pr}} \tag{11-9}$$

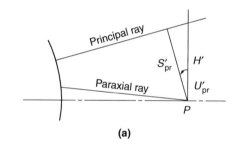

(a)

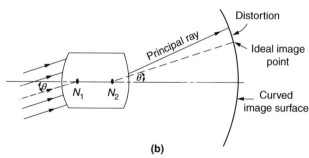

(b)

Figure 11.17 Distortion diagrams. (a) Basic geometry for distortion computation in image plane and (b) distortion when image surface is curved.

Note that the two parts of this formula are similar in magnitude, the first being caused by the difference in slope of the principal ray as it enters and leaves the system, the second being derived from the lens surface contributions.

To verify the accuracy of this formula, we take the much-used cemented doublet of Section 2.5 and trace a principal ray entering at 8° through the anterior focal point to form an almost perfectly *telecentric system* (see Table 11.2).[20]

The agreement in the results of this calculation between the direct measure of the image height and the sum of the various contributions is excellent. For the distortion itself we first calculate

$$h'_{\text{Lagrangian}}\left(\frac{\cos U_1}{\cos U'_k} - 1\right) = -0.0163489$$

When this is added to the summation value in Table 11.2 we find the distortion to be -0.0618321, again in excellent agreement. The change in slope of the principal ray has contributed about one-third of the distortion, the remainder coming from the lens surfaces themselves.

Unfortunately, the quantities under the summation sign are not really "contributions" that have merely to be added together to give the distortion. Each lens surface, to be sure, provides an amount to be summed, but it also

Table 11.2

Calculation of Distortion Contributions

c		0.1353271		-0.1931098	-0.0616427	
d			1.05	0.4		
n			1.517	1.649		
			Paraxial			
ϕ		0.0699641		-0.0254905	0.0400061	
$-d/n$			0.6921556	-0.2425713		$l' = 11.285857$
y		1		0.9515740	0.9404865	$f' = 12.00002$
nu	0		-0.0699641	-0.0457080	-0.0833332	
u	0		-0.0461200	-0.0277186	-0.0833332	
$(yc+u)=i$		0.1353271		-0.2298783	-0.0856927	
			8° Principal ray, with $L_1 = -11.76$			
Q		1.6527600		1.7050560	1.7263990	
Q'		1.6947263		1.7117212	1.7233187	
U	8°		0.56367°	2.10291°	$-0.50119°$	
			Distortion contributions			
$(Q-Q')_{\text{pr}}$		-0.0419663		-0.0066652	0.0030803	
ni		0.1353271		-0.3487254	-0.1413073	
$1/u'_k \cos U'_k$		12.000478		12.000478	12.000478	
Product		-0.0681528		0.0278930	-0.0052234	$\sum = -0.0454832$

Hence $H' = 1.6701438 - 0.0454832 = 1.6246606$

contributes to the slope of the emergent ray in the first term of the distortion expression. The relation just given is therefore mainly of theoretical interest; however, the Buchdahl coefficients for distortion discussed in Chapter 4 do not suffer the above issue.

11.4.3 Distortion When the Image Surface Is Curved

If a lens is designed to form its image on a curved surface, the meaning of distortion must be clearly defined. As always, distortion is the radial distance from the ideal image point to the crossing point of the principal ray; but now the ideal image is represented by the point of intersection of a line drawn through the second nodal point at the same slope as that of a corresponding ray entering through the first nodal point (Figure 11.17b). Then

$$\text{distortion} = [(Y_2 - Y_1)^2 + (Z_2 - Z_1)^2]^{1/2}$$

where subscript 2 refers to the traced principal ray and subscript 1 to the ideal ray through the nodal points.

11.5 LATERAL COLOR

Lateral color is similar to distortion in that it is calculated by finding the height of intercept of principal rays at the image plane, but now we must compare two principal rays in two different wavelengths, typically the C and F lines of hydrogen, although, of course, any other specified lines can be used if desired. Then

$$\text{lateral color} = H'_F - H'_C$$

Lateral color can be resolved into a power series, but now there is a first-order term that does not appear in distortion (the first-order term in distortion is the Gaussian image height; see Figure 4.5):

$$\text{lateral color} = aH' + bH'^3 + cH'^5 + \ldots$$

Some people consider that only the first term represents lateral color, all the others being merely the chromatic variation of distortion. No matter how it is regarded, lateral color causes a radial chromatic blurring at image points located away from the lens axis. Of course, both distortion and lateral color vanish at the center of the field.

11.5.1 Primary Lateral Color

The first term of this series, representing the primary lateral color, can be calculated by a method similar to the calculation of the OSC, except that now we trace paraxial rays in C and F light instead of tracing a marginal and a paraxial ray in brightest light. Thus, writing paraxial data in F in place of the original marginal ray data, and paraxial data in C in place of the original paraxial ray data, our formula Eq. (9-4) becomes

$$CDM = \frac{\text{lateral color}}{\text{image height}} = \frac{u'_C}{u'_F}\left(\frac{l'_C - l'_{pr}}{l'_F l'_{pr}}\right) - 1 \quad \text{for a near object} \quad (11\text{-}10)$$

$$= \frac{\Delta f'}{f'} - \frac{\Delta l'}{l' - l'_{pr}} \quad \text{for a distant object} \quad (11\text{-}11)$$

where $\Delta f' = f'_F - f'_C$ and $\Delta l' = l'_F - l'_C$. The latter is, of course, the ordinary paraxial longitudinal chromatic aberration. The expression CDM is an abbreviation for *chromatic difference of magnification* and it is strictly analogous to OSC.

In a symmetrical lens, or any other lens in which the pupils coincide with the principal planes, $l' - l'_{pr} = f'$, and Eq. (11-11) becomes

$$CDM = (\Delta f' - \Delta l')/f' \quad (11\text{-}12)$$

The numerator of this expression is simply the distance between the second principal planes in C and F light. Thus, if these principal planes coincide, there will be no primary lateral color. This is often a convenient computing device for use in the early stages of a design. Later, of course, it is necessary to trace true principal rays in F and C and calculate the difference in the heights of these rays at the focal plane.

The logic of this last relationship can be understood by the diagram in Figure 11.18, which shows the principal rays in C and F, at small obliquity, emerging from their respective principal points and proceeding to the image plane. It is clear that

$$\text{primary lateral color} = z \tan U'_{pr} = z(h'/f')$$

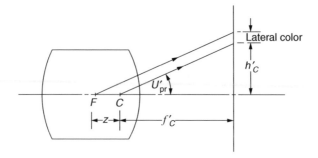

Figure 11.18   Primary lateral color depends on z.

and hence

$$CDM = \text{lateral color}/h' = z/f'.$$

Although six of the cardinal points are wavelength dependent, reference to Section 3.3.7 shows that the seventh cardinal point (optical center) is spatially stationary with wavelength. Just as higher-order lateral color can be thought of as chromatic variation of distortion, all of the comatic and astigmatic aberrations are wavelength dependent (unless the optical system is all reflective). Spherochromatism was covered in some depth in Chapter 7.

11.5.2 Application of the (D − d) Method to an Oblique Pencil

It has been shown by Feder[21] that Conrady's $D - d$ method can be applied to an oblique pencil through a lens. He pointed out that if we calculate $\sum D \, \Delta n$ along each ray of the pencil and $\sum d \, \Delta n$ along the principal ray, then we can plot a graph connecting $\sum (D - d) \, \Delta n$ as ordinate against $\sin U'$ of the ray as abscissa. The interpretation of this graph is that the ordinates represent the longitudinal chromatic aberration of each zone, while the slope of the curve represents the lateral color of that zone.

Typical curves at 0 and 20°, calculated for the $f/2.8$ triplet used in Section 8.4.1, are given in Figure 11.19 for $\Delta n = (n_F - n_C)$. The fact that the axial graph

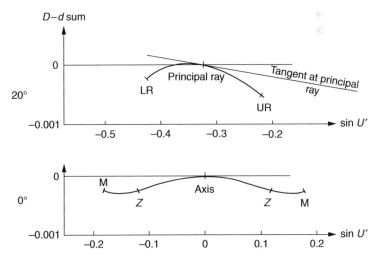

Figure 11.19 Application of the $(D - d)$ method to an axial and an oblique pencil through a triplet objective.

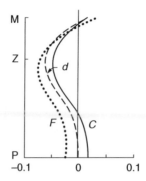

Figure 11.20   Spherochromatism of $f/2.8$ triplet objective.

is not a straight line indicates the presence of spherochromatism; this is shown plotted in the ordinary way in Figure 11.20. The tilt of the 20° curve at the principal-ray point (Figure 11.19) indicates the presence of lateral color, of amount about -0.0018. The lateral color found by actual ray tracing was $H'_F - H'_C = -0.00168$, which is in excellent agreement considering the difficulty in graphically determining the exact tangent to the curve at the principal-ray point.

11.6 THE SYMMETRICAL PRINCIPLE

A fully symmetrical (holosymmetrical) system is one in which each half of the system, including the object and image planes, is identical to the other half, so that if the front half is rotated through 180° about the center of the stop it will coincide exactly with the rear half.

Such a fully symmetrical system has several interesting and valuable properties, notably complete absence of distortion and lateral color, and absence of coma *for one zone* of the lens. These are the three transverse aberrations, with the contributions of the front component being equal and opposite to the contributions of the rear. The two half-systems also contribute identical amounts to each of the longitudinal aberrations, but now the contributions have the same sign and add together instead of canceling out.

The reason for this cancellation of the transverse aberrations can be seen by consideration of Figure 11.21a. Any principal ray in any wavelength starting out from the center of the stop and traveling both ways to the object and image planes will intersect those planes at the same height above and below the axis, giving a magnification of exactly -1.0 over the entire field. Thus distortion and lateral color are automatically absent.

To demonstrate the absence of coma, we must trace a pair of upper and lower oblique rays in the stop both ways until they intersect each other at P

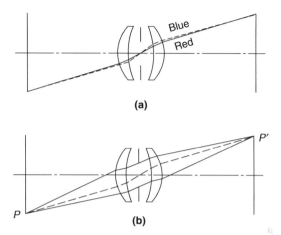

Figure 11.21 Transverse aberrations of a holosymmetrical system. (a) Distortion and lateral color and (b) coma.

and P' (Figure 11.21b). We then add a principal ray through the center of the stop at such a slope that it passes through P. Symmetry will then dictate that it will also pass exactly through P'. Hence, this one zone of the lens will be coma-free, although one cannot draw any similar conclusion for other zones of the lens. It should be noted that if there is any coma in each half of the lens, the principal ray in the stop will not be parallel to the parallel upper and lower oblique rays initially placed there. The symmetry principle is a powerful tool for the lens designer, but its limitations must be kept in mind.

DESIGNER NOTE

If the lens is symmetrical but the conjugates are not equal, then the distortion will be corrected only if the entrance and exit pupils, where the entering and emerging portions of the principal ray cross the axis, are fixed points for all possible obliquity angles.[22]

Similarly, lateral color will be absent if the entrance and exit pupils are fixed points for all wavelengths of light. These two conditions are often referred to as the Bow–Sutton condition. No corresponding conclusions can be drawn for coma, but it is generally found that coma is greatly reduced by symmetry, even though the conjugate distances are not equal. The point to notice is that if distortion and lateral color must be well corrected over a wide range of magnifications, as in a process lens used to copy maps, then the designer must concentrate on correcting the spherical and chromatic aberrations of the principal rays rather than on correcting the primary image, stopping the lens down if necessary to maintain the image quality. Stopping the lens down, of course, has no effect on the aberrations of the principal ray.

11.7 COMPUTATION OF THE SEIDEL ABERRATIONS

In some designs it is advantageous to determine the contributions of the various surfaces, or thin lens elements, to the seven primary or Seidel aberrations. This procedure has the advantage of indicating where each aberration arises in the system, and the computation is rapid enough to permit an approximate design to be reached in a short time before any real ray tracing is attempted.

To calculate the surface contributions, we first trace a regular paraxial ray from object to image, and also a paraxial principal ray through the center of the stop. The entering values of the (y, u) of the paraxial ray and the (y_{pr}, u_{pr}) of the paraxial principal ray must correspond to the desired values for the real lens, so that the y is equal to the true Y at the first surface, and the u_{pr} is equal to the tan U_{pr} of the angular field for which the primary aberrations are desired. In this notation, the Lagrange invariant can be written

$$hnu = n(u_{pr}y - y_{pr}u)$$

11.7.1 Surface Contributions

Both Conrady[23] and Feder[24] have given simple formulas by which the surface contributions to the Seidel aberrations can be rapidly computed. We calculate the following equations in order, noting that subscript 0 in u_0' and h_0' refers to the final image, while other symbols refer to the surface in question. Having traced the paraxial ray and the paraxial principal ray, we calculate their angles of incidence by the usual relation $(i = yc + u)$, where c is the surface curvature. Then

$$K = yn\left(\frac{n}{n'} - 1\right)(i + u')/2u_0'^2$$

$$SC = Ki^2, \quad CC = Kii_{pr}u_0', \quad AC = Ki_{pr}^2$$

$$PC = -\frac{1}{2}h_0'^2 c\left(\frac{n' - n}{nn'}\right)$$

$$DC = (PC + AC)(u_0'i_{pr}/i) = CC_{pr} + \tfrac{1}{2}h_0'(u_{pr}'^2 - u_{pr}')$$

$$L = yn\left(\frac{\Delta n}{n} - \frac{\Delta n'}{n'}\right)/u_0'^2$$

$$L_{ch}C = Li, \quad T_{ch}C = Li_{pr}u_0'$$

(11-13)

Here c is the surface curvature, as usual; for the aberrations, SC is the contribution to longitudinal spherical aberration, CC to the sagittal coma, AC to sagittal astigmatism (i.e., the longitudinal distance from the Petzval surface to the sagittal focal line), PC to the sag of the Petzval surface, and DC to the distortion; $L_{ch} C$ and $T_{ch} C$ are the surface contributions to the longitudinal chromatic aberration and lateral color, respectively. Of the two expressions for the distortion contribution, the first is easier to compute by hand, but it fails if the object is at the center of curvature of the surface, for then $AC = -PC$ and DC becomes indeterminate.

The second alternative expression requires the calculation of CC for the principal ray, and it is therefore recommended that a subroutine be written for CC that can be applied successively to the paraxial ray and the principal ray data. Alternative equations for the surface contributions were presented in Section 4.4, as were the scaling factors for presenting the aberrations as transverse, longitudinal, or wave aberrations.

11.7.2 Thin-Lens Contributions

In some thin-lens predesigns it is convenient to reduce the system to a succession of thin elements separated by finite air spaces. We trace the same two paraxial rays through the system, using Eqs. (3-17) for this purpose. We then list the values of $Q = (y_{pr}/y)$ at each thin lens. The computation now falls into two parts: first, the calculation of each contribution as if the stop were at the thin lens, and then the modification of each contribution to place the stop in its true position. The second stage makes use of the Q at each lens.

The equations for the first stage are

$$SC = -\frac{y^4}{u_0'^2} \sum \left(G_1 c^3 - G_2 c^2 c_1 + G_3 c^2 v_1 \right)$$

$$+ G_4 cc_1^2 - G_5 cc_1 v_1 + G_6 cv_1^2$$

$$CC = -y^2 h_0' \sum \left(\tfrac{1}{4} G_5 cc_1 - G_7 cv_1 - G_8 c^2 \right)$$

$$AC = -\tfrac{1}{2} h_0'^2 \sum (1/f) \tag{11-14}$$

$$PC = -\tfrac{1}{2} h_0'^2 \sum (1/nf)$$

$$DC = 0, \quad T_{ch} C = 0$$

$$L_{ch} C = -\frac{y^2}{u_0'^2} \sum \left(\frac{1}{Vf} \right)$$

The summations in these expressions are used only if the thin component is compound, such as a thin doublet or a thin triplet; they are not required for a single thin element. The formulas for the G terms are in Sections 6.3.2 and 9.3.4.

We next apply the calculated Q factors ($Q = y_{pr}/y$) to place the stop in its correct position. The true contributions, marked with asterisks, are found in the following way:

$$SC^* = SC, \quad PC^* = PC, \quad L_{ch}C^* = L_{ch}C$$
$$CC^* = CC + SC(Qu_0')$$
$$AC^* = AC + CC(2Q/u_0') + SCQ^2 \qquad (11\text{-}15)$$
$$DC^* = (PC + 3AC)Qu_0' + 3CCQ^2 + SC(Q^3 u_0')$$
$$T_{ch}C^* = L_{ch}C(Qu_0')$$

These expressions are generally known as the *stop-shift formulas*. It should be observed that for the stop shift to affect the value of coma, residual spherical aberration must be present in an adequate amount. In a like manner, spherical aberration and/or coma must be present for the stop shift to affect the value of astigmatism. Distortion can be affected by a stop shift if spherical aberration and/or coma and/or astigmatism exist.

11.7.3 Aspheric Surface Corrections

If we are computing the Seidel surface contributions and encounter an aspheric surface, we first calculate the contributions, assuming that the surface is a sphere with the vertex curvature c, and then add a set of correcting terms depending on the asphericity.

The aspheric surface is assumed to be of the form

$$Z = \tfrac{1}{2}cS^2 + j_4 S^4 + j_6 S^6 + \ldots$$

where $S^2 = y^2 + z^2$ and the j values are the aspheric coefficients. Then

$$\text{addition to } SC = 4j_4 \left(\frac{n - n'}{u_0'^2} \right) y^4$$

$$\text{addition to } CC = 4j_4 \left(\frac{n - n'}{u_0'^2} \right) y^3 y_{pr}$$

$$\text{addition to } AC = 4j_4 \left(\frac{n - n'}{u_0'^2} \right) y^2 y_{pr}^2 \qquad (11\text{-}16)$$

$$\text{addition to } DC = 4j_4 \left(\frac{n - n'}{u_0'^2} \right) y y_{pr}^3$$

It should be noted that only the j_4 coefficient appears in the primary aberrations, since the higher aspheric terms affect only the higher-order aberrations.[25] *Also, if the stop is located at an aspheric surface, the y_{pr} there will be zero, and the only aberration to be affected by the asphericity is the spherical aberration.*

11.7.4 A Thin Lens in the Plane of an Image

This case is exemplified by a field lens or a field flattener (see also Section 11.2.2). We cannot now use the thin-lens contribution formulas already given because both the stop and the image cannot lie in the same plane. Consequently, we have to return to the surface contribution formulas and add them up for the case in which $y = 0$. When this is done, we find that for a thin lens situated in an image plane

$$SC = CC = AC = 0, \quad L_{ch}C = T_{ch}C = 0$$

The Petzval sag PC has its usual value of $-\frac{1}{2}h_0'^2/f'N$, where N is the index of the glass. The distortion must be carefully evaluated. It turns out to be

$$DC = \frac{1}{2}h_0'^2 u_0' \left(\frac{y_{pr}}{Nf'u_1}\right)\left[\frac{1}{r_1} + \frac{N}{r_2} - \frac{1}{l_{pr1}}\right]$$

where f' is the focal length of the thin lens. The distortion contribution depends on the shape of the thin lens in addition to its focal length and refractive index.

In practice, the lens designer may decide that it is preferable to flatten the tangential field rather than to minimize the Petzval sum (see Section 11.2.1). Figure 11.22a illustrates a simple concave-plano lens, with $N = 1.5$, used as a field flattener for an optical system having $f'=10$, $f/10$, and FOV of $10°$. The vertical dashed line represents the location of the original paraxial image plane. The finite axial thickness of the field flattener necessarily shifts the image plane farther from the optical system in a manner similar to a parallel plate (see Section 6.4); however, the shift is about 15% larger than $t(N-1)/N$ due to the negative power of the front surface. Figure 11.22b shows the inward curving fields of the original optical system and the excellent tangential field correction by use of the field flattener. A consequence of using this field flattener form is the introduction of distortion, as shown in Figure 11.22c. The original optical system had essentially no distortion. Potentially, this can be corrected by adjusting the optical system to have compensating distortion.

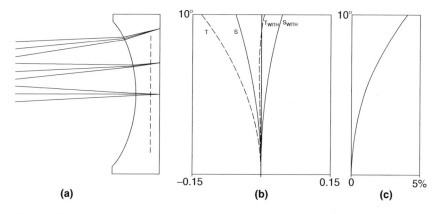

Figure 11.22 (a) Basic field flattener showing image shift from original image plane, (b) astigmatic field curves with (T_{WITH} and S_{WITH}) and without (T and S) the field flattener, and (c) distortion caused by field flattener.

ENDNOTES

[1] H. Coddington, *A Treatise on the Reflexion and Refraction of Light*, p. 66. Simpkin and Marshall, London (1829).

[2] Coddington's book was digitized by the Google Books Project and is freely available to read at *http://books.google.com/books/download/A_Treatise_on_the_Reflection_and_Refract. pdf?id=WI45AAAAcAAJ&output=pdf&sig=ACfU3U0eX3OvIkczIHZL5iJSyLEFEJza-A*. This 1829 treatise likely was read by Gauss and Petzval, the contributions by both occurring over a decade later. Those interested in understanding the fundamentals of optics should read Coddington's book. Coddington credits the works of Professor Airy and Mr. Herschel's invaluable (*sic*) article on light in the *Encyclopedia Metropolitana*.

[3] A. E. Conrady, p. 588.

[4] T. Smith, "The contributions of Thomas Young to geometrical optics," *Proc. Phys. Soc.*, 62B:619 (1949).

[5] Not to be confused with the entrance pupil coordinate.

[6] A. E. Conrady, p. 739.

[7] A. E. Conrady, pp. 290-294.

[8] B. K. Johnson, *Optical Design and Lens Computation*, p. 118, Hatton Press, London (1948).

[9] H. H. Emsley, *Aberrations of Thin Lenses*, p. 194, Constable, London, (1956).

[10] B. K. Johnson, pp. 93-118.

[11] R. Barry Johnson, "Balancing the astigmatic fields when all other aberrations are absent," *Appl. Opt.*, 32(19):3494-3496 (1993).

[12] Spot sizes are about equal when the field angle is $\sqrt{2/3}$ ($\approx 80\%$) of the full field angle.

[13] Seymour Rosin, "Concentric Lenses," *JOSA*, 49(9):862-864 (1959).

[14] R. Barry Johnson, Texas Instruments, Dallas, Texas.

[15] Allen Mann, *Infrared Optics and Zoom Lenses, Second Edition*, pp. 60-61, SPIE Press, Bellingham, WA (2009).

[16] In addition to spherical aberration discussed in Section 6.4, coma and astigmatism are also present.

[17] See also Section 7.4.3.

[18] Donald C. O'Shea, *Elements of Modern Optical Design*, pp. 214-215, Wiley, New York (1985).

[19] U.S. Patent 404,506 (1888).

[20] A stop is located 11.75 before the first surface of the doublet lens, which is the location of the front or anterior focal point. The stop is also the entrance pupil. The principal ray passes through the front focal point and emerges from the lens parallel to the optical axis at the corresponding image height. This is known as the *telecentric condition*.

[21] D. P. Feder, "Conrady's chromatic condition," *J. Res. Nat. Bur. Std.*, 52:47 (1954); Res. Paper 2471.

[22] For at least the monochromatic case, the entrance and exit pupil will be located at the nodal points if the aperture stop is placed at the optical center of the lens. The limitations of the entrance pupil shape and position related to large field angles still apply. See Section 3.3.7.

[23] A. E. Conrady, pp. 314, 751.

[24] D. P. Feder, "Optical calculations with automatic computing machinery," *J. Opt. Soc. Am.*, 41:633 (1951).

[25] It should be realized that the j_4 term and the conic constant both affect the primary aberrations; however, they do not create the same surface contour and should not in general ever be used at the same time on a specific surface. It is most common to use the conic constant. Any aspheric coefficient affects its own order level and those higher but not lower.

Chapter 12

Lenses in Which Stop Position Is a Degree of Freedom

It is obvious that, depending on its position in a lens system, a stop selects some rays from an oblique pencil and rejects others. Thus, if the stop is moved along the axis (or for that matter, if it is displaced sideways, but that case will not be considered here), some of the former useful rays will be excluded while other previously rejected rays are now included in the image-forming beam. Consequently, unless the lens happens to be perfect, a longitudinal stop shift changes all the oblique aberrations in a lens. It will not affect the axial aberrations provided the aperture diameter is changed as necessary to maintain a constant f-number.

12.1 THE $H' - L$ PLOT

The results of a stop shift can be readily studied by tracing a number of meridional rays at some given obliquity through the lens, and plotting a graph connecting the intersection length L of each ray from the front lens vertex as an abscissa, with the intersection height H' of that ray at the paraxial image plane as ordinate. This graph (Figure 12.1a) is similar to the meridional ray plot in Figure 8.7 discussed in Section 8.2, except that the abscissas have reversed signs, so that the upper rim ray of the beam now appears at the left end of the graph while the lower rim ray falls at the right, as illustrated in Figure 12.1b. Locating a stop in any position selects a portion of the graph and rejects the rest of it. The ray passing though the center of the stop is, of course, the principal ray of the useful beam.

Figure 12.1b shows the lens with the stop in front. The dashed line extending from the top of the stop to the plot in Figure 12.1a indicates the height of the principal ray in the Gaussian image plane. The diameter of the beam of light from the left is limited by the stop. If the obliquity angle is increased, it is evident that the lower rim ray is determined by the stop while the upper rim ray will be vignetted by the top of the lens. The outer vertical dashed lines from the axial crossing points of the upper and lower rim rays bound the portion of the $H' - L$ plot that corresponds to the aforementioned meridional ray fan plot. Assuming that the

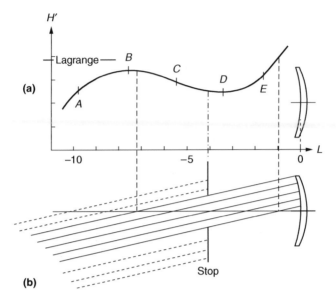

Figure 12.1   A typical $H' - L$ graph for a meniscus lens.

stop, having diameter D, is limiting the beam diameter at the obliquity θ under study, the distance along the axis having intersections with the upper and lower rays defining the beam is simply $D/\tan\theta$. In Figure 12.1 these axial intersections are shown by the vertical dashed lines. This distance is centered on the principal ray intersection.

This graph tells us a great deal about the aberrations in the image and how they will be changed when the stop is shifted along the axis. It should be understood that this technique can be used with a lens of any complexity, not just a simple singlet. Let us now explore how to interpret $H' - L$ plots.

12.1.1 Distortion

The height of the graph at the principal-ray point above or below the Lagrangian image height is a direct measure of the distortion. As shown in Figure 12.1a, the principal-ray point is below the Lagrangian image height which means the distortion is negative.

12.1.2 Tangential Field Curvature

The first derivative or slope of the graph at the principal-ray point is a measure of the sag of the tangential field Z'_t, a quantity that is ordinarily determined by the Coddington equations. If the slope is upward from left to right it indicates an inward-curving field because the upper rim ray strikes the image plane lower than the lower rim ray. Conversely, if the slope is downward from left to right, it indicates a backward-curving field because the upper rim ray strikes the image plane higher than the lower rim ray. A graph that is horizontal at the principal-ray point indicates a flat tangential field.

12.1.3 Coma

The second derivative or curvature of the graph at the principal-ray point is a measure of the tangential coma present in the lens. If the ends of the portion of the graph used are above the principal ray, this indicates positive coma. The coma is clearly zero at a point of inflection where the graph is momentarily a straight line. It is possible that a stop position can be found where both the slope and the curvature are zero; however, this requires that spherical aberration be present.

12.1.4 Spherical Aberration

The presence of spherical aberration is indicated by a cubic or S-shaped curve, undercorrection giving a graph in which the line joining the ends of the curve is more uphill than the tangent line at the principal-ray point. If the line joining the ends of the curve is more downhill, then the spherical aberration is overcorrected.

All of these phenomena are illustrated in the typical $H' - L$ graph shown in Figure 12.1. If the principal ray falls at A, the field will be drastically inward-curving. At B the field is flat but there is strong negative coma. At C the coma is zero but the field is now backward-curving. At D the field is once more flat but now the coma is positive, while at E the field is once more drastically inward-curving. The overall S shape of the curve indicates the presence of considerable undercorrected spherical aberration.

Thus we reach the important conclusion that we can eliminate coma by a suitable choice of stop position if there is spherical aberration present; indeed, this result is implicit in the OSC formulas in Section 9.3. Furthermore, we can flatten the tangential field by a suitable choice of stop position if there is a sufficiently large amount of coma or spherical aberration or both. In terms of the primary or Seidel aberrations, these conclusions are in agreement with the stop-shift formulas given in Eq. (11-15) in Section 11.7.2.

12.2 SIMPLE LANDSCAPE LENSES

It is instructive to plot the 20° $H' - L$ curves for a single lens bent into a variety of shapes, as in Figure 12.2. The focal length is everywhere 10.0, the thickness 0.15, and the refractive index 1.523. In these graphs the abscissa values

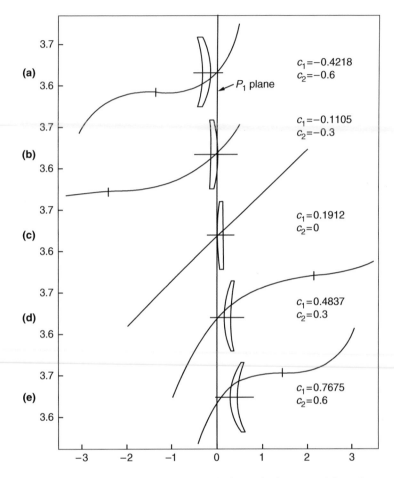

Figure 12.2 Bending a meniscus lens (20°). The abscissa value is measured from the anterior principal point in each case.

are measured from the front or anterior principal point in each case. The reference point for the parameter L can be the vertex of the first lens surface, the anterior principal point, or any other point the designer may select. In curve (a) the lens is shown bent into a strongly meniscus shape, concave to the front where parallel light enters. There is a large amount of spherical undercorrection, leading to an S-shaped cubic curve, and the interesting region containing the maximum, inflection, and minimum points lies close to the lens. Placing the stop at the location denoted by a "tick mark" on the $H' - L$ plot results in a flat tangential field (slope is zero) and no coma (zero curvature or inflection point). The distortion is negative since the Gaussian image height is 3.64.

In curve (b) of Figure 12.2, the lens is bent into such a weak meniscus shape that there is very little spherical aberration, with no maxima or minima. With the stop at the tick mark, the astigmatism is inward curving and the coma is positive. In curve (c) of the figure, a plano-convex lens with its curved face to the front is shown. There is now no coma and very little spherical aberration so that the curve is practically a straight line. The tangential astigmatic field is strongly inward curving. In the remaining graphs the lens is a meniscus with a convex side to the front, and now the interesting region has moved behind the lens, still on the concave side of the lens. Curve (d) in Figure 12.2 shows spherical aberration with no coma (inflection point) and some inward curving tangential astigmatic field. The final curve is for a stronger bending and shows stronger spherical aberration and a slight inward curving field. It will be noticed that all the graphs have about the same slope at $L = 0$. This bears out the well-known fact that *any reasonably thin lens with a stop in contact has a fixed amount of inward-curving field independent of the structure of the lens.*

As a simple meniscus lens has only two degrees of freedom, namely, the lens bending and the stop position, it is clear that only two aberrations can be corrected. Invariably the two aberrations chosen are coma and tangential field curvature. The axial aberrations, spherical and chromatic, can be reduced as far as necessary by stopping the lens down to a small aperture; $f/15$ is common although some cameras with short focal lengths have been opened up as far as $f/11$. The remaining aberrations, lateral color, distortion, and Petzval sum, must be tolerated since there is no way to correct them in such a simple lens. Changes in thickness and refractive index have very little effect on the aberrations.

DESIGNER NOTE

In designing a landscape lens, one should choose a bending such that the $H' - L$ curve is a horizontal line at the inflection point. This will ensure that the coma is corrected and the tangential field will be flat at whatever field angle was chosen for plotting the $H' - L$ curve. Of course, the field may turn in or out at other obliquities.

12.2.1 Simple Rear Landscape Lenses

To meet the specified conditions, it is found that by interpolating between the examples shown in Figure 12.2 for a simple rear landscape lens, a front surface curvature of about –0.28 is required. With the thickness and refractive index used here, there is very little latitude. Solving the rear curvature to give a focal length of 10.0, we arrive at the 25° $H' - L$ curve shown in Figure 12.3. This curve indicates that the stop must be at B, a distance of 1.40 in front of the lens. At $f/15$ the stop diameter will be 0.667, and to cover a field of up to 30° the lens diameter

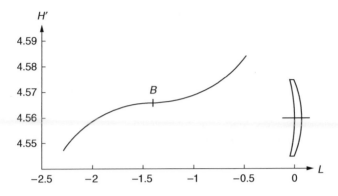

Figure 12.3 The $H' - L$ graph of a rear meniscus lens having a flat coma-free field (25°).

must be about 1.80. Actually, because of excessive astigmatism, it is unlikely that this lens would be usable beyond about 25° from the axis. The lens system is

c	d	n
−0.28		
	0.15	1.523
−0.4645		

with $f' = 10.0003$, $l' = 10.1445$, LA' ($f/15$) = -0.2725, and Petzval sum = 0.0634. The astigmatism is shown in Figure 12.4 and has the values presented in Table 12.1.

If a flatter form were used, the spherical aberration would be slightly reduced and the tangential field would be inward curving. This would reduce the astigmatism, but the sagittal field is already seriously inward curving and flattening the lens would make it even worse. It therefore appears that the present design is about as good as could be expected with such a simple lens.

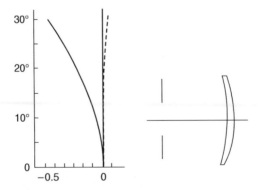

Figure 12.4 Astigmatism of a simple rear meniscus lens. The sagittal field is the solid curve and the tangential field is the dashed curve.

Table 12.1

Astigmatism and Distortion for Lens Shown in Figure 12.4

Field (deg)	X'_s	X'_t	Distortion (%)
30	−0.584	0.044	−3.34
20	−0.260	0.003	−1.41

For a second example, consider a similar lens having a focal length of 10, operating at $f/10$, and a $20°$ semi-field-of-view. Since the f-number is significantly lower (larger aperture) than in the prior example, we increase the lens thickness to 0.40 to maintain reasonable thickness at the trim diameter. Using Eqs. (4-6) and (4-8), with the entrance pupil coordinate ρ being a tenth of the pupil radius (same as stop in this case), we have real-ray definitions for astigmatism and coma. Using these equations for defect optimization in a lens design program produced the following design.

c	d	n
−0.192009		
	0.40	1.523
−0.373366		

This lens is a bit flatter as a consequence of the somewhat smaller field of view and lower f-number. The stop to first vertex distance is 1.592 or about 14% greater than for the prior case. The Seidel coefficients are $\sigma_1 = -0.002047$, $\sigma_2 = 0.000000$, $\sigma_3 = -0.000659$, $\sigma_4 = 0.002063$, and $\sigma_5 = -0.007043$. The primary tangential astigmatism is $3\sigma_3 + \sigma_4 = 0.000086$ or essentially flat, while the sagittal astigmatism remains the same as the preceding lens. However, the distortion increases to -1.9% from -1.4% for the previous lens.

12.2.2 A Simple Front Landscape Lens

Quite by chance, the curve for lens (e) in Figure 12.2 has a horizontal inflection, and it therefore meets the requirements for a landscape lens. Its structure is

c	d	n
0.7675		
	0.15	1.523
0.60		

with $f' = 9.99918$, $l' = 9.60387$, LA' ($f/15$) $= -0.4729$, stop distance $= 0.8641$, stop diameter $= 0.5830$, and Petzval sum $= 0.0575$. The results are shown in Table 12.2.

Table 12.2

Astigmatism and Distortion for Lens Shown in Figure 12.5

Field (deg)	Z'_s	Z'_t	Distortion (%)
30	−0.603	+0.074	+3.50
20	−0.246	+0.005	+1.41

The lens diameter should be about 1.6 (Figure 12.5) so that the lens will not vignette the rays entering at the maximum obliquity. Note that the surfaces on this lens are much stronger than those of the rear meniscus and the aberration residuals are generally worse. The spherical aberration particularly is much greater than in the system shown in Figure 12.4. Nevertheless, front landscape lenses are usually preferred because the camera can then be made shorter for the same focal length, and the large front lens acts as an effective shield to prevent the entry of dirt into the shutter mechanism. In an effort to reduce the spherical aberration a flatter lens is often employed, but the resulting inward-curving field must then be offset in part by the use of a cylindrically curved film gate. The compensation is not very good, however, because a cylinder does not fit very well on a spherically curved image.

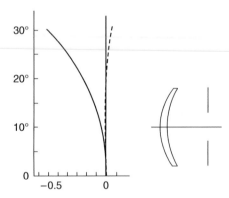

Figure 12.5 Astigmatism of a front-meniscus lens. The sagittal field is the solid curve and the tangential field is the dashed curve.

DESIGNER NOTE

It is worth noting that the remaining spherical and chromatic aberration residuals in a simple landscape lens have the effect of greatly increasing the depth of field, so that if the film is correctly located relative to the lens, any object from, say, 6 ft to infinity will be sharply imaged on the film in some particular wavelength and some particular lens zone. The other wavelengths and other zones will be more or less out of focus. Thus we have a sharp image superposed on a slightly blurred image of objects at all distances (within limits), and if the exposure is kept on the short side, very acceptable photographs can be obtained without the necessity for any focusing mechanism on the camera.

12.3 A PERISCOPIC LENS

It was found empirically very early in the development of photography that placing two identical landscape lenses symmetrically about a central stop removed the distortion and lateral color thereby giving a better image than could be obtained by the use of a simple meniscus lens alone. Such a lens was called *periscopic*.

To design the rear half of a symmetrical lens, we assume that there will be parallel light in the stop space, and now we can evidently ignore coma since it will be corrected automatically by the symmetry (in at least a single zone). Therefore we have to consider only the tangential field curvature, and by the $H' - L$ curve we can select a stop position to flatten the field, provided the lens bending is equal to or stronger than that used for a landscape lens, but *we cannot use stop position to flatten the field if the bending is weaker than that of a landscape lens*. This is because the $H' - L$ curve doesn't contain a place where the slope is zero. Also, the steeper the bending the closer the stop will be to the lens, resulting in a more compact system.

Using the thickness and refractive index employed in our previous designs, we will try a rear-meniscus lens with $c_1 = -0.8$. The structure is

c	d	n
−0.8		
	0.15	1.523
−0.95198		

with $f' = 9.99975$, $l' = 10.41182$. The $H' - L$ curve for a 20° obliquity is shown in Figure 12.6.

This graph tells us that our lens will have a flat tangential field if the stop is placed at a distance of −0.85 or −0.23 from the front lens surface. Naturally, we choose the nearer position, and we mount two similar lenses about a central stop located 0.23 from each of the facing surface vertices. The focal length now

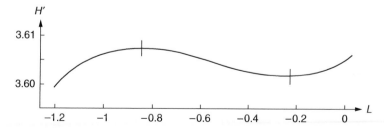

Figure 12.6   The $H' - L$ curve of the rear component of a periscopic lens (20°).

drops to 5.3874, and so we scale up the combined system to a focal length of 10 (scale factor = 10/5.3874; remember that the radius is scaled by this value, not the curvature). The resulting system is shown in Figure 12.7a.

c	d	n
0.51287		
	0.278	1.523
0.431		
	$\dfrac{0.427}{0.427}$	
−0.431		
	0.278	1.523
−0.51287		

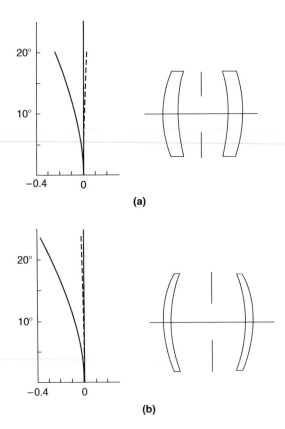

(a)

(b)

Figure 12.7   Two periscopic designs. The sagittal field is the solid curve and the tangential field is the dashed curve.

Table 12.3

Astigmatism and Distortion for Lens Shown in Figure 12.7a

Field (deg)	Z'_s	Z'_t	Distortion (%)
19.8	−0.231	+0.019	+0.04

with $f' = 10.00414$, $l' = 9.32841$, stop diameter ($f/15$) = 0.620, LA' ($f/15$) = –0.2959, Petzval sum = 0.0562. Astigmatism and distortion are presented in Table 12.3.

The spherical aberration and field curvature calculated here are for parallel light entering the left-hand end of the system, but it was designed on the assumption that there would be parallel light in the stop. It is actually rather surprising that the aberrations for a distant object resemble so closely the aberrations of the rear half alone. It is clear from examination of Figure 12.7a that the tangential field is slightly too far backward, and it is therefore desirable to *reduce the central air space slightly to flatten the field.* Also, the scaling-up process has made the lens elements unnecessarily thick, and it would be worth going back to the beginning and redesigning the system with much thinner lenses.

It is of interest to compare this design with the original Steinheil "Periskop" lens, which was of this type. According to von Rohr,[1] the specification was

c	d	n
0.5645		
	0.1316	1.5233
0.4749		
	0.6484	
	0.6484	
−0.4749		
	0.1316	1.5233
−0.5645		

with $f' = 10$, $l' = 9.2035$, stop diameter ($f/15$) = 0.627, LA' ($f/15$) = –0.355, Petzval sum = 0.0615. Table 12.4 presents the astigmatism and distortion. The modified lens is shown in Figure 12.7b.

Table 12.4

Astigmatism and Distortion for Lens Shown in Figure 12.7b

Field (deg)	Z'_s	Z'_t	Distortion (%)
23.4	−0.364	−0.010	+0.07

12.4 ACHROMATIC LANDSCAPE LENSES

12.4.1 The Chevalier Type

In this type, a flint-in-front lens of slightly meniscus shape is used, with stop in front and the concave side facing the distant object.

As an example we will use the following glasses:

(a) Flint: $n_d = 1.62360$, $V = 36.75$, $\Delta n = 0.01697$
(b) Crown: $n_d = 1.52122$, $V = 62.72$, $\Delta n = 0.00831$

For a focal length of 10, we find using Eq. (5-4)

$$c_a = -0.2269, \quad c_b = +0.4634$$

Assuming an equiconcave flint as a starter and establishing suitable thicknesses (actually those used here were too thick), we solve the last radius by the $D - d$ method and find the focal length to be 10.515. After scaling down to a focal length of 10 we have

c	d	n
−0.1189		
	0.28	1.62360
0.1189		
	0.56	1.52122
−0.3424		

with $f' = 10.00$, $l' = 10.4510$, LA' ($f/15$) $= -0.162$, Petzval sum $= 0.0667$.

We next trace a set of oblique rays through the upper half of the lens at 20° to locate the stop position for zero coma (Figure 12.8). This gives the values shown in the table on the next page.

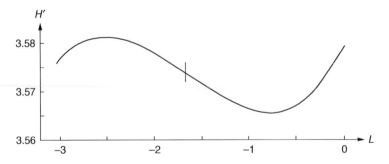

Figure 12.8   The $H' - L$ curve of a Chevalier achromat (20°).

L	H'
0	3.579278
−1.0	3.566382
−2.0	3.577830
−3.0	3.576493

The inflection point of this graph is at $L = -1.67$, and because the graph is S-shaped, the tangential field will obviously be backward-curving (see Section 12.1.2). Performing Coddington traces at several obliquities produces data for Figure 12.9 (see Table 12.5).

Unfortunately, it is not possible to flatten the tangential field in a lens of this type at the same time as eliminating the coma. The concave front face and the dispersive interface both contribute overcorrected astigmatism of about the same amount, and bending the lens merely increases one contribution while reducing the other. Using modern barium crown glass with a flint of the same index, one could make an achromatic lens that would behave like a simple landscape lens so far as the monochromatic aberrations are concerned, with the interface then being merely a buried surface. Another possibility is to depart

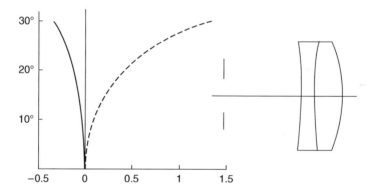

Figure 12.9 Astigmatism of a Chevalier achromat. The sagittal field is the solid curve and the tangential field is the dashed curve.

Table 12.5

Astigmatism and Distortion for Lens Shown in Figure 12.9

Field (deg)	X'_s	X'_t	Distortion (%)
30	−0.341	1.325	−4.18
20	−0.134	0.394	−1.84
10	−0.043	0.079	−0.46

from strict achromatism by weakening the cemented interface, but the high cost of such a lens over that of a single element would be scarcely justified.

12.4.2 The Grubb Type

In 1857 Thomas Grubb[2] made a lens that he called the *aplanat*, consisting of a meniscus-shaped crown-in-front achromat. The spherical aberration was virtually corrected by the strong cemented interface, and as a result the user had to accept either the coma or the field curvature since both could not be corrected together. The Grubb lens eventually led to the "Rapid Rectilinear" design discussed in Section 12.5.1.

12.4.3 A"New Achromat" Landscape Lens

Since the cemented interface in the "old" Chevalier achromat has the effect of overcorrecting the astigmatism at high obliquities, it is evident that we could reverse the effect if we were to use a crown glass of higher refractive index than the flint glass (a "new achromat"). Furthermore, this combination of refractive indices has the effect of reducing the Petzval sum, but it will be accompanied by a large increase in the spherical aberration.

The design procedure for a new achromat is entirely different from that for an old achromat, because now we leave the achromatizing to the end and solve the outside radii of curvature for Petzval sum and focal length. We select refractive indices such that there are a variety of dispersive powers available for achromatizing after the design is completed. Two typical refractive indices meeting this requirement are

(a) Flint: 1.5348 (available V numbers from 45.7 to 48.7)
(b) Crown: 1.6156 (available V numbers from 54.9 to 58.8)

As a first guess we will aim for a Petzval sum of 0.03 on a focal length of 10. We must also guess at a likely interface radius and lens thicknesses. This gives the following as a starting system.

c	d	n
−0.551		
	0.1	1.5348
0.164		
	0.4	1.6156
−0.5687		

with $f' = 9.9998$, $l' = 10.8865$, Petzval sum $= 0.030$. The large thickness helps to reduce the Petzval sum without using very strong elements (see "A Thick Meniscus" in Section 11.2.2).

In plotting the $H' - L$ graph, we use a larger obliquity angle than before because new achromats tend to cover an exceptionally wide field. The graph shown in Figure 12.10 represents the curve for 25°, and we see that the inflection falls at $L = -0.326$. The astigmatic field curves are also shown and indicate that they suffer higher-order astigmatic and Petzval terms. As was mentioned in Section 11.2.1, the longitudinal distance from the Petzval surface to the tangential focal line is three times as great as the corresponding sagittal astigmatism when considering primary (Seidel) aberrations.

In a like manner, when the secondary or fifth-order Petzval occurs, the longitudinal distance from the secondary Petzval surface to the tangential focal line is five times as great as the corresponding sagittal astigmatism.[3,4,5] The field focus locations can be written as

$$z_t = [(3AST3 + PTZ3)H^2 + (5AST5 + PTZ5)H^4]/u'$$

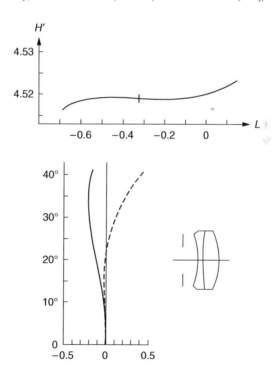

Figure 12.10 Tentative design of a new-achromat lens. The sagittal field is the solid curve and the tangential field is the dashed curve.

and

$$z_s = [(AST3 + PTZ3)H^2 + (AST5 + PTZ5)H^4]/u'.$$

It is evident that the collective interface should be made stronger to move the field inward, and a smaller Petzval sum would also be desirable. Hence for our next attempt we try $c_2 = 0.25$ and Petzval sum $= 0.027$. The changes listed in the following table give the system that is shown in Figure 12.11.

c	d	n
−0.5777		
	0.1	1.5348
0.25		
	0.4	1.6156
−0.57795		

with $f' = 9.99996$, $l' = 10.91516$, Petzval sum $= 0.027$, LA' $(f/15) = -0.50$, and stop position $= -0.102$. The astigmatic and distortion behavior are shown in Table 12.6.

Assuming that this is acceptable, the last step is the selection of real glasses for achromatism. A few trials, using the $D - d$ method, indicate that the following Schott glasses would be excellent:

(a) LLF1: $n_e = 1.55099$, $\Delta n = 0.01198$, $V = 45.47$
(b) N-SK4: $n_e = 1.61521$, $\Delta n = 0.01046$, $V = 58.37$

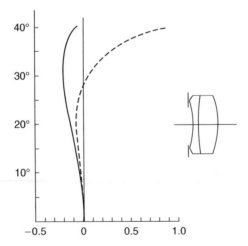

Figure 12.11 Astigmatism of a new achromat, later form. The sagittal field is the solid curve and the tangential field is the dashed curve.

Table 12.6

Astigmatism and Distortion for Lens Shown in Figure 12.11

Field (deg)	Z'_s	Z'_t	Distortion (%)
40	−0.075	0.869	−9.23
30	−0.207	0.027	−4.68
20	−0.130	−0.083	−2.04
10	−0.041	−0.041	−0.46

Of course, the design should be finalized using these specific glasses since the n_e and V values are a bit different.

It is perhaps not obvious which of these two designs would be the better. For a narrow field such as ±22°, the lens in Figure 12.10 is to be preferred, while for a wider field such as ±33° the lens in Figure 12.11 would obviously be better. It is interesting to see how the small changes in the design have made such a large difference to the tangential field at the wider field angles.

The large spherical aberration is a definite disadvantage of the new-achromat form. This was corrected by Paul Rudolph in his Protar design which will be discussed in Section 14.4.

12.5 ACHROMATIC DOUBLE LENSES

12.5.1 The Rapid Rectilinear

The *Rapid Rectilinear*, or aplanat lens is one of the most popular photographic lenses ever made. The lens is symmetrical, and the rear half is spherically corrected and has a flat field. In order to keep the lens compact, a large amount of positive coma is required in the rear component. This implies that a graph of spherical aberration against bending should rise high above the zero line, much higher than is usual for telescope objectives. To achieve this, the V difference between the old-type crown and flint glasses should be small, but a large index difference is helpful. The exact V difference depends on the aperture and field required. For a normal lens of $f/6$ or $f/8$ aperture, a V difference of about 7.0 is satisfactory. A smaller V difference can be used for a wide-angle lens of $f/16$ aperture, while a larger V difference leads to a longer lens of higher aperture, suitable for portraiture applications.

All three of these variations have been used by different manufacturers. At first, two flint glasses were utilized, but after about 1890 it was common to find an ordinary crown in combination with a light barium flint (see "A New-Achromat Combination" in Section 11.2.2).

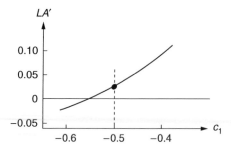

Figure 12.12 Bending curve for the rear component of a Rapid Rectilinear.

To initiate the design procedure, we will select the following glasses:

(a) Light Flint: $n_e = 1.57628$, $\Delta n = n_F - n_C = 0.01343$, $V = 42.91$
(b) Flint: $n_e = 1.63003$, $\Delta n = n_F - n_C = 0.01756$, $V = 35.87$

The Abbe number difference is $V_a - V_b = 7.04$. In designing the rear component, the procedure already described for telescope doublets is followed, except that because of the strongly meniscus shape of the lenses, the preliminary G-sum analysis is not very helpful and will be omitted.

Using these glasses for a focal length of 10, the (c_a, c_b) formulas give,

$$c_a = 1.0577, \quad c_b = -0.8089$$

Assuming that c_1 will be about one-half c_a with negative sign, we make a drawing of the lens at a diameter of about one-tenth the focal length, enabling us to set the thicknesses at 0.3 for the crown and 0.1 for the flint.

Taking a few bendings and solving each for perfect achromatism by the $D - d$ method on a traced $f/16$ ray, we can plot the graph in Figure 12.12. Recalling Figures 7.2 and 9.4, it should be evident that we want to select a value for c_1 in the neighborhood of the left-hand solution, where the coma is positive; the stop position will be in front of the rear component. The right-hand solution with negative coma is useless since it would require the stop to be behind the lens to flatten the field. Since this is a photographic lens, we desire a small amount of spherical overcorrection to offset the zonal undercorrection shown in Figure 12.13a, which suggests that we try $c_1 = -0.5$ for further study.

This lens has a focal length of 10.806, $LA'_m = +0.026$, and $LZA = -0.0178$. To find the stop position for a flat tangential field, we plot the $H' - L$ graph at 20° for a succession of L values as illustrated in Figure 12.13b. Remember that such plots are easily generated using an optical design program by filling the lens aperture (assuming the lens is the temporary stop) with meridional rays and then viewing the tangential ray fan plot for that obliquity. As already stated, the abscissa is reversed between the two plots. We now observe that the minimum point falls at $L = -0.2$, which is the distance from the stop to the front (concave) surface.

We now assemble two of these lenses together about a central stop, as illustrated in Figure 12.14a, and find that the focal length is 5.6676. It is best to scale this immediately to a focal length of 10.0, yielding the prescription in the following table.

c	d	n
0.3974		
	0.1764	1.63003
0.8828		
	0.5293	1.57628
0.2834		
	0.3529	
	0.3529	
−0.2834		
	0.5293	1.57628
−0.8828		
	0.1764	1.63003
−0.3974		

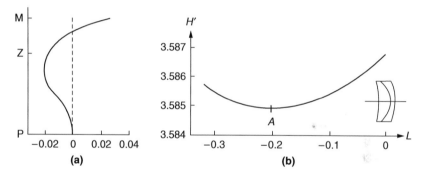

Figure 12.13 Aberrations of the rear component of a Rapid Rectilinear: (a) spherical aberration; (b) the $H' - L$ curve at 20° obliquity.

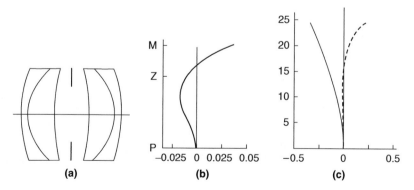

Figure 12.14 The final Rapid Rectilinear design: (a) layout; (b) longitudinal spherical aberration; (c) astigmatic field curves where the sagittal field is the solid curve and the tangential field is the dashed curve.

In the table at the top of 341, we have $f' = 10.00$, $l' = 9.0658$, lens diameter = 1.8, and Petzval sum = 0.0630. The $f/8$ axial ray from infinity gives $LA' = 0.0350$, and it also tells us that the $f/8$ stop diameter must be 1.110. An $f/11.3$ zonal ray gives $LZA = -0.0108$, enabling us to plot the spherical aberration graph in Figure 12.14b.

To plot the fields, we now add two other principal rays having slope angles in the stop space of 28° and 12°, respectively. The principal-ray slope angles in the stop space between the lenses are generally somewhat different than the entering or outside slope angle (see Section 12.5.2). The sagittal and tangential fields traced along these principal rays are shown in Table 12.7.

The fields are plotted in Figure 12.14c and closely resemble those of the rear half-system. Both the spherical aberration and the astigmatism are thus very stable in this type of lens for changes in the object distance, which was one of the reasons for its great popularity.

Table 12.7

Astigmatism and Distortion for Lens Shown in Figure 12.14a

Outside angle (deg)	Angle at stop (deg)	Z'_s	Z'_t	Distortion (%)
24.4	28	−0.3411	0.2050	0.09
17.5	20	−0.2013	0.0044	0.04
10.6	12	−0.0789	−0.0196	0.01

12.5.2. A Flint-in-Front Symmetrical Achromatic Doublet

There is, of course, a companion system to the Rapid Rectilinear in which the rear component is a flint-in-front spherically corrected achromat. To design such a lens we may use the same glasses as for the Rapid Rectilinear, and we plot a graph of spherical aberration at $f/16$ against bending, of course in the region of the left-hand solution where the coma is positive (Figure 12.15). For each plotted point the last radius is solved for strict achromatism by the $D - d$ method, and the curvatures are scaled to a focal length of 10, keeping the thicknesses at 0.1 and 0.3 as before.

We recall that when we were designing a telescope objective, we found that the left-hand solution for a flint-in-front doublet has a much smaller zonal residual than the left-hand crown-in-front doublet (Section 7.2). Consequently we shall plan the present design to be a "portrait" lens with an aperture of $f/4.5$ and covering a somewhat narrower field than the Rapid Rectilinear.

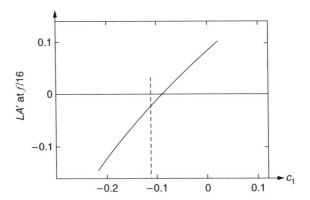

Figure 12.15 Spherical aberration vs. bending for a flint-in-front doublet.

The rear half of the new lens will therefore have to work at $f/9$, and since the graph in Figure 12.15 represents the $f/16$ aberration, we must select a bending having a small residual of undercorrected aberration, at say $c_1 = -0.11$. This gives the following rear half-system:

	c	d	n
	-0.11		
		0.1	1.63003
	0.69		
		0.3	1.57628
$(D - d)$	-0.3489		

with $f' = 10.0542$, $l' = 10.3008$, Petzval sum $= 0.0706$, LA' $(f/9) = -0.0336$, LA' $(f/11.4) = -0.0365$, and LA' $(f/16) = -0.0254$. The residual aberration at $f/9$ was deliberately made negative since it was found that mounting two similar components about a central stop tended to overcorrect the aberration. The last radius was determined, of course, by the $D - d$ method as usual.

To locate the stop, we trace several rays at $20°$, giving the $H' - L$ curve shown in Figure 12.16. The minimum falls at $L = -0.50$ for a flat tangential field. Mounting two of these lenses about a central stop as depicted in Figure 12.17 and scaling to $f' = 10$ gives the prescription shown in the table on the next page.

c	d	n
0.19450		
	0.5382	1.57628
−0.38462		
	0.1794	1.63003
0.06132		
	0.8970	
	0.8970	
−0.06132		
	0.1794	1.63003
0.38462		
	0.5382	1.57628
−0.19450		

with $f' = 10.0$, $l' = 8.4795$, Petzval sum $= 0.0787$, LA' ($f/4.5$) $= +0.0181$, and LA' ($f/5.6$) $= -0.0069$. The astigmatism and distortion are shown in Table 12.8.

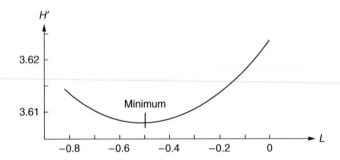

Figure 12.16   The $H' - L$ graph of the rear component of a flint-in-front double lens (20°).

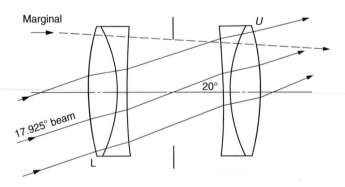

Figure 12.17   Completed $f/4.5$ symmetrical portrait lens.

Table 12.8

Astigmatism and Distortion for Lens Shown in Figure 12.17

Angle in object space	Angle in stop	Z'_s	Z'_t	Distortion (%)
24.956°	28°	−0.496	+0.543	+0.21
17.925°	20°	−0.294	−0.021	+0.10
10.798°	12°	−0.115	−0.055	+0.03

Plotting the fields and aberrations of this lens makes an interesting comparison with the comparable data for the Rapid Rectilinear (Figure 12.18). The reasons for regarding this as a portrait lens are evident.

As a final check, we will trace a family of rays at 17.925° to complement the 20° principal ray already traced, and we plot the (H − tan U) curve shown in Figure 12.19. As mentioned previously, the slope angle of the entering oblique bundle of parallel rays is slightly different than slope angle in the space between the lenses. The ends of this curve represent rays passing through the extreme top and bottom of the diaphragm, and as can be seen, the lower ray is very bad and should be vignetted off. It is customary in lenses of this kind to limit every surface to a clear aperture equal to the entering aperture of the marginal ray, which in this case is $Y = 1.1111$. This limitation cuts off the lower rays drastically, placing the true lower rim ray at the point marked L on the graph in Figure 12.19 and the lens drawing in Figure 12.17. It also somewhat reduces the upper part of the aperture to a limiting rim ray marked U in both figures.

It is clear that the remaining aberration of the lens is a small residual of negative coma (see Eq. (4-8)) of magnitude

$$\text{Coma}_t = \tfrac{1}{2}(H'_U + H'_L) - H'_{pr} = -0.0182$$

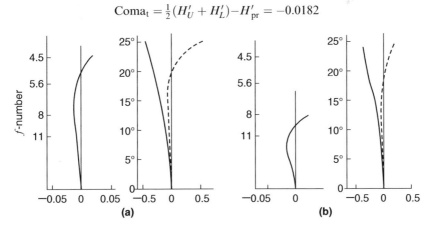

Figure 12.18 Comparison of Rapid Rectilinear (a) flint-in-front and (b) crown-in-front forms. (Spherical aberration and astigmatism curves for $f' = 10$.) The sagittal field is indicated by the solid curve and the tangential field is indicated by the dashed curve.

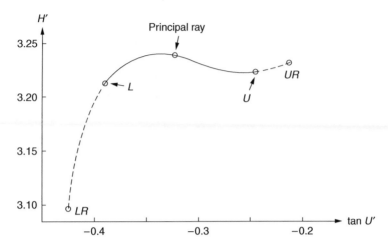

Figure 12.19   The meridional ray plot for the final system at 17.9° field angle.

Assuming that the sagittal coma is one-third the tangential coma, the equivalent *OSC* becomes –0.00096, which is small enough to be neglected, especially since a lens of this type is unlikely to be used at a field as wide as 17.9°. See Sections 4.3.4 and 9.3.

Although we have regarded this as a portrait lens, it has seldom been used in this way, but it could very well be used at or near unit magnification as a relay lens in a telescope, in which case the coma and distortion would automatically vanish due to symmetry (Section 11.6).

12.5.3 Long Telescopic Relay Lenses

In many types of telescopes and periscopes, an erector system working at or near unit magnification is inserted between the objective and eyepiece to give an erect image. This *erector* often consists of two identical spherically corrected doublets mounted symmetrically about a central stop, the stop position being chosen to give a flat field exactly as in the Rapid Rectilinear lens, except that now we often need a long system rather than a short one.

As was pointed out in connection with the design of the Rapid Rectilinear in the beginning of Section 12.5.1, the greater the amount of coma in the rear lens the smaller the stop shift required to give a flat tangential field, and the shorter the relay will be. For a long relay, we therefore need a spherically corrected lens with a very small amount of coma, and hence we select a design in which the graph of spherical aberration against bending rises only a little way above the abscissa line. Furthermore, whatever lens we use for the rear component of our relay must have positive coma in order that the flat-field stop position will be in front of the lens.

Referring to the bending curve in Section 9.3.5 for a normal cemented doublet, we see that the left-hand solution has positive coma, and it is therefore suitable for the rear of a telescopic relay. We locate the stop position for a flat field as we did for the Rapid Rectilinear by tracing several oblique parallel rays through the upper half of the lens, a suitable obliquity being now about 4°. The left-hand flint-in-front solution is much preferable to the crown-in-front form since it has only about a third of the zonal aberration, and we will continue with that design here. The graph connecting H' with L for this lens is shown in Figure 12.20, and since the minimum falls at $L = -3.2$, that will be the stop position in this case. The computed astigmatism at 4° when the stop is at that position is found to be $Z'_s = -0.0117$ and $X'_t = +0.0006$, representing the desired flat tangential field. Figure 12.21 shows that two of these lenses mounted together about a central stop would make an excellent relay.

If a still longer relay is required, the spherical aberration graph must be lowered still further, and the left-hand solution for the near aplanat discussed in Section 10.3 can be used. In this case the stop position, calculated at a very small obliquity such as 2°, falls at a distance of 9.2 in front of the lens which is close to the anterior focal point thereby making the system nearly telecentric in the image space. The combination of two such systems forms a 1:1 afocal *telecentric relay*, which has been used in contour projectors to give a longer working distance, and in borescopes, where up to four relays can be assembled in sequence without any need for field lenses at the intermediate real images.

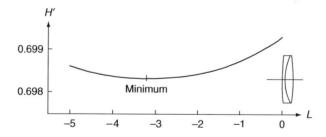

Figure 12.20   The $H' - L$ curve of a flint-in-front telescope objective at 4° obliquity.

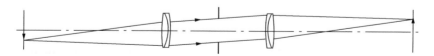

Figure 12.21   A 1:1 telescopic relay consisting of two flint-in-front objectives with a stop at the flat-field position.

This is important since, when a field lens is employed, it must have positive power and will adversely impact the field curvature.

The main relay in a submarine periscope consists of a pair of highly corrected aplanatic objectives spaced apart by a distance equal to two or three times their focal length, the field angle being then less than 1°. In this case the astigmatism is negligible so long as the tangential field is flat. Coma is corrected by the symmetry in the usual way.

12.5.4 The Ross "Concentric" Lens

This is the classic example of a symmetrical objective consisting of two deeply curved new achromats surrounding a central stop. It was patented by Schroeder in 1889, with the structure of the rear half, after scaling to a focal length of 10, according to von Rohr[6] as follows:

c	d	n	V
	0.194	(air)	
−1.94125			
	0.020	1.5366	48.69
0			
	0.071	1.6040	55.31
−1.78358			

with $f' = 10$, $l' = 10.5961$, $L_{pp} = +0.6166$, Petzval sum $= -0.00618$. The glasses are assumed to be light flint No. 26 and dense barium crown No. 20 in Schott's catalog of 1886. The nearest "modern" Schott glass for No. 26 is LLF-6 and for No. 20 is SK-8.

Tracing a fan of rays entering at −20° gives the $H' - L$ curve shown in Figure 12.22a. Clearly the stop position for a flat tangential field should be at about $L = -0.237$ (point A), but von Rohr's specification places it at B, where $L = -0.194$, resulting in a slightly backward-curving field. (Incidentally, measurements made on an actual Concentric lens did not agree with this specification in any respect.) The front principal point is at C.

After combining two of these lenses together and scaling to a focal length of 10, the spherical aberration at $f/15$ was found to be an unacceptable value of −0.65, so that the lens should not be used at any aperture greater than about $f/20$ (spherical aberration decreases to about −0.27; see Eq. (6-12)). The fields with the preferred air space (0.868), and with that given by von Rohr (0.768), are also shown in Figure 12.22b. The unusual backward-curving sagittal field is, of course, due to the Petzval sum being negative. It is remarkable how great an effect a small change in the central air space has on the two fields.

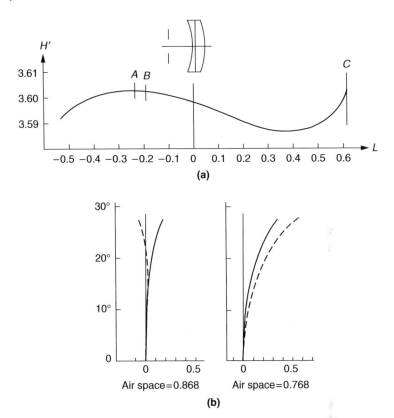

Figure 12.22   Ross Concentric lens: (a) $H' - L$ curve of rear half (20°) and (b) astigmatism of complete lens, where the sagittal field is indicated by the solid curve and the tangential field by the dashed curve.

ENDNOTES

[1] M. von Rohr, *Theorie und Geschichte des Photographischen Objektivs*, p. 288. Springer, Berlin (1899).

[2] T. Grubb, British Patent 2574 (1857).

[3] R. R. Shannon, *The Art and Science of Optical Design*, pp. 256–257. Cambridge University Press, Cambridge (1997).

[4] H. A. Buchdahl, *An Introduction to Hamiltonian Optics*, pp. 66, 132. Cambridge University Press, Cambridge (1970).

[5] H. A. Buchdahl, *Optical Aberration Coefficients*, p. 76, Dover, New York (1968). Using the relationships between polar and nonpolar coefficients, Eqs. (31.8), it can be shown using Eq. (41.21) that the secondary Petzval curvature is $(5\mu_{11} - \mu_{10})/4$.

[6] M. von Rohr, p. 234. Also see U.S. Patent 404,506.

Chapter 13

Symmetrical Double Anastigmats with Fixed Stop

13.1 THE DESIGN OF A DAGOR LENS

For 25 years after the introduction of the Rapid Rectilinear lens, designers tried unsuccessfully to modify it in such a way as to reduce the Petzval sum, and so remove the astigmatism that limited the performance of the Rectilinear in the outer parts of the field. In 1892 the German designer von Höegh[1] made three useful suggestions to mitigate this limitation: (1) to insert a collective interface convex to the stop in the flint element of the Rapid Rectilinear, thus turning the half-system from a doublet into a triplet; (2) to use progressively increasing refractive indices outward from the stop; and (3) to use almost equal outside radii of curvature and to thicken the lens sufficiently to give the desired focal length and Petzval sum. In this way he created the famous "Double Anastigmat Goerz," later renamed the Dagor,[2] which covered a fairly wide anastigmatic field at $f/6.3$. The symmetry, of course, automatically eliminated the three transverse aberrations, leaving the designer only spherical and chromatic aberrations and astigmatism to be corrected in each half.

As an example of the design of such a lens, we first select three refractive indices for which there are many glasses available having different dispersive powers, so that we can achromatize the lens at the end by choosing suitable types of glass from available catalogues. These indices will be 1.517, 1.547, and 1.617, although of course other values could have been chosen that would be equally satisfactory. For a focal length of 10.0, we can start the design of the rear half-system with the radii −1, −0.5, +2, and −1, suitable thicknesses being determined from a scale drawing as 0.14, 0.06, and 0.19, respectively.

These thicknesses are actually somewhat meager, but it is better to keep the lens as thin as possible to reduce vignetting at high field angles. Since the stop position will not be a degree of freedom, because we can correct all the aberrations without its help, we place the stop as close as possible to the lens, at a distance of 0.125, to minimize the diameter of the lens elements.

We shall employ the four radii in the rear component in the following manner. First r_1 is varied to give the desired Petzval sum after solving r_4 for focal length. The two internal surfaces contribute very little to the sum since the refractive index difference across these two surfaces is small. We find that a suitable value for the Petzval sum in a lens of this type is about 0.018 for a focal length of 10, which is about 0.2% of the focal length. Now r_2 is varied to make the marginal spherical aberration approximately equal and opposite to the 0.7 zonal aberration[3]; and we select r_3 to give a flat tangential field on a principal ray passing through the center of the stop at an angle of 30°.

With the tentative initial data given above, we find that for the focal length and Petzval sum desired, c_1 must be –0.78 and c_4 should equal –0.7748. Our starting system prescription is therefore as follows:

c	d	n
	0.125	(air)
−0.78		
	0.14	1.517
−2.00		
	0.06	1.547
0.50		
	0.19	1.617
−0.7748		

with $f' = 10.0$, $l' = 11.057$, Petzval sum = 0.0182, $f/12.5$ stop diameter = 0.8, LA' ($f/12.5$) = 0.160, LZA ($f/17.7$) = –0.150, X'_t at 30° = –0.076.

A scale drawing of the lens is shown in Figure 13.1. By chance the spherical aberration is about right and will be accepted. However, the 30° tangential field is more inward-curving than we would like, so we proceed to reduce c_3 slightly, a suitable value being 0.486. This gives $X'_t = -0.0056$ and $X'_s = -0.0686$. Tracing a few more principal rays at other obliquities enables us to plot the field curves for the rear half of the system in the figure.

Regarding this as a satisfactory rear half, we now assemble two of these lenses together and scale up to an overall focal length of 10.0. This gives the following prescription for the front half with the rear half being symmetrical:

c	d	n
0.4464		
	0.3304	1.617
−0.2795		
	0.1043	1.547
1.1502		
	0.2434	1.517
0.4486		
	0.2173	
Stop	(rear half symmetrical about stop)	

with $f' = 10.0$, $l' = 9.2260$, Petzval sum $= 0.0211$, $f/6.8$ stop diameter $= 1.276$, LA' ($f/6.8$) $= 0.0001$, LZA ($f/9.6$) $= -0.1130$. The spherical aberration is much less overcorrected than for the half-system, and the strong interfaces could be slightly deepened to rectify this.

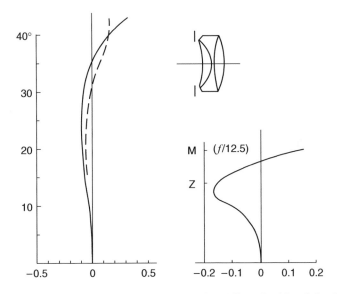

Figure 13.1 Aberrations of rear half of a Dagor lens ($f' = 10$) with c_3 being 0.486.

It will be noticed that the zonal aberration here is greater than in the corresponding Rapid Rectilinear lens; this is the major problem in lenses of the Dagor type. The fields are shown plotted in Figure 13.2, and it will be seen that they are not greatly different from those of the half-system. As should be expected, there is a minute amount of distortion, about 0.13% at 30°, which can be ignored.

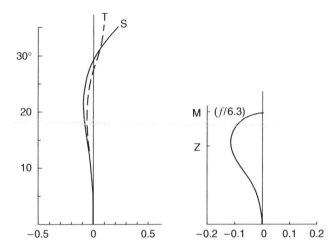

Figure 13.2   Aberrations of a complete Dagor ($f' = 10$).

The final step in the design is the selection of glasses for achromatism. We return to the marginal ray trace through the rear half-system, and compute the value of $D - d$ in each of the three lens elements. At an aperture of $f/12.5$ these are, respectively, −0.27299, +0.50191, and −0.25560. Our problem is therefore to find three glasses with the approximate indices used in this design, having Δn values such that $\sum (D - d) \Delta n = 0$. A brief search in the Schott catalog suggests that the following would make an achromatic combination:

(a) KF-3: $n_e = 1.51678$, $\Delta n = 0.00950$, $V_e = 54.40$
(b) KF-1: $n_e = 1.54294$, $\Delta n = 0.01079$, $V_e = 50.65$
(c) SK-6: $n_e = 1.61635$, $\Delta n = 0.01100$, $V_e = 56.08$

The refractive indices are close but not exactly equal to those assumed in the design. It would therefore be necessary to repeat the whole procedure using the exact index data for the real glasses, to obtain the final formula.

DESIGNER NOTE

This lens is a good example of how the lens designer can gain an understanding of how specific parameters such as radii and thicknesses control the various aberrations. It is unlikely that these parameters will be independent or orthogonal to one another, but very often the coupling is relatively small. This is why the "doubling graphing" technique used often in examples already presented illustrates reasonably linear behavior. When the parameters are more tightly coupled or correlated, they behave in a nonlinear manner such as will be seen in Figure 14.10.

Although an automatic lens design program can design a Dagor-type lens, it is an interesting exercise to set up the program to perform such a design and compare the results with the manual method taught above. Quite often one will find that the program will take an unexpected path toward a solution as it does not have the inherent understanding of the lens designer. Frequently, the program will attempt to use the thicknesses that are more than desired to control the aberrations, so in the early stages, it is typically a good idea to not allow the glass thicknesses to vary and to appropriately limit the element spacings.

13.2 THE DESIGN OF AN AIR-SPACED DIALYTE LENS

This name is given to symmetrical systems containing four separated lens elements, as illustrated in Figure 13.3. This type was originated by von Höegh,[4] who called it the "Double Anastigmat Goerz type B," this name being later changed to Celor. The rear separated achromat contains five degrees of freedom, namely, two powers, two bendings, and an air space, with which it is possible to obtain the desired focal length and correct four aberrations: Petzval sum, spherical aberration, chromatic aberration, and astigmatism. If we then mount two of these components symmetrically about a central stop, we correct in addition the three transverse aberrations: coma, distortion, and lateral color. The stop position is not a degree of freedom since we have sufficient variables without it. However, as the lens will generally be used with a distant object, we may have to depart slightly from perfect symmetry to remove any residuals of coma that may appear.

We can save a good deal of time by first determining the two powers and the separation of the rear component to yield the desired lens power, chromatic, and Petzval values, assuming thin lenses and using the Seidel contribution formulas given in Eq. (11-14), Section 11.7.2. The thin-lens predesign requires

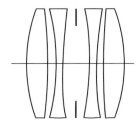

Figure 13.3 The dialyte objective.

solution of the following three equations for ϕ_a, ϕ_b, and d:

$$\sum y\phi = y_a\phi_a + y_b\phi_b = \Phi y_a \quad \text{(power)} \tag{13-1}$$

$$\sum \frac{y^2\phi}{V} = \left(\frac{y_a^2}{V_a}\right)\phi_a + \left(\frac{y_b^2}{V_b}\right)\phi_b = -L'_{ch}u_0'^2 \quad \text{(chromatic)} \tag{13-2}$$

$$\sum \frac{\phi}{n} = \left(\frac{1}{n_a}\right)\phi_a + \left(\frac{1}{n_b}\right)\phi_b = \text{Ptz} \quad \text{(Petzval)} \tag{13-3}$$

From Eq. (13-1) we express y_b as a function of y_a, by $y_b = y_a(\Phi - \phi_a)/\phi_b$. Inserting this in Eq. (13-2) gives

$$\frac{y_a^2\phi_a}{V_a} + \frac{y_a^2(\Phi - \phi_a)^2}{\phi_b V_b} = -L'_{ch}u_0'^2 \tag{13-4}$$

However, using Eq. (13-3),

$$\phi_b = n_b(\text{Ptz} - \phi_a/n_a) \tag{13-5}$$

and putting this into Eq. (13-4) gives a quadratic for ϕ_a:

$$\phi_a^2[V_a - V_b n_b/n_a] + \phi_a[\text{Ptz } n_b \ V_b - 2\Phi V_a - L'_{ch}\Phi^2 V_a V_b n_b/n_a]$$
$$+ \Phi^2 V_a[1 + L'_{ch} \text{ Ptz } n_b V_b] = 0 \tag{13-6}$$

Thus we obtain ϕ_a by Eq. (13-6), ϕ_b by Eq. (13-5), and finally the separation d by

$$d = (\phi_a + \phi_b - \Phi)/\phi_a\phi_b \tag{13-7}$$

As an illustration of the design of such a lens, we will first solve the rear component for a focal length of 10, a Petzval sum of 0.030 (0.3% of f'), and zero chromatic aberration. *The selection of glasses must, however, be made with some care for if the V difference is too great the negative lens will be too weak to enable us to correct the other aberrations.* A reasonable glass choice is

(a) Barium flint: $n_D = 1.6053$, $n_F = 1.61518$, $n_C = 1.60130$, $V = 43.61$
(b) Dense barium crown: $n_D = 1.6109$, $n_F = 1.61843$, $n_C = 1.60775$, $V = 57.20$

with V difference = 13.59. The above algebraic solution gives

$$\phi_a = -0.4958, \quad \phi_b = 0.5458$$

and since $\phi = c/(n - 1)$, we find that

$$c_a = -0.8191, \quad c_b = 0.8934, \quad d = 0.1848$$

This completes the predesign of the thin-lens powers and separation.

We could, of course, determine the bendings of the two thin elements in the rear half-system for spherical and astigmatic correction, but it is best to insert thicknesses first and assemble the two components before doing this. To assign thickness we must decide on the relative aperture of the finished lens, and $f/6$ is a good value to adopt. This makes the aperture of the rear component about $f/12$, and a diameter of 1.0 is suitable. The thicknesses of the two lenses will then be 0.06 for the flint and 0.20 for the crown.

For our starting bendings we may assign 40% of the total flint curvature to the front face of the rear negative element, and 25% of the crown curvature to the front face of the rear positive element. However, because of the finite thicknesses, we must scale each thick element to restore its ideal thin-lens power, and the air space must be adjusted to maintain the ideal separation between adjacent principal points. With the stop at a distance of 0.12 from the flint element, the whole lens becomes

c	d	r	n
$c_1 = -c_8 = 0.6788$		(1.473)	
	0.2		flint
$c_2 = -c_7 = -0.2263$		(−4.418)	
	0.0756		air
$c_3 = -c_6 = -0.4893$		(−2.043)	
	0.06		crown
$c_4 = -c_5 = 0.3262$		(3.065)	
	0.12		air
(stop)			

The lens in its present state is drawn to scale as was shown in Figure 13.3. The focal length is 5.6496 and the Petzval sum (for $f' = 10$) is 0.04039. Tracing $f/8.5$ rays in F and C light, with $Y_1 = 0.3323$, gives the zonal chromatic aberration as 0.03312. The increase in Petzval sum is due to the finite thicknesses of the lenses.

We must now restore the desired values of Petzval sum and chromatic aberration by changing the power of the two crown elements and the two outer air spaces, maintaining symmetry about the stop and letting the focal length go. Of course we could equally as well vary the power of the flint elements, but we must adopt a fixed procedure or we shall never reach a satisfactory solution. A double graph is a great convenience here, plotting the zonal chromatic aberration as ordinate and the Petzval sum for $f' = 10$ as abscissa. The starting point will be (0.0404, 0.0331) and the aim point will be (0.034, 0). A trial change of the outer air spaces by 0.05 gives zonal chromatic aberration = −0.0016 and

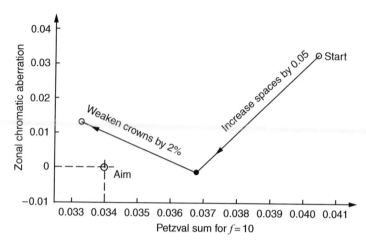

Figure 13.4   Double graph for chromatic aberration and Petzval sum.

$Ptz = 0.0368$, and then weakening both surfaces of the crown elements by 2% gives that the zonal chromatic aberration $= 0.0133$ and $Ptz = 0.0331$. The graph shown in Figure 13.4 tells us that we should have increased the original air spaces by 0.061 and weakened the crown lens surfaces by 1.24%. These changes give zero chromatic aberration and Petzval sum $= 0.0339$.

At this stage we find

$$LA' = \text{marginal spherical aberration} = 0.1335, \quad X'_s \text{ at } 22° = 0.1088$$
$$LZA = \text{zonal spherical aberration} = 0.0429, \quad X'_t \text{ at } 22° = 0.4216$$

We desire to have the marginal and zonal spherical aberrations equal and opposite, or $LA' + LZA = 0$, and we would also like to have $X'_t = 0$ for a flat tangential field. We proceed to accomplish this by bending both crowns and both flints in such a way as to maintain symmetry.

In the double graph of Figure 13.5, we see that at the start $X'_t = 0.4216$ and $LA' + LZA = 0.1764$. The aim point is (0, 0). Bending the crowns by $\Delta c_1 = -0.02$ toward a more nearly equiconvex form gives $X'_t = 0.1344$ and $LA' + LZA = 0.1742$. Then bending the flints by $\Delta c_3 = +0.02$ toward a more nearly equiconcave form gives $X'_t = 0.0254$ and $LA' + LZA = 0.0494$. The graph indicates that we should have used $\Delta c_1 = -0.0190$ and $\Delta c_3 = +0.0282$. These changes gave $X'_t = -0.0043$ and $LA' + LZA = 0.0022$, both of which are acceptable. The Petzval sum for $f' = 10$ is now 0.0341 and the zonal chromatic aberration is -0.0010; hence both are virtually unaffected by the small bendings that we have applied to the lenses.

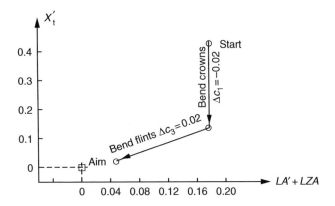

Figure 13.5 Double graph for spherical aberration and field curvature.

At this point the symmetrical system has the construction

c	d	n
$c_1 = -c_8 = 0.6514$		
	0.20	flint
$c_2 = -c_7 = -0.2425$		
	0.1366	air
$c_3 = -c_6 = -0.4611$		
	0.06	crown
$c_4 = -c_5 = 0.3544$		
	0.12	air
(stop)		

The four longitudinal aberrations are at the desired values, and we must now investigate the transverse aberrations to see how well they have been removed by the lens symmetry.

Tracing the 22° principal rays in *C*, *D*, and *F* light tells us that the distortion is 0.474% and the transverse chromatic aberration is 0.000362. Since these are both positive, we can improve both at once by shifting a small amount of power from the front to the back. Weakening both surfaces of the front crown element by 2% and strengthening the rear crown surfaces by 2% lowers the distortion to 0.190% and reduces the transverse chromatic aberration to –0.00017, both of which are now acceptable. However, this change has slightly affected the other corrections, which are now

$$\text{focal length} = 5.4122$$

$$\text{zonal chromatic aberration} = -0.00022$$

$$\text{Petzval sum} = 0.0341, \text{ for } f' = 10$$

$$LA' + LZA = -0.0158$$

$$X'_s = 0, \quad X'_t = 0.0304$$

Since the last change has slightly altered the spherical aberration and the field curvature, we return to the graph in Figure 13.5 and apply $\Delta c_1 = -0.0034$ and $\Delta c_3 = -0.0027$ to restore these. The system is now as follows:

c	d	n_D
0.6350		
	0.2	1.6109
−0.2411		
	0.1366	
−0.4638		
	0.06	1.6053
0.3517		
	$\dfrac{0.12}{0.12}$	
−0.3517		
	0.06	1.6053
0.4638		
	0.1366	
0.2508		
	0.2	1.6109
−0.6610		

with focal length for half lens = 5.4212, zonal chromatic aberration = 0.00011, Petzval sum ($f' = 10$) = 0.0342, $LA' + LZA = -0.0022$; for 22°: $X'_s = -0.0088$, $X'_t = -0.0005$, distortion = 0.189%, lateral color = −0.00017; stop diameter for $f/6 = 0.7896$. Everything is thus known except coma, which we must now investigate.

The easiest way to evaluate the coma is to trace several oblique rays and draw the meridional ray plots at two or three obliquities and look for a parabolic trend, although in general this will be mixed with a cubic tendency due to oblique spherical aberration, and a general slope caused by inward or backward tangential field curvature. If the parabolic trend is not particularly noticeable, the amount of coma is probably negligible in view of the other aberration residuals that are unavoidably present. However, if *coma is the dominant aberration it is necessary to reduce it by bending the two crown elements in the same direction, and not symmetrically about the stop* as was done previously to correct the spherical aberration and field curvature.

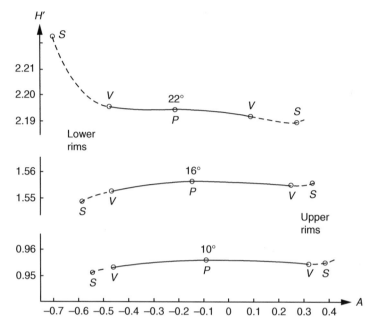

Figure 13.6 Meridional ray plots for dialyte ($f' = 5.42$).

In the present example, meridional ray fans were traced at the three obliquities: 10°, 16°, and 22°, as shown in Figure 13.6. The abscissa is the A values of the rays, that is, the height of the incidence of each of the rays at the tangent plane to the front surface.[5] To locate the endpoints of the curves we must decide which clear aperture we should allow at the front and the rear of the complete lens.

It is customary in a short lens such as this to give all eight lens surfaces the initial aperture of the marginal beam, which in our case is $f/6$ or 0.904. The limiting rays at each obliquity are then found by trial such that the lower rim rays meet r_1 at a height of –0.452, and the upper rim rays meet r_8 at a height of +0.452. These limiting rays are marked V in Figure 13.6, their paths being shown in Figure 13.7. If there were no vignetting, the upper and lower rays would be limited only by the diaphragm, these rays being marked S in Figure 13.6. It is clear that the vignetting has proved to be very beneficial, especially for the lower rays at 22°; these would cause a bad one-sided haze due to higher order coma if they were not vignetted out in this way.

A glance at this graph reveals that there is a small residual of negative coma, requiring a small negative bending of the front and rear crowns to remove it. A Δc of –0.005 was found to be sufficient to make all three curves quite straight. To gild the lily, trifling bendings were applied to remove small residuals of

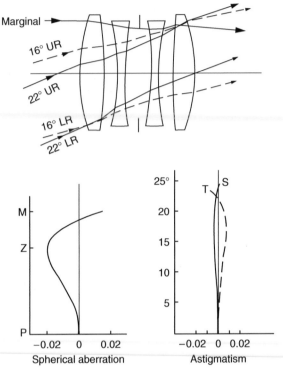

Figure 13.7   An *f*/6 dialyte objective.

LA' and X'_t, namely, $\Delta c_1 = -0.0005$, and $\Delta c_3 = -0.0025$. The final lens was then
scaled to a focal length of 10, with the following specification:

c	d	n_D
0.34138		
	0.369	1.6109
−0.13373		
	0.252	
−0.25288		
	0.111	1.6053
0.18937		
Stop	$\dfrac{0.221}{0.221}$	
−0.18937		
	0.111	1.6053
0.25288		
	0.252	
0.13357		
	0.369	1.6109
−0.36091		

with $f' = 10.0$, $l' = 9.1734$, Petzval sum $(10) = 0.0342$, LA' $(f/6) = 0.0143$, LZA $(f/8.5) = -0.0193$; for $22°$: $X'_s = -0.0155$, $X'_t = 0$, distortion $= 0.218$ %, lateral color $= -0.0004$; zonal chromatic aberration $= 0.0001$.

The aberrations of this final system were shown in Figure 13.7. One interesting point here is that after all the changes in powers and bendings that have been made, radii r_2 and r_7 are almost identical. It might be a significant manufacturing economy to make them identical by a further trifling bending to one or both of the positive elements.

Lenses of this dialyte type perform admirably and can be designed with apertures up to about $f/3.5$; the field, however, is limited to about $22°$ to $24°$ from the axis.

13.3 A DOUBLE-GAUSS–TYPE LENS

The mathematician Gauss once suggested that a telescope objective could be made with two meniscus-shaped elements, the advantage being that such a system would be free from spherochromatism. However, this arrangement has other serious disadvantages and it has not been used in any large telescope. Alvan G. Clark tried to use it with no success, but with considerable insight he recognized that two such objectives mounted symmetrically about a central stop might make a good photographic lens. He patented[6] the idea in 1888, and a lens of this type called the Alvan G. Clark lens was offered for sale by Bausch and Lomb from 1890 to 1898. The same type was also used in the Ross Homocentric, the Busch Omnar, and the Meyer Aristostigmat. An unsymmetrical version was later used by Kodak in their Wide Field Ektar lenses.

The design is suitable for a low-aperture wide-angle objective, the design procedure following closely the design of the dialyte just described. However, the glasses must be much further apart on the $V - n$ chart than before, possible types being

(a) Dense flint: $n_D = 1.6170$, $V = 36.60$, $n_F = 1.62904$, $n_C = 1.61218$
(b) Dense barium crown: $n_D = 1.6109$, $V = 57.20$, $n_F = 1.61843$, $n_C = 1.60775$

Using these glasses, the formulas given in Eqs. (13-5) and (13-6) can be solved for the two lens powers in the rear component, assuming zero L'_{ch} and a smaller Petzval sum such as 0.028 for a focal length of 10. The powers are much smaller than before and the air space much larger:

$$\phi_a = -0.2937, \quad c_a = -0.4760$$
$$\phi_b = 0.3376, \quad c_b = 0.5526$$
$$d = 0.5657$$

Since the lens elements are to be meniscus in shape, we can start by selecting bendings having $c_1 = 1.9c_a$ and $c_3 = -0.17c_b$. For a half-system of aperture $f/16$ we could try thicknesses of 0.1 for the flint element having a diameter of 0.9, and 0.3 for the crown element of diameter 1.9. (This is actually thicker than necessary, and 0.23 would have been better.) As before, after inserting the thicknesses we scale each element back to its original power, and we calculate the air space required to restore the separation between the adjacent principal planes. The stop is placed conveniently at 0.15 in front of the vertex of the negative element.

Having assembled the double lens, we find that its focal length is 6.255, the zonal chromatic aberration is −0.00398, and the Petzval sum for a focal length of 10 is 0.0249. As before, we proceed to correct the chromatic aberration and Petzval sum by changing the outer spaces and the powers of the positive elements, maintaining symmetry at all times. The double graph for these changes is shown in Figure 13.8.

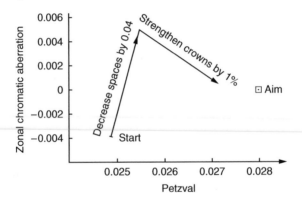

Figure 13.8 Double graph for chromatic aberration and Petzval sum for $f = 10$.

The graph suggests that we should strengthen both crown elements by 1.47% and decrease the outer spaces by 0.0466. These changes give the following front half-system (the rear is identical with the stop centered 0.15 from each half):

c	d	n_D
0.6491		
	0.3	1.6109
0.0952		
	0.1860	
0.4141		
	0.10	1.6170
0.9044		
	0.15	

with $f' = 6.0810$, aperture $= f/8$, zonal chromatic $= 0.00015$, Petzval sum (10) $=$ 0.0279. We regard these residuals as acceptable and proceed to correct the spherical aberration and tangential field curvature by bending the elements in a symmetrical manner. The aberrations of this system were found to be[7]

$$f/8 \text{ spherical aberration} = -0.0652$$
$$f/11.3 \text{ zonal aberration} = -0.0311$$
$$LA' + LZA = -0.0963$$
$$(32°)X'_s = 0.1538, \quad X'_t = 0.0937$$

The results of separately bending the crowns and flints are shown in the double graph in Figure 13.9, and a few trials indicate that we should bend the front crown by 0.0128 and the front flint by 0.0627 to remove both spherical aberration and tangential field curvature simultaneously. These changes give:

$$\text{focal length} = 5.8951 \quad LA' = -0.0020, \quad LZA = -0.0003$$
$$X'_s = 0.1196, \quad X'_t = -0.0049$$

which are acceptable, but now we find that the bendings have upset our previous corrections for Petzval and chromatic aberration, which have become Ptz $= 0.0271$, and zonal chromatic $= -0.0067$. *It is characteristic of meniscus elements that any change in the lens shape affects all aberrations*, an unfortunate property that makes the design of a Gauss-type lens much more difficult and time-consuming than the design of a comparable dialyte lens.

To remove the residual Petzval and chromatic aberrations, we return to the graph in Figure 13.8, which suggests that we should make a further reduction in the air spaces of 0.037, and strengthen the crowns by 0.191%. These changes

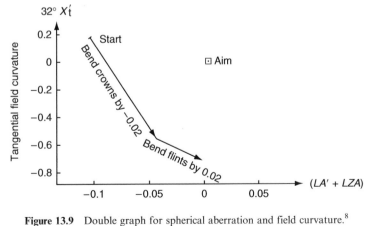

Figure 13.9 Double graph for spherical aberration and field curvature.[8]

in their turn upset the spherical and field corrections, so we have to make further small bendings for these, and so on back and forth until all four aberrations are corrected. The system then is as follows:

c	d	n_D
0.6733		
	0.3	1.6109
0.1183		
	0.145	air
0.4768		
	0.1	1.6170
0.9671		
	0.15	
	(symmetrical)	

with focal length = 5.9394, Ptz(10) = 0.0278; for 32°: $X'_s = 0.0674$, $X'_s = -0.0134$; $LA' = -0.0015$, $LZA = 0.0000$, zonal chromatic aberration = -0.0008.

Finally, we come to the correction of distortion and lateral color:

$$32° \text{ distortion} = 1.28\%, \quad 32° \text{ lateral color} = 0.0014$$

These were reduced by shifting 3% of the power of the front crown element to the rear crown, giving 0.70% distortion and –0.0001 of lateral color. However, since this move upset everything, it was necessary to return to the previous graphs and repeat the whole design process once or twice more. After scaling up to a focal length of 10.0, the final system is as follows:

c	d	n
0.38600		
	0.5083	1.6109
0.06787		
	0.2355	
0.28732		
	0.1694	1.6170
0.57670		
	0.2542	
	0.2542	
−0.57670		
	0.1694	1.6170
−0.28732		
	0.2355	
−0.07201		
	0.5083	1.6109
−0.40990		

with $f' = 10.0$, $l' = 8.9971$, Ptz(10) $= 0.0279$, zonal chromatic aberration $= 0.00030$, LA' ($f/8$) $= 0.00046$, LZA ($f/11$) $= 0.00225$, stop diameter ($f/8$) $= 0.5149$. The results are shown in Table 13.1.

Table 13.1

Astigmatism, Distortion, and Lateral Color for Example Double-Gauss Lens

Field (deg)	X'_s	X'_t	Distortion (%)	Lateral color
32	0.0617	−0.1303	0.660	−0.00034
25	−0.0529	0.0586	0.266	−0.00086
15	−0.0456	0.0239	0.065	−0.00064

A scale drawing of this lens, together with its aberration graphs, is shown in Figure 13.10. It is evident that the zonal aberration is of the unusual over-corrected type and that the crown elements are quite unnecessarily thick.

To complete the work we must determine how much vignetting should be introduced, mainly to cut off the ends of the curves in Figure 13.11. Our procedure will be to decide to accept a maximum departure of the graphs from the principal ray by, say, ±0.025, and cut off everything beyond that limit. The vignetted rays are shown in the lens diagram in Figure 13.10. The limiting rays are marked V on the ray plots (Figure 13.11), and the extreme unvignetted rays that just fill the diaphragm are marked S. It will be seen that the 15° beam is unvignetted. In view of this, the limiting surface apertures shown in Table 13.2 are recommended for this lens. This completes the design.

If a higher aperture than $f/8$ is desired, it is necessary to thicken the negative elements considerably, and introduce achromatizing surfaces into them. The process for the design of the $f/2$ "Opic" lens of this type has been described by H. W. Lee.[9] Using the Buchdahl coefficients $\sigma_3, \sigma_4, \mu_{10}$, and μ_{11}, Hopkins[10] has shown that it is possible to calculate the image height, relative to the Gaussian image height, where the sagittal and tangential field curves intersect one another. This of course assumes that higher-order astigmatic terms are negligible.

Beyond this intersection height, the two field curves rapidly diverge from one another, as can be observed in Figures 13.2, 13.7, and 13.10, and is shown in Figure 13.14 in the next section (page 372). Attempting to use the lens beyond this image height will be unfruitful. This image height, H_n is derived by equating the linear terms in aperture of Eqs. (4-6) and (4-7) through fifth order, such that

$$[(3\sigma_3 + \sigma_4)H_n^2 + \mu_{10}H_n^4]\rho = [(\sigma_3 + \sigma_4)H_n^2 + \mu_{11}H_n^4]\rho$$

$$2\sigma_3 H_n^2 + \mu_{10}H_n^4 = \mu_{11}H_n^4$$

$$H_n = \sqrt{\frac{2\sigma_3}{\mu_{11} - \mu_{10}}}$$

$$(13\text{-}8)$$

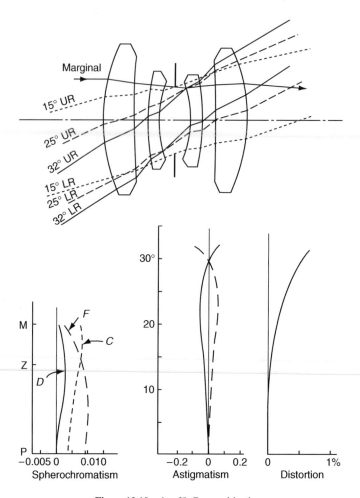

Figure 13.10   An *f*/8 Gauss objective.

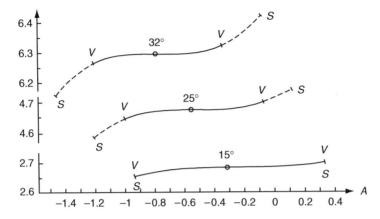

Figure 13.11 Meridional ray plots for *f*/8 Gauss objective.

Table 13.2

Limiting Surface Apertures for
***f*/8 Gauss Objective**

Surface	Clear aperture
1	1.061
2	0.890
3	0.603
4	0.520
Stop	0.515
5	0.510
6	0.596
7	0.859
8	1.023

13.4 DOUBLE-GAUSS LENS WITH CEMENTED TRIPLETS

Consider now that the two negative elements in Figure 13.10 are replaced with cemented meniscus triplets (see Section 10.4). In the late 1950s, Altman and Kingslake developed such a 100-mm focal-length Double-Gauss lens, shown in Figure 13.12, that could be, for example, used as a 1:1 relay or erector lens in a sighting telescope.[11] In the case to be examined, light in the central air space is collimated, with each half of the lens operating at *f*/7.6; the entire lens therefore operates at *f*/3.8.

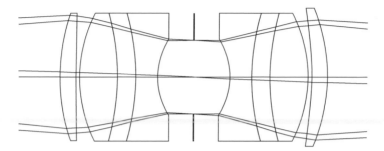

Figure 13.12 Unity magnification of the Double-Gauss lens with negative cemented meniscus triplets.

An important objective of their design was to attain highly corrected zonal spherical aberration and spherochromatism so that the lens gives decisive definition of a modest field-of-view when used as a process lens. When used as an erector or relay lens in a telescope, the observer's eye can move about without causing the image to shift or flutter as it does in the presence of even slight spherical aberration. Rather than following the more common approach of splitting the elements, they replaced the negative lenses with a cemented triplet. The triplet has a weak meniscus element ($1.43 < n < 1.60$) cemented between a biconcave element and a biconvex element, with both having a refractive index at least 0.08 greater than the weak meniscus element. The thickness of the weak meniscus can be used to control the zonal spherical aberration, with an increase in the thickness tending to overcorrect the zonal with respect to the marginal spherical aberration. They selected glasses for the meniscus and the biconvex elements that have about the same Abbe number so that the lens designer will have a *chromatically ineffective surface* that can be varied without appreciably affecting the color (see Buried Surface, Section 10.5).

The marginal spherical aberration is primarily controlled by adjusting (1) the concave surfaces facing the stop, which also strongly affect the Petzval sum, and (2) the bendings of the positive elements. The longitudinal color is corrected by adjusting the cemented surface between the biconcave element and the meniscus element. By varying the aforementioned chromatically ineffective surface, zonal spherical aberration can be corrected along with adjusting the bendings of the positive elements to maintain marginal spherical aberration correction. It was observed that the marginal spherical aberration varies more rapidly than the zonal spherical aberration as the chromatically ineffective surface is changed, with both aberrations becoming more overcorrected as this surface is strengthened.

In addition, when the marginal spherical aberration is restored by changing the bendings of the positive elements, both the marginal and zonal spherical aberrations change at about the same rate. Thus, the net effect is to change the zonal spherical aberration toward overcorrection. The field curvature is controlled primarily by varying the central air space while the coma, distortion, and lateral color are controlled in the usual way by varying the front half of the objective with respect to its rear half, so the cemented meniscus triplets are the same while the positive elements differ. It was observed that this lens structure has good spherochromatism. An example structure is as follows:

r	d	n_d	V_d
36.02			
	3.1	1.517	64.5
418.3			
	0.7		
24.59			
	7.4	1.611	58.8
−45.33			
	3.5	1.523	58.6
−44.52			
	4.3	1.617	36.6
13.42			
Stop	$\dfrac{6.900}{6.900}$		
−13.42			
	4.3	1.617	36.6
44.52			
	3.5	1.523	58.6
45.33			
	7.4	1.611	58.8
−24.59			
	0.7		
−74.42			
	3.1	1.720	29.3
−32.20			

Figure 13.13 shows the longitudinal aberrations while Figure 13.14 presents the field curvature and distortion. As one would expect, the distortion is small (0.012%) and the field is quite flat and slightly backward curving. The ray fans in Figure 13.15a illustrate again that the spherochromatism is well corrected, although the axial color in C light is slightly overcorrected. The presence of some lateral color is seen in Figure 13.15b. Linear and higher-order coma and spheroastigmatism are observable in Figure 13.15b and Figure 13.15c.

This example lens was designed for use as a relay in a sighting scope comprising a typical objective lens having two cemented doublets, reticle, a prism system of light barium crown glass, a relay lens, and an eyepiece. The lens

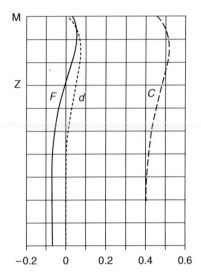

Figure 13.13 Longitudinal aberration.

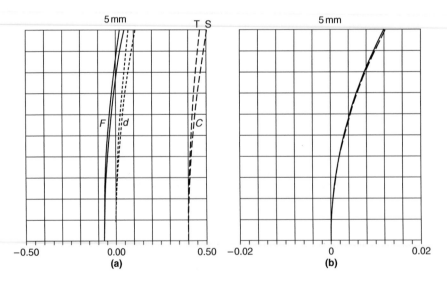

Figure 13.14 Field curvature and distortion: (a) The tangential field curve lies to the left of the sagittal field curve for each of the colors and the abscissa is in lens units. (b) Distortion is in percentage.

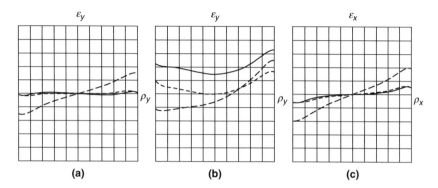

Figure 13.15 Ray fans (a) for on-axis and (b) and (c) for 5 mm off-axis. Ordinate scales are ±0.1 mm. Solid curves = F light, short dashed curves = d light, and long dashed curves = C light.

designer is often faced in practice with the design of an entire system rather than just simply a lens. In the example relay lens given by Altman and Kingslake, they designed it as part of an overall system. The lateral color of the system was well corrected for all zones. To accomplish this, the eyepiece was allowed to have moderately large residual undercorrected lateral color, which was matched by opposite lateral color in the rest of the system. The objective lens was well corrected since a color-free image was desired at the reticle.

In this particular system, the prism system needed to be placed between the reticle and the relay lens, which made correction more difficult than if the prism system had been placed following the relay lens. To achieve a balance in the lateral color, they found it necessary to make the rear element of the relay lens from a very-high dispersion dense flint glass and the front lens from a very-low dispersion crown glass. It also required the refractive index of these outer lenses to be markedly different, which caused serious zonal spherical aberration, spherochromatism, and coma; however, the novelty of the use of the meniscus elements in each negative triplet provided a means to achieve excellent correction. Reading of their patent is encouraged by those interested in further design details.

13.5 DOUBLE-GAUSS LENS WITH AIR-SPACED NEGATIVE DOUBLETS

The basic Gauss lens that was shown in Figure 13.10 can be improved by replacing the negative lenses with air-spaced negative doublets and the rear positive element with a cemented doublet as illustrated in Figure 13.16.[12] This 100-mm focal length lens is well corrected at $f/2$ operating at unity magnification. The purpose of this lens was for printing on a film that is sensitive to a particular wavelength of blue light, say 435.8 nm. Consequently, chromatic correction was

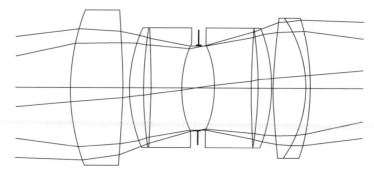

Figure 13.16 Unity-magnification Double-Gauss lens with air-spaced negative doublets.

not of particular importance except that it was desirable to achromatize the lens for blue and green light where the green light was used for alignment of the system. A typical structure is as follows:

r	d	n_d	V_d
66.300			
	17.42	1.75510	47.2
−192.96			
	2.23		
46.049			
	4.7	1.65820	57.2
108.97			
	2.36		
−352.51			
	10.06	1.69873	30.1
33.510			
Stop	$\frac{5.460}{5.460}$		
−33.285			
	12.15	1.61633	31.0
252.60			
	2.18		
−133.01			
	4.63	1.69680	56.2
−50.695			
	0.51		
138.25			
	10.95	1.68235	48.2
−37.751			
	1.41	1.62032	60.3
−79.381			

The principal invention of this lens structure is the use of a pair of negative doublets, located about a central stop, with each having a strong negative air lens (see Section 7.4.3). All of the elements in this lens use high refractive index glasses and large thicknesses to simplify correction of aberrations by using weaker surfaces. Examination of Figure 13.17 shows that the spherical aberration is undercorrected and that the axial image quality can benefit by moving the image plane slightly toward the lens by an amount of 85 μm.

The astigmatic field curves in Figure 13.18a show that they intersect at 5, which implies that this is essentially the limit of the useful field-of-view. These field curves are also inward curving, which is advantageous to enhance the off-axis resolution since the axial refocus is inward toward the lens. Figure 13.19a presents the axial ray fan after refocus. Inspection of this plot shows that the spherical aberration contains at least third-, fifth-, seventh-, and ninth-order spherical aberration.

The off-axis ray fans, when refocused is invoked, are shown in Figure 13.19b and Figure 13.19c. Figure 13.18b illustrates that the distortion is triflingly small. The patent suggests that this lens can resolve over 400 lines per mm

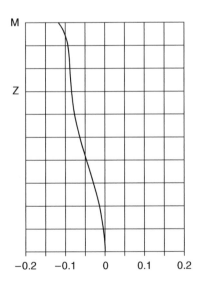

Figure 13.17 Longitudinal aberration focused at the paraxial focal point; the abscissa is in lens units.

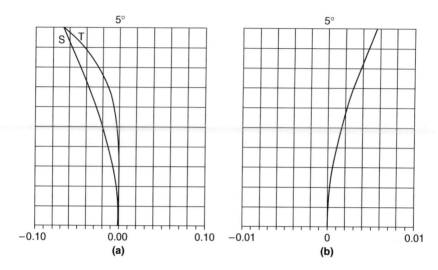

(a)   **(b)**

Figure 13.18 Field curvature and distortion. The abscissa for the field curve (a) is in lens units and distortion (b) is in percentage.

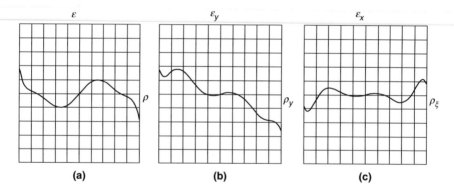

(a)   **(b)**   **(c)**

Figure 13.19 Ray fans (a) for on-axis and (b) and (c) for 5° off-axis when lens has been refocused by −0.085 mm with respect to the paraxial focus. Ordinate scales are ±0.01 mm.

(200 line-pairs per mm). Figure 13.20 shows the MTF for a diffraction-limited $f/2$ lens, the axial MTF, and the MTF for an object 5° off axis. It is evident that the lens is nearly diffraction-limited on-axis with excellent sagittal off-axis performance and somewhat degraded tangential off-axis performance at the edge of the field of view.

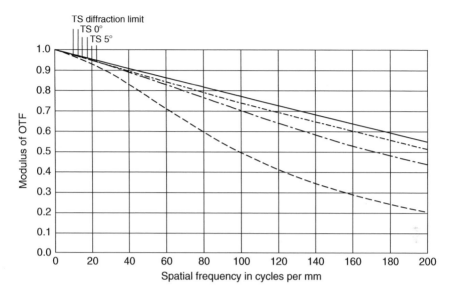

Figure 13.20 MTF for the axial and the 5° off-axis objects when lens has been refocused by −0.085 mm with respect to the paraxial focus.

ENDNOTES

[1] C. P. Goerz and E. von Höegh, U.S. Patent 528,155, filed February (1893).

[2] A light-hearted account of the creation of the Dagor has been given by R. Schwalberg, *Pop. Phot.,* 70:56 (1972).

[3] A technique to approximate axial image refocusing while remaining in the paraxial image plane—see discussion in Section 6.2.

[4] C. P. Goerz and E. von Höegh, U.S. Patent 635,472, filed July (1898).

[5] These meridional ray fans are similar to the more typical ray fans where the abscissa is the ray intercept value in the entrance pupil. In this case, P designates the location of the principal ray for each fan plot. A reason to use the first tangent plane for the incident ray coordinate A is that it aids in selecting the diameter of the lenses.

[6] A. G. Clark, U.S. Patent 399,499, filed October (1888).

[7] The maximum speed of this lens is $f/8$. The 0.7 zone corresponds to $f/11.3$.

[8] The spherical aberration measured on the abscissa is the sum of the marginal and zonal spherical aberration. The goal is for them to have equal and opposite values so that a value of zero is the aim point. The reason for this was explained in Section 13.1.

[9] H. W. Lee, "The Taylor–Hobson $f/2$ anastigmat, *Trans. Opt. Soc.,* 25:240 (1924).

[10] Robert E. Hopkins, "Third-order and fifth-order analysis of the triplet," *JOSA*, 57(4): 389–394 (1962). In this paper, the derivation is not provided for the H_n equation, which contains a typographical error, with σ_4 rather than σ_3 being shown.

[11] Fred E. Altman and Rudolph Kingslake, U. S. Patent 2,823,583 (1958).

[12] Rudolf Kingslake, U.S. Patent 3,537,774 (1970).

Chapter 14

Unsymmetrical Photographic Objectives

14.1 THE PETZVAL PORTRAIT LENS

This ancient lens was the first photographic objective to be deliberately designed rather than being put together by an empirical selection of lenses out of a box. It consists of two fairly thin achromats spaced widely apart with a central stop.[1] It has excellent correction for spherical aberration and coma, but because the Petzval sum is uncorrected, the angular field is limited by astigmatism to about 12° to 15° from the axis. Modified forms of the Petzval lens are still used, mainly for the projection of 16 and 8 mm movie films and other projection devices, although if a negative field flattener is added close to the image plane the lens becomes a true anastigmat, and in this form it has been used as a long-focal-length lens for aerial reconnaissance purposes.

The front component of the original Petzval design of 1839 was an ordinary $f/5$ telescope doublet. It is possible that Petzval attempted to assemble two identical lenses symmetrically about a central stop, in order to raise the aperture to $f/3.5$ for use with the slow daguerreotype plates of the time, but the aberrations were so bad that he had to separate the two elements in the rear component and bend them independently to correct the spherical aberration and coma. Later, in 1860, J. H. Dallmeyer turned the rear component around,[2] with the crown element leading, and he thus obtained a lens that was better than the Petzval design near the middle of the field, but the inevitable uncorrected astigmatism was so great that the two designs are virtually indistinguishable. In 1878 F. von Voigtländer[3] found that by suitably bending the front component of the Dallmeyer type he could cement the rear component also, and it is this last arrangement that is used today as a small projection lens of high aperture.

14.1.1 The Petzval Design

In designing a Petzval portrait lens it is customary to make both doublets of the same diameter and to mount the stop approximately midway between them. If the front doublet consists of the familiar form with an equiconvex crown, this stop position has the effect of making the tangential field of the front component somewhat backward-curving, and to correct this requires a positive rear component somewhat weaker than the front component. To correct the spherical aberration as well as the OSC and to flatten the tangential field, we find that we must select glass types having a rather large V difference; with the refractive indices used by Petzval, 1.51 and 1.57, a V difference of at least 18 is required. In the present examples the following Schott glasses are used:

(a) Crown: K-1, $n_e = 1.51173$, $n_F - n_C = 0.00824$, $V_e = 62.10$
(b) Flint: LF-6, $n_e = 1.57046$, $n_F - n_C = 0.01325$, $V_e = 43.05$

The V difference is 19.05.

The Front Component

For the front component we adopt a thin-lens focal length of 10 and a clear aperture of 1.8. This aperture may have to be adjusted later after the actual focal length of the system has been determined. For this front lens, the thin-lens formulas give $c_a = 0.63706$ and $c_b = -0.30618$. Assuming an equiconvex crown, our front component is as follows:

c	d	n
0.31853		
	0.4	1.51173
−0.31853		
	0.12	1.57046
$(D - d)$ 0.086680		

Assuming an air space of 2.6, the 10° principal ray enters at $L_{pr} = 2.054$ and crosses the axis midway between the two lenses.

The Petzval Rear Component

For a Petzval-type rear component, we may start with the arbitrary Setup that follows:

	c	d	n
	0.25		
		0.12	1.57046
	0.6		
		0.025516	
	0.55		
		0.4	1.51173
$(D-d)$	−0.017292		

with $f' = 6.1898$, $l' = 3.9286$, LA' ($f/3.44$) = 0.0005, OSC ($f/3.44$) = 0.001944. The focal length and aberration data given here are calculated for the complete system. The space between the two rear elements was determined so that they would be in edge contact at a diameter of 1.8. As the design proceeds this separation must be recalculated for each Setup to maintain the edge–contact condition.

The best way to correct the spherical aberration and coma is to bend the two rear elements separately and plot a double graph as shown in Figure 14.1. The graph data are

(a) Original Setup A: $LA' = 0.000449$, $OSC = 0.001944$
(b) Bend flint by 0.02 for Setup B: $LA' = 0.024885$, $OSC = 0.004688$
(c) From Setup B, bend crown by 0.02 to obtain Setup C: $LA' = 0.010455$, $OSC = 0.001965$

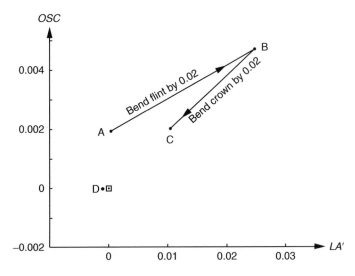

Figure 14.1   Double graph for rear component of Petzval portrait lens ($f' = 6.2$).

Extrapolating in the usual way, and because the graphs are remarkably straight, we quickly reach the aplanatic form (Setup D):

	c	d	n
	0.27		
		0.12	1.57046
	0.62		
		0.018802	
	0.5841		
		0.40	1.51173
$(D-d)$	0.0220382		

with $f' = 6.2206$, $l' = 3.9233$, LA' ($f/3.46$) $= -0.0009$, OSC ($f/3.46$) $= -0.00003$. The fields along the computed $10°$ principal ray were $X_s' = -0.0597$, $X_t' = -0.0123$.

To move the fields backward, we must weaken the entire rear component. A few trials indicate that c_c should be reduced by 0.025, and after recorrecting the spherical and chromatic aberrations and the OSC we obtain the following solution (Setup E):

	c	d	n
	0.27		
		0.12	1.57046
	0.595		
		0.023158	
	0.5495		
		0.40	1.51173
$(D-d)$	0.0287696		

with $f' = 6.4012$, $l' = 4.0408$, LA' ($f/3.56$) $= 0.0030$, LZA ($f/5$) $= -0.0021$, OSC ($f/3.56$) $= -0.00002$, Ptz (10) $= 0.0811$. The results are shown in Table 14.1. These aberrations are plotted in Figure 14.2.

The final check on our system is made by drawing a meridional ray plot at $10°$ obliquity, which is shown in Figure 14.3a. The abscissas are the height of each ray at the stop with the height of the marginal ray at the stop being shown on the graph ordinate. However, because of vignetting at the front and rear surfaces,

Table 14.1

Astigmatism and Distortion for Setup E

Field (deg)	X_s'	X_t'	Distortion (%)
15	−0.1034	0.1551	0.32
10	−0.0571	0.0007	0.11

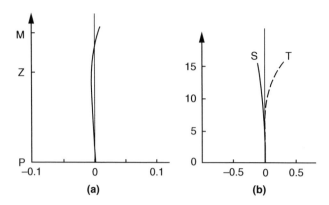

Figure 14.2 Aberrations of Setup E ($f' = 6.4$): (a) longitudinal spherical aberration and (b) astigmatism.

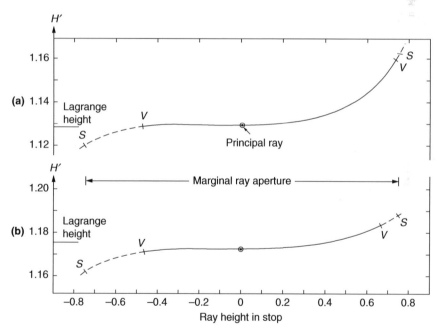

Figure 14.3 Ray plots for Petzval objectives at 10°: (a) Ray plot with rear elements in close contact; (b) Ray plot with air-spaced rear elements.

which are assumed to have a free aperture of 1.8, only a part of the graph is valid. The upper and lower vignetted rays are indicated by VV, whereas the limiting rays through the top and bottom of the stop are marked SS on this graph. It should be noted particularly that the middle of the curve is straight and level as a result of the good correction of OSC and the flat tangential field at 10°, but the upper end of the curve rises precipitously because of the extremely high values of the angles of incidence in the rear air space. Furthermore, the tangential field at 15° becomes rapidly more backward-curving for the same reason.

The best way to improve both these conditions is to increase the air space between the two rear elements. We will try a fixed air space of 0.15 and repeat the entire design. Following the same procedure, and plotting the usual graphs, gives us this final solution for the rear component, using the same front component and central air space as before:

	c	d	n
	0.25		
		0.12	1.57046
	0.54		
		0.15	
	0.468		
		0.40	1.51173
$(D-d)$	0.0028107		

with $f' = 6.6685$, $l' = 4.2468$, LA' $(f/3.70) = 0.0012$, LZA $(f/5.2) = -0.0031$, OSC $(f/3.70) = 0.00001$, Ptz $(10) = 0.0804$. The results are shown in Table 14.2.

The 10° meridional ray plot for this lens, to the same scale as before, is shown in Figure 14.3b. It will be seen that this design is much better than the previous one, and indeed almost all of the Petzval portrait lenses made since 1840 have had a wide space between the two rear elements. A sectional drawing of this system and its aberration graphs are shown in Figure 14.4. The solid curve is the sagittal field and the dashed curve is the tangential field.

Table 14.2

Astigmatism and Distortion for the Final Petzval-type Lens Design

Field (deg)	X'_s	X'_t	Distortion (%)	Lateral color
15	-0.1105	0.0157	-0.95	
10	-0.0553	-0.0002	-0.28	-0.00049

14.1.2 The Dallmeyer Design

To design a lens of the Dallmeyer type, we can start by merely turning around the rear component of the last system, recomputing the last radius by the $D - d$ method, and tracing enough rays to evaluate the system. The crown element was made slightly thinner as it appeared to be too thick before. It was found that the spherical aberration had become decidedly undercorrected

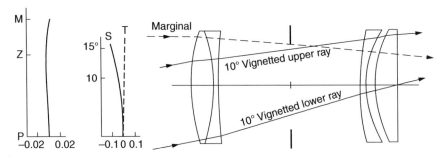

Figure 14.4 The final Petzval-type lens.

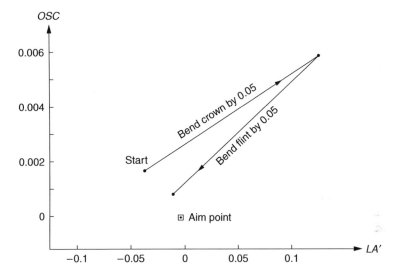

Figure 14.5 Double graph for the Dallmeyer lens.

and the *OSC* overcorrected, and so a double graph was plotted by which these aberrations could be corrected, using suitable bendings of both rear elements. This graph is shown in Figure 14.5, and it led us to the following rear system:

	c	d	n
	0.0722		
		0.35	1.51173
	−0.3930		
		0.15	
	−0.4600		
		0.12	1.57046
(D − d)	−0.1408571		

with $f' = 6.9991$, $l' = 4.3100$, LA' ($f/3.89$) = -0.0125, $OSC = -0.00006$; for 10°: $X'_s = -0.0703$, $X'_t = -0.0423$.

In an attempt to correct the inward-curving tangential field, the third element was weakened by 0.1, and since this had the effect of reducing the relative aperture of the system, the front clear aperture was increased at the same time from 1.8 to 2.0. This required a recomputation of c_3 for achromatism by the $D - d$ method. It was found that the same double graph could be used, and a few trials gave the following rear component:

	c	d	n
	−0.0741		
		0.35	1.51173
	−0.4393		
		0.15	
	−0.5283		
		0.12	1.57046
$(D - d)$	−0.2880611		

with $f' = 7.3340$, $l' = 4.6524$, LA' ($f/3.67$) = 0.0018, $OSC =$ zero; for 10°: $X'_s = -0.0481$, $X'_t = 0.0358$. Obviously we have gone too far in our weakening of the rear component, so we decided to strike a compromise and repeat the design. The final complete system then became as follows:

	c	d	n
	0.31853		
		0.40	1.51173
	−0.31853		
		0.12	1.57046
$(D - d)$	0.0847414		
		2.6	
	0		
		0.35	1.51173
	−0.411		
		0.15	
	−0.4884		
		0.12	1.57046
$(D - d)$	−0.2114208		

with $f' = 7.1831$, $l' = 4.4796$, LA' ($f/3.59$) = -0.0014, LZA ($f/5.1$) = -0.0136, OSC ($f/3.59$) = -0.00007, Petzval (10) = 0.0774. The results are shown in Table 14.3.

Table 14.3

Astigmatism, Distortion, and Lateral Color for Dallmeyer-type Portrait Lens

Field (deg)	X'_s	X'_t	Distortion (%)	Lateral color
15	−0.1278	0.0359	1.54	
10	−0.0599	−0.0049	0.18	0.000692

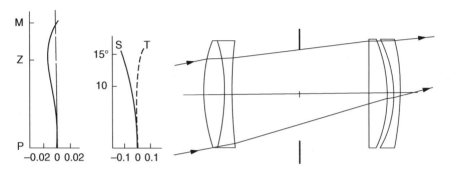

Figure 14.6 A Dallmeyer-type portrait lens. Plots for spherical aberration and astigmatism (solid curve is sagittal; dashed curve is tangential).

A section of the lens is shown in Figure 14.6 along with the graphs of the aberrations. A meridional ray plot is shown in Figure 14.7, where it is seen to be somewhat flatter than the better of the two preceding Petzval designs. However, the large astigmatism would swamp this slight improvement. The zonal spherical aberration, although still small, is about four times as great as for the Petzval form (recall that the Dallmeyer-type portrait lens is considered to be of the Petzval form).

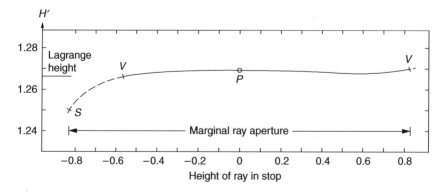

Figure 14.7 Ray plot for Dallmeyer portrait lens at 10°.

14.2 THE DESIGN OF A TELEPHOTO LENS

A telephoto lens is one in which the "total length" from front vertex to focal plane is less than the focal length; telephoto lenses are used wherever the length of the lens is a serious consideration.

Most telephoto objectives contain a positive achromat in front and a negative achromat behind, the lens powers being calculable when the focal length F, the total length kF, and the lens separation d are all given (Figure 14.8). The factor k is known as the *telephoto ratio*, and its value is ordinarily about 0.8.

In terms of thin lenses, the ratio

$$\frac{y_b}{y_a} = \frac{f_a - d}{f_a} = \frac{kF - d}{F}$$

hence

$$f_a = \frac{Fd}{F(1 - k) + d}$$

For lens (b), we have $l = f_a - d$ and $l' = kF - d$. Therefore,

$$\frac{1}{f_b} = \frac{1}{kF - d} - \frac{1}{f_a - d}$$

from which it follows that

$$f_b = \frac{(f_a - d)(kF - d)}{f_a - kF}$$

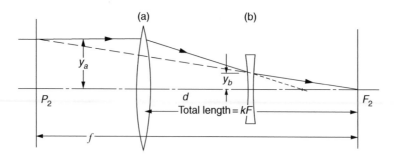

Figure 14.8 Thin-lens layout of a telephoto system with a distant object.

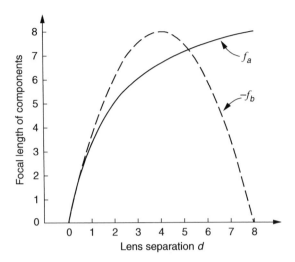

Figure 14.9   Relation between lens powers and separation when $F = 10$ and $k = 0.8$.

As an example, if $F = 10.0$ and $k = 0.8$, we can plot graphs of the focal lengths of the two components against the lens separation d (Figure 14.9). It is clear that as the separation is increased, both the front and rear lenses become weaker. Indeed, the power of the rear negative component reaches its minimum value when that lens lies midway between the front positive component and the focal plane. However, as the separation is increased, the diameters of the lenses must also be increased to reduce the vignetting.

For our present design we will assume a thin-lens separation d equal to 3.0. This will require a positive front component with $f_a = 6.0$ and a negative rear component with focal length $f_b = -7.5$. Since the two chromatic aberrations will be controlled by a suitable choice of glass dispersions at the end, we adopt refractive indices such that a range of dispersions is available. The crown index is therefore set at 1.524, for which there are glasses with V values ranging from about 51 to 65, and the flint at 1.614, for which V values exist between about 37 and 61.

For a start, let us suppose that the chosen glasses are K-3 and F-3:

(a) K-3: $n_e = 1.52031$, $\Delta n = n_F - n_C = 0.00879$, $V_e = 59.19$
(b) F-3: $n_e = 1.61685$, $\Delta n = n_F - n_C = 0.01659$, $V_e = 37.18$

with the V difference $= 22.01$. For the front component, then, $c_a = 0.8615$, while for the rear component $c_c = -0.6892$. We may assume an equiconvex crown for the front and an equiconcave crown in the rear. We assign suitable thicknesses for a clear aperture of 1.8 (i.e., an aperture of $f/5.6$), and we consider an angular semifield of $10°$. For every Setup we calculate the last radii of the front and rear components to yield the desired focal lengths of $+6.0$ and -7.5, respectively, and we determine the central air space so that the separation of adjacent principal points is 3.0. The stop is assumed to be in the middle of the air space. Our starting system is as follows:

	c	d	n
	0.4308		
		0.50	1.524
	−0.4308		
		0.15	1.614
(for f')	0.04155		
		2.517648	
	−0.3446		
		0.15	1.524
	0.3446		
		0.50	1.614
(for f')	−0.023990		

with $f' = 10.0$, $l' = 4.5205$, LA' ($f/5.6$) $= 0.3022$, OSC ($f/5.6$) $= -0.0260$; for $10°$: distortion $= 2.04\%$, $X'_t = -0.0218$.

We now proceed to change the rear component to correct distortion and tangential field curvature, using c_4 and c_5 on a double graph, of course maintaining the thin-lens telephoto conditions by solving for c_6 and the central air space d'_3 at all times. It is found that the graph for changes in c_4 bends back on itself but the graph for c_5 is quite straight (Figure 14.10). The aim point at distortion $= 0.5\%$ and $X'_t = 0$ is nearly reached by the following Setup:

	c	d	n
	0.4308		
		0.50	1.524
(unchanged)	−0.4308		
		0.15	1.614
	0.04155		
		3.058468	
	−0.7446		
		0.15	1.524
	0.4100		
		0.50	1.614
	−0.310175		

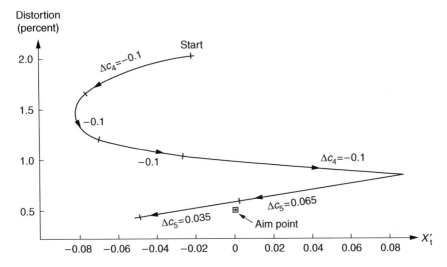

Figure 14.10 Double graph for distortion and field curvature.

with $f' = 10$, $l' = 3.8945$, LA' ($f/5.6$) $= 0.6093$, OSC ($f/5.6$) $= -0.0161$; for 10°: distortion $= 0.580\%$, $X'_t = 0.0022$. These changes have led to considerable overcorrection of the spherical aberration, while the OSC is slightly smaller.

We now move to the front and plot a double graph of spherical aberration and OSC for changes in c_1 and c_2 (Figure 14.11). The closest Setup to the aim point at $LA' = 0.02$ and $OSC = 0$ is as follows:

	c	d	n
	0.4228		
		0.50	1.524
	−0.2888		
		0.15	1.614
	0.0551473		
		3.062479	
	−0.7446		
		0.15	1.524
(unchanged)	0.4100		
		0.50	1.614
	−0.310175		

with $f' = 10$, $l' = 3.8945$, LA' ($f/5.6$) $= 0.1017$, OSC ($f/5.6$) $= -0.00003$; for 10°: distortion $= 0.878\%$, $X'_t = -0.3247$.

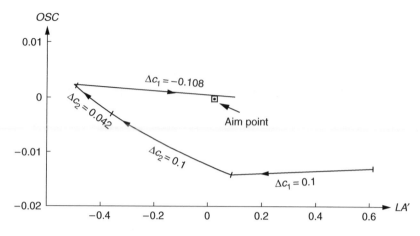

Figure 14.11   Double graph for spherical aberration and *OSC*.

It is clear by now that in this type of lens every change affects every aberration, and it is not very profitable to go back and forth between the two double graphs. Instead, therefore, we will resort to the solution of four simultaneous linear equations in four unknowns, each equation being of the type

$$\Delta \, ab = \sum (\partial \, ab / \partial \, var) \Delta \, var$$

where ab signifies an aberration and var signifies a variable lens parameter (see Section 17.1.3).

DESIGNER NOTE

As lens constructions become more complicated or have significant cross-correlations between parameters and aberrations, it becomes highly desirable to utilize the optimization feature of a lens design program or a computer-based math package. Solving the system of equations given by Δab_i can be simple, difficult, or even impossible. Various techniques in numerical analysis have been applied to this problem where nonlinearities and singularities are common.

The following illustrates the challenge of solving even this rather minimal problem. When one has numerous parameters, aberrations, and constraints to consider in finding a suitable solution to a lens design problem, it is of course necessary to use some form of computer-aided optimization. Much effort has been expended since the 1950s in developing optimization algorithms that include least squares, steepest descent, additively damped least squares, multiplicatively damped least squares, full and pseudo second-derivate damped least squares, orthonormalization, simulated annealing, and many others. The importance of how the lens designer assigns the importance of each aberration (or more generally a system defect), constructs the parameter boundaries, and the design plan followed cannot be overemphasized.

By design plan, we mean the steps the lens designer will take in the design process, as the outcome can be profoundly different! As a simple example, consider one design plan to be just to allow the lens design program to have full control over all of the parameters with the expectation that the program will find an acceptable solution if allowed to run long enough. This is a common approach with many novice lens designers and frequently yields unsatisfactory results. Another design plan might be to attempt to control the higher-order aberrations first and then control the lower-order aberrations since the higher the aberration order, the more stable the abberations are with respect to lens parameters.[4,5] Mastery of the design methods taught in this book can definitely help the lens designer in the quest to develop satisfactory lens designs.

Long before Glatzel[6] and Shafer[7] discussed strain in optical systems, Kingslake often lectured to his students that a well-designed lens will have a pleasing appearance while those that do not will likely not perform well and/or be difficult to manufacture and align. The paper by Shafer provides a useful discussion of stain in optical systems.

The 16 coefficients of the type (∂ ab/∂ var) are found by trial, by applying a small change, say 0.1, to each variable in turn and finding its effect on each of the four aberrations. The coefficients were found to be

Aberration	c_1	c_2	c_4	c_5
LA'	−5.488	−3.180	−0.6211	0.1416
OSC	0.01886	0.11995	−0.03421	−0.00590
Distortion (%)	−3.704	2.357	2.030	−4.021
X'_t	2.807	−2.439	−1.121	−1.274

These four simultaneous equations were solved to give the desired changes in the four aberrations, namely,

$$\Delta LA' = -0.08 \text{ (to yield } +0.02)$$
$$\Delta OSC = 0 \text{ (correct as is)}$$
$$\Delta \text{ distortion} = -0.38 \text{ (to give } +0.5)$$
$$\Delta X'_t = +0.32 \text{ (for zero)}$$

The solution of the equations was

$$\Delta c_1 = 0.0438, \ \Delta c_2 = -0.0333$$
$$\Delta c_4 = -0.0906, \ \Delta c_5 = -0.0111$$

Applying these changes to our lens, and solving as before for the two focal lengths and the thin-lens separation, we get the following prescription:

c	d	n
0.4666		
	0.50	1.524
−0.3221		
	0.15	1.614
(f) 0.092643		
	3.12194	
−0.8352		
	0.15	1.524
0.3989		
	0.50	1.614
(f) −0.372419		

with $f' = 10$, $l' = 3.74711$, LA' ($f/5.6$) = 0.0248, OSC ($f/5.6$) = 0.00026; for 10°: distortion = 0.534%, $X'_t = 0.0307$. These aberrations are almost correct, but a second solution using the same coefficients gave this final system:

c	d	n
0.4664		
	0.50	1.524
−0.3208		
	0.15	1.614
(f) 0.0926424		
	3.11078	
−0.8273		
	0.15	1.524
0.4083		
	0.50	1.614
(f) −0.3660454		

with $f' = 10.0$, $l' = 3.7618$, LA' ($f/5.6$) = 0.0211, LZA ($f/8$) = −0.0108, OSC ($f/5.6$) = −0.00001; for 10°: distortion = 0.50%, $X'_t = -0.0012$, $X'_s = 0.0261$.

We next trace a number of oblique rays at 10° obliquity and draw a meridional ray plot to determine the best stop position (Figure 14.12). The abscissas of this plot are conveniently the Q of each ray at the front surface. Since the lower end of this graph sags downward excessively, we must move the stop closer to the front than the midway position previously assumed. This puts the diaphragm at a distance of 0.5 from surface 3 (rear surface of front achromat), and the limiting upper and lower rays that just fill the stop are shown by SS. When we set the front surface aperture at the diameter of the entering $f/5.6$ axial beam, namely, 1.786, the lower limiting ray is that shown at V_1. However, the graph indicates that we can safely increase the diameter of the rear aperture to 1.94, so that the upper limiting ray is located at V_2.

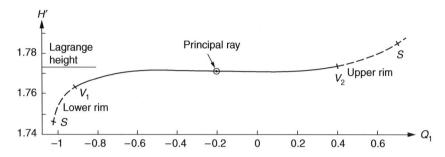

Figure 14.12 Meridional ray plot of telephoto lens at 10° obliquity.

The graph also indicates the presence of a small amount of overcorrected oblique spherical aberration, which is normal in lenses of this type. The principal ray now has a starting Q_1 of –0.2 for the 10° beam, or $L_{pr} = 1.1518$. Keeping this L_{pr} value we can add principal rays at 7° and 12°, giving what is shown in Table 14.4. These results are plotted in Figure 14.13.

Table 14.4

Astigmatism and Distortion for Telephoto Lens

Field (deg)	X'_s	X'_t	Distortion (%)
12	0.0018	−0.1611	0.26
10	0.0322	−0.0241	0.47
7	0.0189	0.0193	0.35

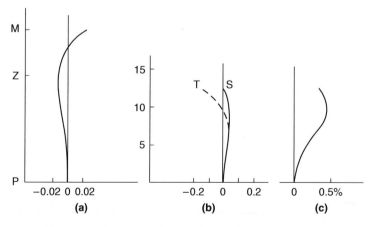

Figure 14.13 Aberrations of a telephoto lens: (a) longitudinal spherical, (b) astigmatism (solid curve is sagittal; dashed curve is tangential), and (c) distortion.

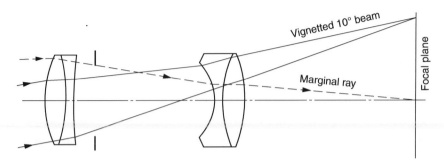

Figure 14.14 Final telephoto design showing $f/5.6$ marginal ray and limiting rays at $10°$.

The lens in its present configuration is shown in Figure 14.14. The telephoto ratio is 0.817, and the lens could be used in a focal length of 120 mm or more on a 35-mm camera. The aperture could be slightly increased, especially if the angular field were less than $10°$.

To complete the design, we must select real glasses for achromatism. The $D - d$ values along the marginal ray in the four lens elements are given, together with the products $(D - d)\, \Delta n$ for a first glass selection as shown in Table 14.5.

Adjusting the catalog indices of these glasses for C and F light by the same amount as the n_e was in error and tracing $10°$ principal rays in C and F gave the lateral color as $H_F' - H_C' = -0.006525$, which was considered excessive. Since the lateral color takes the same sign as the longitudinal color of the rear component and the opposite sign of that of the front component, it is clear that the $\sum$ $(D - d)\, \Delta n$ of the rear should be more positive, while that of the front component should be more negative.

A second glass selection is shown in Table 14.6. Now the lateral color is observed to be $+0.00088$, which is much better. No further improvement is possible using Schott glasses, and so the design is considered complete (unless glasses from other manufacturers are investigated). It is, of course, necessary to repeat the final stages with the actual refractive indices of the chosen glasses, the procedure being to trace a paraxial ray with the true

Table 14.5

Initial Glass Selection for Telephoto Lens

Lens element	a	b	c	d
$D - d$	−0.314961	0.156459	0.073780	−0.045341
Glass type	BK-8	F-3	KF-7	F-3
$n_F - n_C = \Delta n$	0.00818	0.01659	0.01021	0.01659
Product $(D - d)\, \Delta n$	−0.0025764	0.0025957	0.0007533	−0.0007522
	0.0000193		0.0000011	$\sum = 0.0000204$

Table 14.6

Second Glass Selection for Telephoto Lens

Lens element	a	b	c	d
Glass type	BaLK-3	F-3	K-4	BaF-5
n_e	1.52040	1.61685	1.52110	1.61022
V	60.58	37.19	57.58	49.49
Δn	0.00859	0.01659	0.00905	0.01233
$(D - d)\,\Delta n$	−0.0027055	0.0025957	0.0006677	−0.0005591
	−0.0001098		0.0001086	$\sum = -0.0000012$

indices, and to adjust the curvature of each surface to maintain the ray-slope angle after each surface at its former value. Any small aberration residuals that appear can be removed by solving the four simultaneous equations again, assuming that the former 16 rate-of-change coefficients are still valid.

There is, of course, no magic in the initial choice of refractive indices, and it is possible that a better design could be obtained by a different choice.

14.3 LENSES TO CHANGE MAGNIFICATION

14.3.1 Barlow Lens

In 1834, an English engineer and mathematician named Peter Barlow discovered a means to increase the magnification of a telescope (or a microscope), often called a telephoto adapter. He accomplished this by placing a negative power lens between the objective lens and its focal point. Figures 3.19, 5.12, and 14.8 illustrate conceptual configurations and, as shown in Section 5.7, the system focal length F' is given by

$$\frac{1}{F'} = \frac{1}{f_a'} + \frac{1}{f_b'} - \frac{d}{f_a' f_b'}$$

where f_a' is the objective focal length, f_b' is the Barlow lens focal length, and d is their separation. The back focal length of this system was shown in Section 3.4.8 to be given by

$$bfl = F' \left(\frac{f_a' - d}{f_a'} \right)$$

(or $bfl = l' = kF - d$ as explained in Section 14.2) so the shift in the back focal length with and without the Barlow lens is

$$d + bfl - f_a'.$$

In general, Barlow lenses are used to produce a change in magnification of typically two (2X) but rarely more than four (4X). The lens is frequently an achromatic doublet where the lens is corrected for aberrations with the object located to the right of the Barlow lens at a distance of $f'_a - d$ and an image distance of bfl also located to the right of the Barlow lens. It should be evident that the use of a Barlow lens increases the *f*-number of the system by the ratio F'/f'_a.

14.3.2 Bravais Lens

Simply put, a Bravais system is a lens or combination of lenses that forms an image in the same plane in which the object is located. A single Bravais lens is the aplanatic hemispherical magnifier that provides a magnification of n assuming the magnifier is in air.[8] It should be noted that the similar Amici aplanatic hyperhemispherical magnifier is not of the Bravais type since the object and image planes are not colocated. A more general use of a Bravais lens is to change the magnification (or focal length) of an existing optical system without disturbing the original system's conjugate points. An example is to consider a commercial photographic printer that has been adjusted for a certain fixed magnification. To allow the capability of producing a different size print from the same negative, the magnification can be changed by the insertion of a Bravais lens without the need for refocusing.

Bravais likely was the first to publish that each lens system has two object positions where the image and object planes are coincident.[9] These positions are called the Bravais points. An existing optical system can be either operating at infinite or finite conjugates to work with a Bravais lens. The existing optical system's image plane becomes the virtual object plane of the Bravais lens whose image plane is at the same location. The Bravais lens can be thought of as a relay lens having magnification that can be positive or negative, and greater or less than unity.[10] Johnson, Harvey, and Kingslake developed a Bravais lens having at least one Bravais point outside of the lens and determined formulas for its position[11] which are

$$p = \frac{Z + \sqrt{Z^2 + 4Zf'}}{2}$$

$$p' = p - Z$$

$$m = \frac{p}{p'}$$

where p and p' are the distances from the Bravais point to the respective principal points, Z is the distance between the principal points, and f' is the focal length of the lens. The powers of the two lens elements comprising the Bravais lens can be found using the equations in Section 3.4.8 by setting the object-to-image distance s equal to zero. One of several well-corrected Bravais lenses (both single and double components) presented in the patent by Johnson et al. is as follows:

r	d	n	V
108.5			
	25	1.617	54.9
−44.3			
	4.9	1.689	30.9
−106.4			

where the magnification is 0.7, $f' = 100$, the Bravais point is 17.5 behind r_3, and both principal points are inside the lens (first principal point is 8.085 behind r_1 and second principal point is 24.022 in front of r_3).

Figure 14.15 shows this lens and the paths of the marginal ray from the existing optical system with and without the Bravais lens in place, and illustrates that the image plane remains at the same position when the Bravais lens is used. The final slope angle increases by about a factor of 1.4 with the Bravais lens, which means that the f-number will decrease by a factor of about 0.7 (shorter system focal length). An apochromatic Bravais lens for Gaussian beams such as used in printers has been investigated by Griffith.[12]

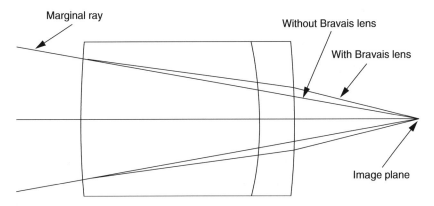

Figure 14.15   Bravais lens having magnification of 0.7.

DESIGNER NOTE

When using a computer-based lens design program to design such a lens as a Bravais, an easy method to create a "virtual" object, that is, a converging beam, is to use an ideal lens focused at the location of the virtual object. Most lens design programs provide such an ideal lens and may be called an ideal lens, perfect lens, paraxial lens, and so on. Should your program not include such a feature, you can use either of the Section 6.1.8 aspheric planoconvex lenses since they are free from spherical aberration or perhaps a parabolic mirror or elliptical (source at one focus and image at other) mirror. Of course if the lens designer has the objective lens available, that can be included instead. It should be noted that, in general, the objective optics and Bravais lens are designed separately since the Bravais lens most often is moved in and out of the system as needed. Consequently, image quality of the objective optics must be acceptable without the Bravais lens; the Bravais lens must not degrade the image quality.

14.4 THE PROTAR LENS

In 1890 Paul Rudolph of Zeiss[13] had the idea of correcting the spherical aberration of a new-achromat landscape lens by adding a front component resembling the front component of a Rapid Rectilinear but with very little power. The thought was that the strong cemented interface in the front component could be used to correct the spherical aberration, and that it would have little effect on field curvature because the principal rays would be almost perpendicular to it. The cemented interface in the rear component would be used to flatten the field as in the new-achromat landscape lens. It is noted that in the Protar patent, one form of the rear component was a cemented triplet although the primary component form was a cemented doublet.

This leaves us with four other radii to be determined. The fourth and sixth radii can be used for Petzval sum and focal length, as in the design of a new achromat, leaving the first and third radii for coma and distortion correction. The two chromatic aberrations are controlled by the final selection of glass dispersions.

As an example, we will first select suitable refractive indices. The two doublets comprise four elements that we denote as (a) to (d). For the outer elements (a) and (d) we may assume that $n_e = 1.6135$. In the Schott catalog we find many glasses having n_e lying close to this figure, with values of $V_e = (n_e - 1)/(n_F - n_C)$ lying between 37.2 and 59.1. For the inner elements (b) and (c) we choose similarly $n_e = 1.5146$ for which V_e values are available between 51.2 and 63.6. Suitable thicknesses are, respectively, 0.25 and 0.4 for the front component and 0.1 and 0.4 for the rear component; the center space is set at 0.4 with the diaphragm midway. We do not use diaphragm position as a degree of freedom since we have enough degrees of freedom already; it is, however, advisable to keep the center space small to reduce vignetting.

For a first trial we may choose $c_1 = 0.5$, $c_2 = 1.2$, and $c_5 = 0.5$. We solve c_3 to make the front component afocal, and we determine c_4 and c_6 by trial and error to make the focal length equal to 10 and the Petzval sum 0.025. The Setup A is as follows:

c	d	n
0.5		
	0.25	1.6135
1.2		
	0.4	1.5146
0.417646		
Stop	$\dfrac{0.2}{0.2}$	
−0.626156		
	0.1	1.5146
0.5		
	0.4	1.6135
−0.572960		

with $f' = 10$, $l' = 9.8120$, Ptz = 0.025, trim diameter = 1.5. A scale drawing of this lens is shown in Figure 14.16. Tracing an $f/8$ marginal ray from infinity and a principal ray passing through the center of the stop at a slope of $-20°$ gives these starting aberrations:

$$LA' \text{ at } f/8 = -0.09026, \quad \left.\begin{array}{l} X'_s = -0.0610 \\ X'_t = +0.0169 \end{array}\right\} \text{ at } U_{pr} = -17.90°$$

Since the main function of radius r_2 is to control the spherical aberration and the main function of r_5 is to flatten the field, we next proceed to vary c_2 and c_5 in turn by 0.05 and plot a double graph by means of which the spherical aberration and the tangential field curvature can be corrected. We assume that the desired values of these aberrations are $LA' = +0.15$ and $X'_t = 0$. The double graph in Figure 14.17 indicates that we should make the following changes from the original Setup A:

1. $\Delta c_2 = 0.034$. But c_2 was 1.2, so therefore try new $c_2 = 1.234$.
2. $\Delta c_5 = 0.009$. But c_5 was 0.5, so therefore try new $c_5 = 0.509$.

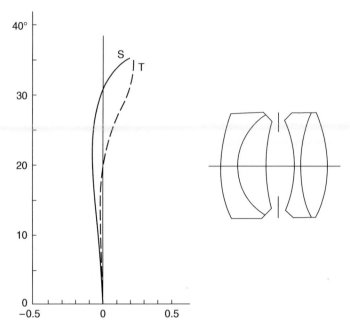

Figure 14.16   Astigmatism of Protar lens, Setup B.

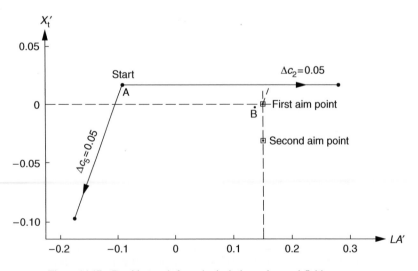

Figure 14.17   Double graph for spherical aberration and field curvature.

These changes give Setup B, the thicknesses and refractive indices remaining as before:

c	d
0.5	
	0.25
1.234	
	0.4
0.410355	
Stop	$\dfrac{0.2}{0.2}$
−0.637781	
	0.1
0.509	
	0.4
−0.579493	

with $f' = 10$, Ptz $= 0.025$, LA' $(f/8) = 0.1361$, LZA $(f/11) = -0.0724$; for $17.91°$: $X'_s = -0.0668$, $X'_t = -0.0023$.

Before making any further changes in LA' and X'_t, we must decide whether the aim point that we have chosen is the best. Certainly the zonal spherical aberration is about right, and so we will maintain our aim for spherical aberration at $+0.15$. However, to study the field requirements, it is necessary to trace several more principal rays at higher obliquities and plot the astigmatism curves. These rays give the tabulation shown in Table 14.7. A plot of these field curves (Figure 14.16) indicates at once that a much better aim point for the sag of the tangential field X'_t would be at -0.03, and this value will be used from now on (second aim point in Figure 14.17).

We next proceed to correct the coma and distortion. The OSC of Setup B is found to be -0.00399 at $f/8$, and both the coma and distortion are clearly excessive. Our free variables are now c_1 and the power of the front component. By using $y = 1$ for the paraxial ray, the power of the front component is given directly by u'_3, and any desired value of this angle can be obtained by solving for c_3. Assuming for a start that both the OSC and the distortion should be

Table 14.7

Astigmatism and Distortion for Setup B

Field angle at object (deg)	Field angle in stop (deg)	X'_s	X'_t	Distortion (%)
−35.00	−40	+0.17590	+0.24096	−2.52
−26.62	−30	−0.05159	+0.10503	−1.25
−17.91	−20	−0.06677	−0.00230	−0.51

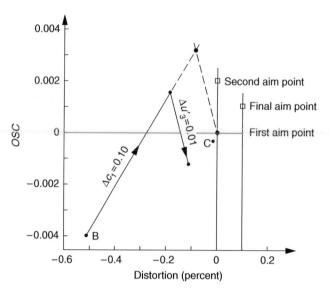

Figure 14.18 Double graph for coma and distortion.

zero, we plot a second double graph (Figure 14.18), changing c_1 and u_3'. This
graph indicates that we should make the following changes in the system:

1. $\Delta c_1 = 0.1227$. But c_1 is 0.5, so therefore we should try $c_1 = 0.6227$.
2. $\Delta u_3' = 0.0117$. But u_3' is zero, so therefore we should try $u_3' = 0.0117$.

With these changes our lens becomes Setup C:

c	d	n
0.6227		
	0.25	1.6135
1.234		
	0.4	1.5146
0.570553		
Stop	$\dfrac{0.2}{0.2}$	
−0.473548		
	0.1	1.5146
0.509		
	0.4	1.6135
−0.453187		

with $f' = 10$, Ptz = 0.025, power of front = $+0.0117$, OSC ($f/8$) = -0.000402,
distortion ($18°$) = -0.017%.

We will assume for the present that zero is a good aim point for the distortion, but we must investigate the coma further. To do this we trace a family of rays entering the lens at $-17.23°$, and plot a graph connecting the height of incidence of each ray at the stop against the height H' of the ray at the paraxial focal plane. This graph, Figure 14.19, indicates the presence of some negative primary coma with an upturn at both ends of the curve due to positive higher-order coma. Since the ends of the curve will probably be cut off by vignetting, it might be better to aim at, say, $+0.002$ of OSC at $f/8$ instead of zero. This will represent the aim point for OSC on future double graphs.

The spherical aberration of Setup C is $+0.0908$ and the field curvature is given by $X'_t = +0.1531$. Reference to the first double graph, using the new aim point, indicates these changes:

1. $\Delta c_2 = +0.026$. But $c_2 = 1.234$, so therefore we should try $c_2 = 1.260$.
2. $\Delta c_5 = +0.0916$. But $c_5 = 0.509$, so therefore we should try $c_5 = 0.6006$.

These changes give us Setup D:

c
0.6227
1.260
0.564736
−0.466835
0.6006
−0.435009

with $f' = 10$, Ptz $= 0.025$, front power $= 0.0117$, LA' ($f/8$) $= 0.1422$; for 17.24°: $X'_s = -0.0704$, $X'_t = -0.0331$.

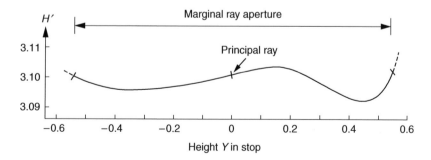

Figure 14.19 Meridional ray plot for Setup C (17°).

The spherical aberration and field curvature of this system are acceptable. However, we find that the OSC at $f/8$ has now become -0.00313 and the $17°$ distortion -0.009%. Reference to the second double graph enables us to remove these residuals, and we then return to the first graph for spherical aberration and field curvature, and so on back and forth several times until all four aberrations are acceptable. The final Setup is E:

c	d	n_e
0.6445		
	0.25	1.6135
1.2466		
	0.4	1.5146
0.628369		
Stop	$\dfrac{0.2}{0.2}$	
-0.383337		
	0.1	1.5146
0.5856		
	0.4	1.6135
-0.395628		

with $f' = 10$, Ptz $= 0.025$, power of front $= 0$, LA' $(f/8) = 0.1529$, LZA $(f/11) = -0.0487$, OSC $(f/8) = 0.00204$. The astigmatism and distortion values are as shown in Table 14.8.

These aberrations are plotted in Figure 14.20, as well as the $17°$ meridional ray plot for the study of coma. It is clear from these graphs that we should have aimed at about $+0.2\%$ of distortion at $17°$ and about $+0.001$ of OSC at $f/8$. These values should be adopted in any future changes. The field and spherical aberrations are just about right.

Table 14.8

Astigmatism and Distortion for Setup E

Field at object (deg)	Field in stop (deg)	X'_s	X'_t	Distortion (%)
-33.63	-40	$+0.08847$	-0.27048	-0.676
-29.62	-35	-0.02411	-0.07838	-0.311
-25.54	-30	-0.07381	-0.03227	-0.119
-21.39	-25	-0.08298	-0.03218	-0.028
-17.18	-20	-0.06921	-0.03188	$+0.007$

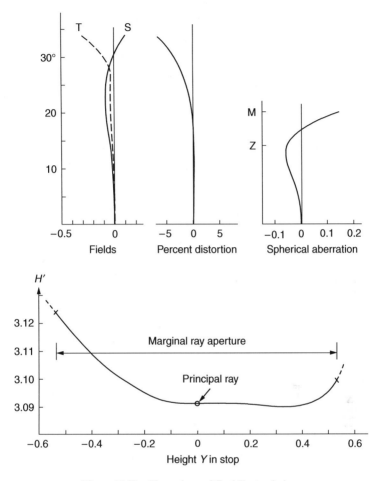

Figure 14.20 Aberrations of final Protar design.

Our next task is to investigate the correction of the two chromatic aberrations by choice of glass. We first attempt to correct the $(D - d)\, \Delta n$ sum of each component separately, since this is the proper thin-lens solution to the problem. For this we use the true $\Delta n = n_F - n_C$ of each likely glass, ignoring the fact that the catalog refractive indices are not quite equal to those we have assumed so far. This gives what is shown in Table 14.9.

No suitable glasses were found by which we could have reduced the negative $(D - d)\, \Delta n$ sum in the rear component.

Table 14.9

Axial Chromatic Error for Setup E

Lens	Glass	n_e	Δn	$D - d$ for $f/8$ ray	$(D - d) \Delta n$	Sum
1	SK-8	1.61377	0.01095	0.108126	0.00118398 ⎫	+0.00003299
2	K-1	1.51173	0.00824	−0.139683	−0.00115099 ⎭	
3	KF-8	1.51354	0.01004	0.151056	0.00151660 ⎫	−0.00012649
4	SK-3	1.61128	0.01034	−0.158906	−0.00164309 ⎭	
					Total	−0.00009350

To calculate the lateral color at 17.18°, we apply the same arithmetic error to the individual F and C indices as we have assumed for the e indices. This gives the numbers shown in Table 14.10.

From these data we find that $H'_F - H'_C = +0.001086$. Now the lateral color in any lens takes the same sign as the longitudinal color of the rear component and the opposite sign to the longitudinal color of the front component. Hence to improve both the longitudinal and lateral color aberrations simultaneously we must make the $(D - d)\,\Delta n$ sum of the front component more positive. To do this we need a glass in lens (1) having a lower V number or a glass in lens (2) having a higher V number. Inspection of the chart enclosed with the glass catalog indicates that the only possible choice is to use BK–1 in place of K–1 in lens (2), because all other possible glasses have a refractive index differing too much from the e indices we assumed for the aberration calculations. This glass has $\Delta n = 0.00805$ giving a value of $\sum (D - d)\,\Delta n$ equal to $+0.00005953$ in the front component, or -0.00006696 for the whole system, and a lateral color of $H'_F - H'_C = +0.000790$. We must accept these residuals in the absence of other more extreme glass types.

Of course, the final stage is to repeat the design using the true n_e refractive indices, and then to adjust the clear apertures to give the desired degree of vignetting.

Table 14.10

Lateral Chromatic Error for Setup E

Lens	Glass	Nominal indices			H'_C	H'_e	H'_F
		n_e	n_C	n_F			
1	SK-8	1.6135	1.60758	1.61853			
2	K-1	1.5146	1.51012	1.51836			
					3.090647	3.091227	3.091733
3	KF-8	1.5146	1.50920	1.51924			
4	SK-3	1.6135	1.60789	1.61823			

14.5 DESIGN OF A TESSAR LENS

The Tessar[14] resembles the Protar in that the rear component is a new-achromat cemented doublet, but the front component is now an air-spaced doublet rather than a cemented old-achromat. The cemented interface in the front component of the Protar was very strong, leading to a large zonal aberration, but the separated doublet in the Tessar gives so much less zonal aberration that an aperture of $f/4.5$ or higher is perfectly feasible. From another point of view, the Tessar can be regarded as a triplet with a strong collective interface in the rear element; this interface has a threefold function: It reduces the zonal aberration, it reduces the overcorrected oblique spherical aberration, and it brings the sagittal and tangential field curves closer together at intermediate field angles. Although sometimes it is mentioned that the Tessar was derived from the Cooke Triplet lens (Section 14.6) invented by Taylor, its actual genesis is the Protar being that Rudolph invented both the Protar (1890) and Tessar (1902) as explained in the Tessar patent specification.

14.5.1 Choice of Glass

It is customary to use dense barium crown for the first and fourth elements, a medium flint for the second, and a light flint for the third element. Possible starting values are therefore as shown in Table 14.11.

Table 14.11

Initial Selection of Glasses for Tessar Design

Lens	Type	n_e	$\Delta n = n_F - n_C$	$V_e = (n_e - 1)/\Delta n$
a	SK-3	1.61128	0.01034	59.12
b	LF-1	1.57628	0.01343	42.91
c	KF-8	1.51354	0.01004	51.15
d	SK-3	1.61128	0.01034	59.12

14.5.2 Available Degrees of Freedom

Because of the importance of the cemented interface in the rear component, it is best to establish it at some particular value, say 0.45, and leave it there throughout the design. Since there is no symmetry to help us, we must correct every one of the seven aberrations, and also hold the focal length, by a suitable choice of the available degrees of freedom; this makes the design decidedly laborious, especially if it is performed by hand on a pocket calculator or even by a computer program.

In the front component we have the two powers, two bendings, and one air space. The second air space is held constant to reduce vignetting, while in the rear component we have only the two outer surface curvatures to be determined. We thus have seven degrees of freedom with which to correct six aberrations and hold the focal length. We must therefore use choice of glass to correct the seventh aberration.

Many possible ways of utilizing the various degrees of freedom could be tried. In this chapter we shall assign the available freedoms in the following way:

1. The power of lens (a) and the dispersion of the glass in lens (d) will be used to control the two chromatic aberrations.
2. The power of lens (b) will be solved to maintain the power of the front component at, say, -0.05 (a focal length of -20) for distortion correction.
3. The curvature of the last surface, c_7, will be solved to make the overall focal length equal to 10.
4. The front air space will in all cases be adjusted to make the Petzval sum equal to, say, 0.025.
5. The spherical aberration will be corrected in all cases by a suitable choice of c_5.
6. This leaves the bendings of lenses (a) and (b) to be used to correct the OSC and the tangential field curvature X_t'.

Our starting System A will be arbitrarily set as follows:

	c	d	n_e
	0.4		
		0.40	1.61128
	0		
		0.3518 (Ptz)	(air)
	-0.2		
		0.18	1.57628
(u_4')	0.406891		
	Stop	$\dfrac{0.37}{0.13}$	(air)
	-0.05		
		0.18	1.51354
	0.45		
		0.62	1.61128
(u_7')	-0.247928		

with $f' = 10$, Ptz $= 0.025$.

Table 14.12

Setup A Chromatic Aberration Contributions

Element	(*a*)	(*b*)	(*c*)	(*d*)
$(D - d) \Delta n$	−0.00266942	0.00404225	0.00270365	−0.00387980
	0.00137283		−0.00117615	
			$\sum = 0.00019668$	

14.5.3 Chromatic Correction

Assuming that this is a reasonable starting system, we next trace an $f/4.5$ marginal ray in e light and find the $(D - d)$ Δn contribution of each lens element, as shown in Table 14.12. We could, of course, adjust the two components to make both totals separately zero, but it is then found that the lateral color $H'_F - H'_C$, calculated by tracing principal rays at 17°, is strongly positive. Since lateral color takes the same sign as the longitudinal color of the rear component, we must have a considerable amount of negative $D - d$ sum in the rear and an equal positive sum in the front component.

We will therefore try to increase the negative sum in the rear component by choosing a glass for element (*d*) with a higher dispersive power, that is, a lower V number. Such a glass is SK-8 with $n_e = 1.61377$, $\Delta n = n_F - n_C = 0.01095$, and $V_e = 56.05$. The slight alteration in refractive index requires a small adjustment of the system, giving Setup B:

c	d	n_e
0.4		
	0.4	1.61128
0		
	0.3421	(air)
−0.2		
	0.18	1.57628
0.4051605		
Stop	$\dfrac{0.37}{0.13}$	(air)
−0.05		
	0.18	1.51354
0.45		
	0.62	1.61377
−0.2444831		

with $f' = 10$, $l' = 8.851896$, Ptz $= 0.025$. An $f/4.5$ marginal ray gives $LA' = +0.30981$ and the $D - d$ values shown in Table 14.13.

This sum is quite acceptable so far as longitudinal chromatic aberration is concerned. We must next check for lateral color. Tracing principal rays at $17°$ in F and C light tells us that the lateral color is $+0.000179$, which is also acceptable, so that now both chromatic aberrations are under control. Fortunately chromatic errors change so slowly with bendings that our future efforts at correcting spherical aberration, coma, and field curvature by bending the three components do not greatly affect the chromatic corrections.

Table 14.13

Setup B Chromatic Aberration Contributions

Element	(a)	(b)	(c)	(d)
$(D - d)\,\Delta n$	−0.00266942	0.00405667	0.00272259	−0.00411584
		0.00138725		−0.00139325
				$\Sigma = -0.00000600$

14.5.4 Spherical Correction

Because elements (a) and (b) are working at about the minimum aberration positions, we cannot hope to correct spherical aberration by bending them. Thus we are obliged to control spherical aberration by bending the rear component, that is, by changing c_5. This will be done by a series of trials at every Setup from now on.

We arbitrarily require LA' to be about $+0.098$; this will yield a zonal residual of about half that amount, giving excellent definition when the lens is stopped down, as it will almost always be in regular use. We then vary c_1 and c_3 to correct the OSC and X'_t by means of a double graph (Figure 14.21). The aim point will be at zero for both these aberrations.

Correcting the spherical aberration of Setup B by adjusting c_5 gives Setup C, shown in the following table:

c	d	n_e
0.4		
	0.4	1.61128
0		
	0.2949	(air)
−0.2		
	0.18	1.57628
0.3968783		
Stop	$\dfrac{0.37}{0.13}$	(air)
−0.080		
	0.18	1.51354
0.45		
	0.62	1.61377
−0.2632877		

with $f' = 10$, $l' = 9.035900$, Ptz = 0.02499, $LA' = 0.09724$, $OSC = -0.01778$. A principal ray traced at 20° through the stop emerges from the front of the lens at 17.3070°, with the following fields:

Angle: 17.31°, X'_s: 0.0716 X'_t: 0.4337, distortion: 0.213%

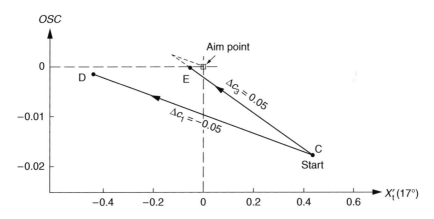

Figure 14.21 This double graph shows the effects of bending the first two elements of a Tessar at 17° field angle. (In each case the Petzval sum and spherical aberration have been corrected first.)

The distortion is negligible, showing that the choice we made of $u'_4 = -0.05$ is about right, and we will continue with that value in what follows. The negative OSC is, however, much too large, and the tangential field is much too backward-curving.

14.5.5 Correction of Coma and Field

To plot a double graph, we make a trial change of $\Delta c_1 = -0.05$ from System C and then restore everything to its original value (i.e., Setup D). We then return to Setup C and now change c_3 by 0.05, which gives Setup E. These changes are shown in Figure 14.21. Following the usual procedure with a double graph, and making several small adjustments, we finally come up with Setup F:

c	d	n_e
0.4126		
	0.40	1.61128
0.013442		
	0.2927	(air)
−0.1366		
	0.18	1.57628
0.464462		
Stop	$\dfrac{0.37}{0.13}$	(air)
−0.0571		
	0.18	1.51354
0.45		
	0.62	1.61377
−0.247746		

with $f' = 10$, $l' = 8.9344$, LA' ($f/4.5$) = 0.0958, LZA ($f/6.4$) = −0.0258, OSC ($f/4.5$) = 0, Ptz = 0.0250. The results are shown in Table 14.14.

The aberration graphs are shown plotted in Figure 14.22. As a check on the coma we next trace a number of oblique rays entering parallel to the principal ray at 17.19° and draw a meridional ray plot (Figure 14.23). It will be seen that the two ends of this graph sag somewhat, but the middle part of the curve is straight. This is an indication of the presence of negative higher-order coma, and it cannot be usefully corrected by the deliberate introduction of positive OSC. A much better method of removing it is to introduce some vignetting. If we limit the clear aperture of each surface to the diameter of the entering $f/4.5$ axial beam, we shall cut off the ends of the ray plot in Figure 14.23 to the marks

Table 14.14

Astigmatism and Distortion for Setup F

Field angle (deg)	X'_s	X'_t	Distortion (%)
29.74	0.1607	0.1303	−1.42
25.61	0.0102	0.0871	−0.92
21.42	−0.0458	0.0305	−0.56
17.19	−0.0537	−0.0020	−0.32

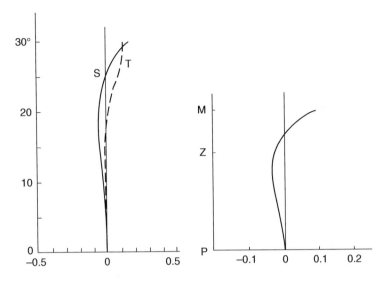

Figure 14.22 Aberrations of Setup F.

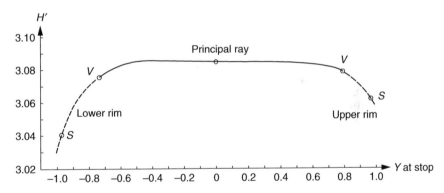

Figure 14.23 Meridional ray plot of Tessar Setup F. Rays *SS* are through top and bottom of the stop. Rays *VV* represent vignetted limiting rays.

VV shown, and we shall thus remove almost the entire higher-order coma without seriously reducing the image illumination. Figure 14.24 shows the lens apertures so reduced and the path of the limiting oblique rays *VV*.

The astigmatic fields shown in Figure 14.22 cross rather too high and the field is a little backward-curving. We shall therefore return to the double graph of Figure 14.21 and establish a new aim point at $OSC = 0$ and $X'_t = -0.04$, which is by chance very close to Setup E. After making several small adjustments in c_1 and c_3, and of course correcting the spherical aberration each time by c_5 and the Petzval sum by d'_2, we arrive at the following solution G:

c	d	n
0.4065		
	0.40	1.61128
0.0069273		
	0.3019	(air)
−0.1421		
	0.18	1.57628
0.4596089		
Stop	$\dfrac{0.37}{0.13}$	(air)
−0.0579		
	0.18	1.51354
0.45		
	0.62	1.61377
−0.2486575		

with $f' = 10$, $l' = 8.925977$, Ptz $= 0.025$, LA' $(f/4.5) = 0.1029$, LZA $(f/6.4)$ $= -0.0216$, OSC $(f/4.5) = 0$, $\sum (D - d)$ $\Delta n = -0.00001096$, lateral color $H'_F - H'_C$ $(17°) = -0.00031$. The results are shown in Table 14.15. The fields and aberration are shown plotted in Figure 14.25.

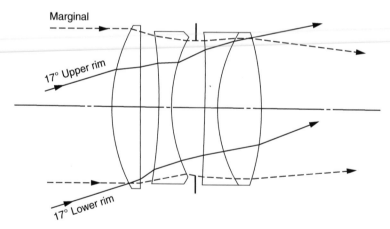

Figure 14.24   Vignetting in Setup F, 17° beam.

Table 14.15

Astigmatism and Distortion for Setup G

Field (deg)	X'_s	X'_t	Distortion (%)
29.64	0.1224	−0.0283	−1.18
25.55	−0.0148	−0.0064	−0.77
21.38	−0.0619	−0.0257	−0.47
17.16	−0.0635	−0.0430	−0.27

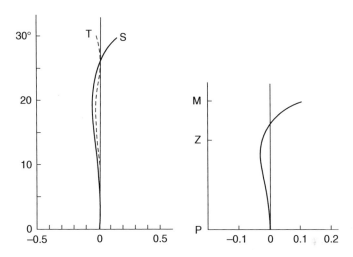

Figure 14.25 Aberrations of Tessar Setup G.

14.5.6 Final Steps

We must now study the effect of changing the cemented interface c_6. This was arbitrarily set at 0.45, and we will next repeat the entire design with $c_6 = 0.325$. The resulting lens is decidedly different from the previous design, as shown in the following table:

c	d	n
0.328		
	0.4	1.61128
−0.0757715		
	0.347	(air)
−0.24		
	0.18	1.57628
0.3564288		
Stop	0.37 / 0.13	(air)
−0.135		
	0.18	1.51354
0.325		
	0.62	1.61377
−0.3216593		

with $f' = 10$, $l' = 9.20712$, LA' ($f/4.5$) $= 0.08714$, LZA ($f/6.4$) $= -0.03475$, OSC ($f/4.5$) $= 0$, $\sum (D - d) \Delta n = -0.0000707$, lateral color ($17°$) $= -0.00121$. The results are shown in Table 14.16.

These aberrations are shown in Figure 14.26. The field is a little narrower than before but quite satisfactory. It should be noted that both of the color aberrations

Table 14.16

Astigmatism and Distortion for Second Tessar System

Field (deg)	X'_s	X'_t	Distortion (%)
25.41	0.0408	−0.0905	+0.12
21.38	−0.0244	0.0198	−0.04
17.22	−0.0413	0.0157	−0.06

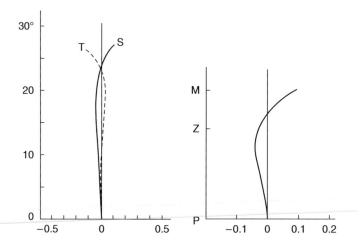

Figure 14.26 Aberrations of second Tessar system.

are negative; to rectify this requires a small increase in the V number of the glass used for the rear crown element, say to SK-1, which has $n_e = 1.61282$ and $V_e = 56.74$, or SK-19 with $n_e = 1.61597$ and $V_e = 57.51$. The lens designer should always be mindful of the impact glass choice can have on a design.

The chief matter requiring study is the meridional ray plot in Figure 14.27, which should be compared with the previous graph in Figure 14.23. It is immediately clear that the change from $c_6 = 0.325$ to 0.45 has had the effect of raising the lower end of the curve and depressing the upper end. That is, strengthening c_6 has introduced some undercorrected oblique spherical aberration to the existing negative higher-order coma, with an improvement in the overall quality of the lens. The lower end of the curve needs cutting off more than the upper end, but obviously we cannot cut it back beyond the marginal ray aperture.

The best way to improve this Tessar is to raise the refractive indices, preferably above 1.6 in all elements. It is doubtful if changing the thicknesses would have any significant effect.

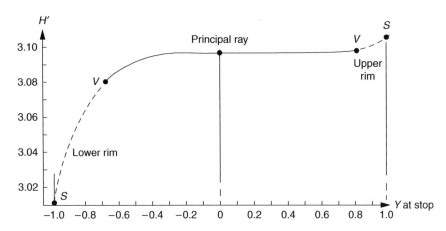

Figure 14.27   Meridional ray plot for Tessar system with $c_6 = 0.325$ (17°).

14.6 THE COOKE TRIPLET LENS

The English designer H. Dennis Taylor was led to this design[15] in 1893 by the simple consideration that if an objective was to consist of a positive lens and a negative lens of equal power and the same refractive index, the Petzval sum would be zero, and the system could be given any desired power by a suitable separation between the lenses. However, he quickly realized that the extreme asymmetry of this arrangement would lead to an intolerable amount of lateral color and distortion, and so he split the positive element into two and mounted the negative element between them, thus making his famous triplet objective (Figure 14.28). He also tried the alternative arrangement of dividing the negative element into two with the positive lens between, but this is much less favorable than the classic arrangement.

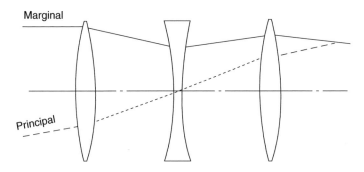

Figure 14.28   The Cooke triplet lens.

The triplet objective is tricky to design because a change in any surface affects every aberration, and the design would be impossibly difficult without a preliminary thin-lens predesign using Seidel aberrations. We assign definite required residuals for each primary aberration, and then by ray tracing determine the actual aberrations of the completed thick-lens system. If any aberration is excessive, we adopt a different value for that primary aberration and repeat the entire predesign. The thin-lens residuals used in the following example are the result of experience with prior designs that result in the final thick system being satisfactory. Of course, in making a design differing from this in any important respect such as aperture, field, or glass selection, we would require a different set of Seidel aberration residuals, which would have to be found by trial.

14.6.1 The Thin-Lens Predesign of the Powers and Separations

If we place the stop at the negative thin element inside the system, we can solve for the powers and separations of the three elements to yield specified values of the overall focal length and primary chromatic aberration, primary lateral color, Petzval sum, and one other condition that will eventually be used for distortion control. This last requirement might be the ratio of the two separations, the ratio of the powers of the outside elements, the ratio of the power of the combination of elements a and b to the power of the system, or some other similar criterion. We thus have five variables (three powers and two separations) with which to solve five conditions, after which we shall have three bendings to correct for the three remaining aberrations: spherical, coma, and astigmatism. Without this convenient division of the aberrations into two groups, those depending only on powers and separations and those depending also on bendings, the entire design process would be hopelessly complicated and almost impossible to accomplish.

The first part of the thin-lens predesign can be performed in several ways, the one employed here having been introduced by K. Schwarzschild around 1904. It uses the formulas for the contributions of a thin element to power, chromatic aberration, and Petzval sum, given in Section 11.7.2. These contributions may be written for each aberration in turn, as follows:

$$(y_a)\phi_a + (y_b)\phi_b + (y_c)\phi_c = (u_0' - u_a) = y_a\Phi \quad \text{if } u_a = 0 \quad \text{(power)}$$

$$(y_a^2/V_a)\phi_a + (y_b^2/V_b)\phi_b + (y_c^2/V_c)\phi_c = -L_{ch}'u_0'^2 \quad \text{(chromatic)}$$

$$(1/n_a)\phi_a + (1/n_b)\phi_b + (1/n_c)\phi_c = \text{Ptz} \quad \text{(Petzval)}$$

These three equations are linear in the three lens powers, and they can be easily solved for the powers once we know the three axial-ray heights y_a, y_b, and y_c. The first of these, y_a, is known when the focal length and f-number are known, but y_b and y_c must be found by trial to satisfy the remaining two conditions, namely, the correction of lateral color and the ratio of the two separations $S_1/S_2 = K$. Reasonable starting values of the other ray heights are $y_b = 0.8y_a$ and $y_c = 0.9y_a$.

As an example, we will proceed to design an objective of focal length 10.0 and aperture $f/4.5$ covering a field of $\pm 20°$. We shall assume that $K = 1$, and use the following types of glass:

$$\text{(a, c) SK-16, } n_D = 1.62031, \ n_F - n_C = 0.01029, \quad V = 60.28$$

$$\text{(b) F-4, } n_D = 1.61644, \ n_F - n_C = 0.01684, \quad V = 36.61$$

In our predesign we shall aim at the following set of thin-lens residuals, hoping that these will give a well-corrected system after suitable thicknesses have been inserted:

$f' = 10$	Petzval sum $= 0.035$
$y_a = 1.111111$	chromatic aberration $= -0.02$
$u_a = 0$	lateral color $= 0$
$u_0' = 0.111111$	spherical aberration $= -0.08$
$u_{pr,a} = -0.364(\tan 20°)$	coma$_s' = +0.0025$
$K = S_1/S_2 = 1.0$	ast$_s' = -0.09$

with $y_a = 1.111111$, $y_b = 0.888888$, and $y_c = 0.999999$. Solving the three Schwarzschild equations for the three powers gives

$$\phi_a = 0.192227, \quad \phi_b = -0.291104, \quad \phi_c = 0.156285$$

The paraxial ray and the paraxial principal ray passing through the middle of the negative lens have the values shown in Table 14.17. Inspection of this table shows that, for the paraxial ray,

$$u_a = 0, \quad u_b = u_a + y_a\phi_a, \quad u_c = u_b + y_b\phi_b$$
$$S_1 = (y_a - y_b)/u_b, \quad S_2 = (y_b - y_c)/u_c$$

Table 14.17

Paraxial Ray Traces for Cooke Triplet Predesign

ϕ		ϕ_a		ϕ_b		ϕ_c	
$-d$			$-S_1$		$-S_2$		
		Paraxial ray					
y		y_a		y_b		y_c	
u	u_a		u_b		u_c		u_0'
		Paraxial principal ray					
y_{pr}		y_{pra}		$y_{prb} = 0$		y_{prc}	
u_{pr}	u_{pra}		u_{prb}		u_{prb}		

Substituting the numerical values of our example gives

$u_a = 0$, $u_b = 0.2135856$, $u_c = -0.0451736$
$S_1 = 1.040436$, $S_2 = 2.459647$

where $K = S_1/S_2 = 0.423002$. Now it is found that K varies almost linearly with y_b, and a couple of trials tells us that $\partial K/\partial y_b = -46.0$. Thus retaining the previous $y_a = 1.111111$ and $y_c = 0.999999$, we find that with $y_b = 0.876380$ we have

$\phi_a = 0.153234$, $\phi_b = -0.296588$, $\phi_c = 0.200775$
$u_b = 0.1702602$, $u_c = -0.0896636$
$S_1 = 1.378661$, $S_2 = 1.378709$, $K = 0.999965$

This is virtually perfect, so we return to the thin-lens ray-trace table and we see that for the paraxial principal ray

$$y_{pra} = \frac{S_1 u_{pra}}{1 - S_1 \phi_a} = -0.636244$$

$$y_{prb} = 0, \; y_{prc} = -y_{pra}/K = +0.636266$$

We can now determine the contribution of each element to the lateral color by the relation

$$T_{ch}C = -yy_{pr}\phi/Vu'_0$$

where

$$T_{ch}C_a = 0.0161736, \quad T_{ch}C_b = 0, \quad T_{ch}C_c = -0.0190729$$

with the total lateral color $= -0.002899$. To correct this, we must change y_c and repeat the whole process.

Omitting all the intermediate steps, we come to the final solution:

$y_a = 1.111111$, $y_b = 0.861555$, $y_c = 0.962510$
$\phi_a = 0.1684127$, $\phi_b = -0.3050578$, $\phi_c = 0.1940862$
$u_b = 0.1871252$, $u_c = -0.0756989$
$S_1 = 1.333632$, $S_2 = 1.333639$, $K = 0.999995$

With $u_{pra} = -0.364$, we find

$$y_{pra} = -0.6260542, \quad y_{prb} = 0, \quad y_{prc} = 0.6260573$$

where

$$T_{ch}C_a = 0.0174910, \quad T_{ch}C_b = 0, \quad T_{ch}C_c = -0.0174616$$

Hence the thin-lens lateral color is $+0.0000294$, which is acceptable.

14.6.2 The Thin-Lens Predesign of the Bendings

The bendings of our three thin-lens elements are defined by c_1, c_3, and c_5, respectively. *Since the stop is assumed to be in contact with lens (b), the astigmatism contribution of that element is independent of its bending.* Our procedure, therefore, is to adopt some arbitrary bending of lens (a) and ascertain its AC^* by the formulas given in Section 11.7.2. We find the AC of lens (b) by $-\frac{1}{2}h_\theta'^2 - \phi_b$ and then solve for the bending of lens (c) that will make the total astigmatism contribution equal to the specified value of -0.09. Having done this, we go to lens (b) and bend it to give the desired value of the sagittal coma, namely, 0.0025. This will not affect the astigmatism in any way. Finally, knowing the bendings of lenses (b) and (c), we can calculate the spherical contributions of all three lens elements, and plot a point on a graph connecting the spherical aberration with the value of c_1. Repeating this process several times with different values of c_1 will enable us to complete the graph and pick off the final solution for any desired value of the thin-lens primary spherical aberration.

The contributions of the thin-lens elements to the three aberrations are given by the formulas in Section 11.7.2 involving the G sums for spherical aberration and coma. These contributions are quadratics in terms of the bending parameters c_1, c_3, and c_5 as follows:

Lens (a)
$$SC^* = -23.227833c_1^2 + 11.968981c_1 - 2.011823$$
$$CC^* = 1.454188c_1^2 - 1.361274c_1 + 0.292417$$
$$AC^* = -7.374247c_1^2 + 10.006298c_1 - 3.442718$$

Lens (b)
$$SC^* = 15.229687c_3^2 + 4.686647c_3 + 1.436793$$
$$CC^* = 0.667069c_3 + 0.095270$$
$$AC^* = 2.020947$$

Lens (c)
$$SC^* = -15.073642c_5^2 + 5.519113c_5 - 0.937665$$
$$CC^* = -1.089393c_5^2 - 0.130340c_5 + 0.030780$$
$$AC^* = -6.377286c_5^2 - 3.861014c_5 - 0.528703$$

Collecting these expressions, we find that with a given c_1, we first solve for c_5 by the quadratic expression

$$c_5^2 + 0.6054322c_5 + (1.15633c_1^2 - 1.569053c_1 + 0.2917344) = 0$$

Only one of the two solutions is useful; the other represents a freakish lens bent drastically to the left that would exhibit huge zonal residuals; that is, it looks odd and is quite strained (see Endnote 7).

Knowing c_1 and c_5, we can solve for c_3 for coma correction by

$$c_3 = -2.179967c_1^2 + 2.040679c_1 + 1.633104c_5^2 + 0.1953917c_5 - 0.6235753$$

Finally, knowing all three parameters c_1, c_3, and c_5, we can calculate the spherical aberration by

$$LA' = -23.227833c_1^2 + 11.968981c_1 + 15.229687c_3^2$$
$$+ 4.686647c_3 - 15.073642c_5^2 + 5.519113c_5 - 1.512695$$

Taking a series of values for c_1 we find what is shown in Table 14.18. Thus all three lenses are bending to the right together. The spherical sums are plotted on a graph (Figure 14.29), from which we can pick off the desired c_1 values for our residual of –0.08. There are obviously two solutions, namely,

$$c_1 = 0.2314 \quad \text{and} \quad c_1 = 0.3780$$

Table 14.18

Primary Spherical Aberration versus Bendings of a Triplet Lens

c_1	c_3	c_5	Primary spherical aberration
0.2	−0.308020	−0.042985	−0.311751
0.25	−0.238049	0.043543	−0.013077
0.3	−0.168838	0.105388	0.044583
0.35	−0.100863	0.152719	0.004574
0.4	−0.060233	0.189738	−0.164065

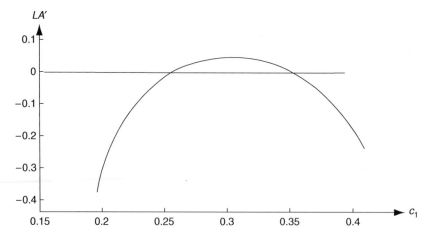

Figure 14.29 Relation between c_1 and primary spherical aberration, after correcting field by c_5 and coma by c_3.

We shall follow up only the left-hand solution since the right-hand solution has more steeply curved surfaces and is likely to exhibit larger zonal residuals. For the left-hand solution, then, we have the following thin-lens curvatures:

$$c_1 = 0.2314 \qquad c_3 = -0.264746 \quad c_5 = 0.015190$$
$$c_2 = -0.040098 \quad c_4 = 0.230124 \quad c_6 = -0.297695$$

14.6.3 Calculation of Real Aberrations

After selecting suitable thicknesses from a scale drawing, scaling the lenses up or down to restore their exact thin-lens powers, and calculating the air spaces to maintain the thin-lens separations between adjacent principal points, we obtain the following thick-lens system:

c	d	n_D
0.2326236		
	0.4	1.62031
−0.04031		
	1.051018	
−0.2617092		
	0.25	1.61644
0.227485		
	0.986946	
0.0152285		
	0.45	1.62031
−0.2984403		

with $f' = 10.00$, $l' = 8.649082$, LA' $(f/4.5) = 0.01267$, LZA $(f/6.3) = -0.01051$, OSC $(f/4.5) = -0.001302$, Ptz $= 0.03801$ (see Table 14.19).

The lateral color correction is evidently about right. Figure 14.30 shows the plotted spherochromatism graphs, which show that both the spherical and chromatic aberrations are also about right. The astigmatic fields are also plotted in

Table 14.19

Astigmatism, Distortion, and Lateral Color for Final Triplet Lens

Field (deg)	X'_s	X'_t	Distortion (%)	Lateral color
24	−0.0386	−0.4338	1.98	0.00195
20	−0.0639	−0.0798	1.09	0.00055
14	−0.0488	+0.0192	0.42	−0.00021

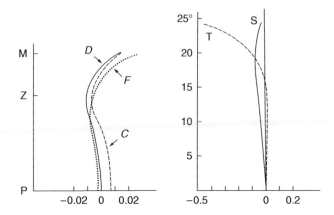

Figure 14.30   Aberrations of final triplet lens (*f*/4.5).

Figure 14.30, where it can be seen that a slight change in the thin-lens sagittal astigmatism in the positive direction might be an improvement.

As far as coma is concerned, we have plotted the graphs of H' against Q_1 for sets of oblique rays at three different obliquities in Figure 14.31. The points VV on each graph represent the limiting vignetted rays, which enter and leave the lens at the initial marginal aperture height of 1.1111, assuming that the front and rear apertures of the lens are limited to that value (Figure 14.32). When the obliquity is increased the vignetting becomes accentuated, and the graphs become shorter at the upper (right-hand) ends. The principal ray along which the astigmatism was calculated is indicated in each case. The slope of the graph at the principal-ray point is, of course, an indication of the X'_t for that obliquity. To improve the distortion, the design could be repeated with a different value of K, say, 0.9 or 1.1.

On the whole, this seems to be a pretty good design, typical of many triplets using these common types of glass. For a good lens at a higher aperture such as *f*/2.8, for example, it would be highly desirable to use glasses with much higher refractive indices, such as a lanthanum crown and a dense flint. A search through the patent files will reveal many triplet designs for use at various apertures and angular fields.

14.6.4 Triplet Lens Improvements

As we have mentioned, Dennis Taylor created the Cooke triplet lens over a hundred years ago and you may wonder why we have spent such effort discussing this lens in this chapter.[16] The reason is that this lens type continues to be of interest for use in various new systems such as low-cost cameras, printers, copiers, and rifle scopes. As odd as it may seem that such a simple lens

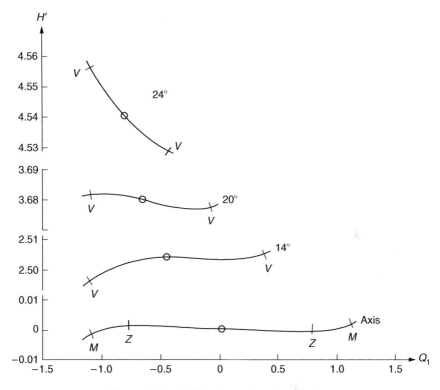

Figure 14.31 Meridional ray plots of triplet lens.

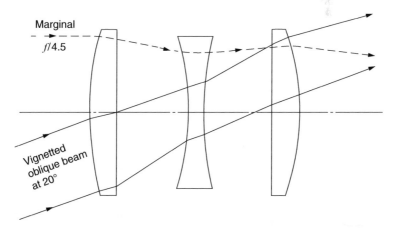

Figure 14.32 Final triplet design, showing $f/4.5$ marginal ray and limiting rays of vignetted $20°$ beam.

configuration continues to receive interest by lens designers, it has because clever lens designers have been able to find solutions that provide better performance and lower manufacturing costs that satisfy certain product requirements. The cost of a lens system includes at least the following considerations:

- Number of elements
- Diameter of elements
- Volume of elements
- Cemented elements
- Tolerance of radii, thicknesses, decenter, tilt, wedge, and so on
- Surface figure and quality
- Cost of glasses (higher index typically costs more; higher production volume glasses typically cost less)
- Mechanical mounting complexity
- Focusing mechanism complexity
- Coatings

Clearly the lens designer is required to consider far more than just the optical configuration and the performance of the design. It is often said that the optical design of the lens is less than half of the lens designer's work in completing a project.

In 1962, Hopkins published a systematic study of a region of triplet solutions in the midst of the infinite number of third-order solutions.[17] His analysis included both third-order and fifth-order aberrations. Hopkins made a number of observations, but perhaps the most significant was that he found that raising the index of the lenses was productive.

Independently in the mid-1950s, Baur and Otzen improved on the triplet photographic objective lens.[18] They pointed out that the known triplet photographic objectives used the highest possible refractive index material for the two positive elements and the lowest possible refractive index material for the negative element in an effort to obtain a low Petzval sum; therefore the field was flattened to achieve a large and useful image area. Their invention was finding structures that provided improved performance where the new lens has greater radii than prior art triplet lenses, requires less glass, and is more economic to manufacture. They found that the refractive index for all three lenses should be in the range of 1.72 to 1.79 for the D spectral line and that the Abbe number should be about 45 for the positive elements and about 28 for the negative element. More particularly, the arithmetic mean of the Abbe numbers of the three elements must obey the relationship,

$$36 < \frac{V_1 + V_2 + V_3}{3} < 41.$$

Baur and Otzen teach in their patent the other relationships that are required to achieve the performance they claim. Their example lenses are $f/2.8$ with field coverage of $26°$ half-angle.

About a decade later, Kingslake[19] disclosed a means to produce a triplet covering a wide field that was remarkably different than that of Baur and Otzen. Ackroyd and Price[20] cofiled a patent application also on wide-angle triplets. Interestingly, their patents were sequentially issued. Kingslake's goals were to markedly increase the useful field-of-view to at least 34° half field-angle with less vignetting than any prior art, to use lower-cost glasses, reduce the size of elements, and minimize the overall length of the lens.

Up to this time, the maximum half field-angle achieved was about 28° and the lens tended toward a *reverse telephoto lens*, which means that the focal length is shorter than the total length from the front lens vertex to the focal plane. He was able to reduce refractive indices used to about 1.61 and Abbe numbers to about 59 for the positive elements and 38 for the negative element. An $f/6.3$ lens having good performance was designed with 35° half field-angle and about 0.30 vignetting. The diameter of the front, middle, and rear elements are greater than 30%, 15%, and 18% of the focal length, respectively. Ackroyd and Price, who also worked at Kodak with Kingslake, improved on Kingslake's design by finding rules for the lens structure that allowed a 34° half field-angle with 0.58 vignetting at $f/6.3$; thus, the diameter of the front, middle, and rear elements are greater than 24%, 16%, and 23% of the focal length, respectively. However, satisfactory results were claimed if diameters were kept above 20%, 13%, and 20%, respectively.

These lens designers made a dramatic leap forward in developing a smaller, lower-cost, and wide-field objective lens suitable for volume production. In both patents, specific and detailed guidance is provided to teach the design procedure for a lens designer to follow. Figure 14.33 illustrates an example lens from their patents. The ray fans in Figure 14.34 show acceptable spherical aberration and chromatic correction on-axis (0°) with some negative coma and slight sagittal astigmatism appearing at 12°. By 24°, oblique spherical aberration is dominating in the meridional plane and sagittal astigmatism continues to increase. A small amount of vignetting of the lower rays can be seen. At the edge of the field (34°), vignetting of both upper and lower rays is evident and necessary to limit the image degradation of the quite strong tangential astigmatism. The sagittal astigmatism has also grown and switched from undercorrected to overcorrected.

The astigmatic field curves in Figure 14.35a show this behavior (see page 432). The tangential astigmatism has been reasonably well controlled to be relatively flat out to about 86% of the field, with the sagittal somewhat less so. At about 29°, the sagittal and tangential curves intersect and then rapidly separate, with the tangential astigmatism becoming more undercorrected and the sagittal astigmatism more overcorrected. This is also another example of using the higher-order astigmatic aberrations to balance against the lower-order terms to achieve a wider field. An estimate of this intersection height is given by Eq. (13-8). The distortion is shown in Figure 14.35b and the 0.25% distortion is quite acceptable for intended application of this lens.

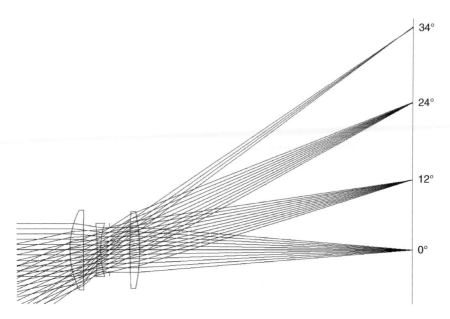

Figure 14.33 Wide-angle triplet lens.

In the early 1990s, Hiroyuki Hirano investigated using the triplet for a low-cost, broad zoom-range lens for a copying system rather than the more typical four-element symmetrical lenses.[21] At that time, available copying lenses had a zooming range of about 0.6X to 1.4X with an *f*-number of 5.6 and total field coverage of about 40°. The new triplet copying lens is shown in Figure 14.36 and is the third of the five examples contained in the patent (see page 433). This lens has a 100-mm focal length, an *f*-number of 6.7, and total field coverage of 46°. The structure is as follows:

r	d	n_d	V_d
23.662			
	8.418	1.58913	61.2
36.430			
	2.003		
−30.805			
	1.033	1.60342	38.0
30.370			
	1.686		
55.410			
	3.238	1.69350	53.2
−29.366			

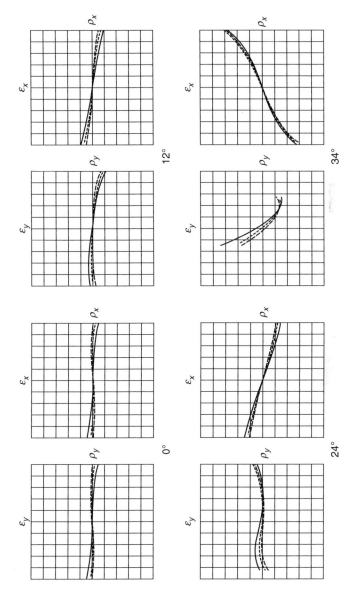

Figure 14.34 Ray fans for wide-angle triplet lens with $f = 100$ and $f/6.3$. Ordinate is ± 0.5 lens units. F light is solid curve, d light is short dashed curve, and C light is long dashed curve.

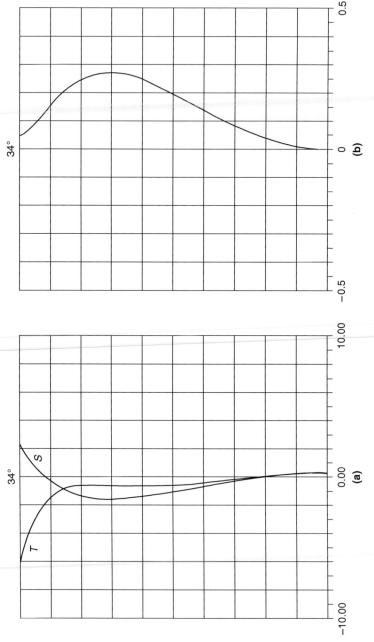

Figure 14.35 Field curves (a) and distortion (b) for wide-angle triplet lens with $f = 100$ and $f/6.3$. Field curves are in lens units and distortion is in percent.

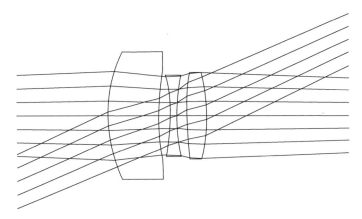

Figure 14.36 Triplet copying lens with $f = 100$, $f/6.7$, and $46°$ total field coverage.

Table 14.20

Alternative Glass Selections for Triplet Copying Lens

Example element	n_e	V_e
1-1	1.51633	64.1
1-2	1.58144	40.7
1-3	1.69100	54.8
2-1	1.49136	57.8
2-2	1.63980	34.5
2-3	1.74400	44.8
3-1	1.58913	61.2
3-2	1.60342	38.0
3-3	1.69350	53.2
4-1	1.58913	61.2
4-2	1.58144	40.7
4-3	1.67790	55.3
5-1	1.51633	64.1
5-2	1.60717	40.3
5-3	1.72916	54.7

An important innovation Hirano made was to use lower refractive index glasses and the minimum number of elements feasible to solve the design goals. Notice that different crown glasses were used for the two positive elements and also it is seen that the refractive index of each of the three lenses is remarkably different, in contrast to typical triplet lenses. The e spectral line refractive index and the Abbe number for each glass used in the five examples are listed in Table 14.20.

Hirano found that if the condition $0.08 < n_3 - n_1$ is met, then additional performance improvement can be realized. In the five examples, $n_3 - n_1$ is observed to range from 0.09 to 0.25. Note also the significant thickness of the first element, which is utilized to control Petzval and to achieve useful distribution of surface powers for aberration control.

Figure 14.37 presents the longitudinal spherical aberration. It is evident that the lens is adequately corrected for spherical aberration and is achromatic since the e and F curves intersect at the 0.7 zone. The ray fans in Figure 14.38 show the existence of high-order aberrations that are controlled nicely by the lens designer. Some tangential coma can be seen starting at 16° and sagittal oblique spherical aberration is marginally acceptable at 23°. Modest vignetting is employed at 23° to control the strong negative tangential coma.

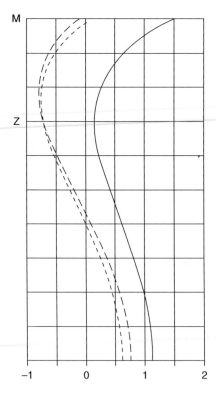

Figure 14.37   Longitudinal spherical aberration for unity magnification of the triplet copying lens. F light is short dashed curve, d light is solid curve, and e light is long dashed curve. The ordinate is in lens units with $f = 100$.

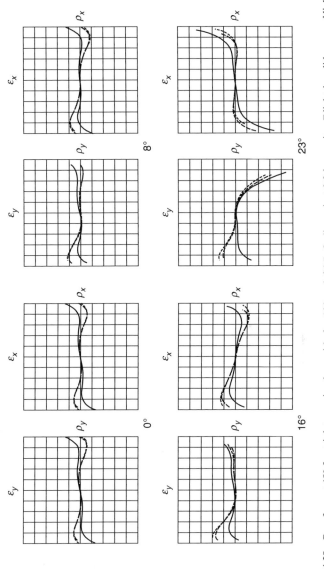

Figure 14.38 Ray fans at 1X for triplet copying lens with $f = 100$ and $f/6.7$. Ordinate is ± 0.2 lens units. F light is solid curve, d light is short dashed curve, and e light is long dashed curve.

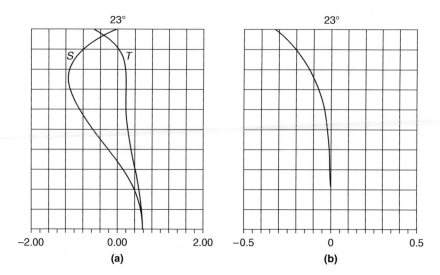

Figure 14.39 Field curves (a) and distortion (b) for triplet copying lens at unity magnification with $f = 100$ and $f/6.7$. Field curves are in lens units and distortion is in percent.

An estimate of the intersection height, as given by Eq. (13-8), of the S and T curves in Figure 14.39a is about 30°, with the Buchdahl coefficients being

$$\sigma_3 = 0.0691, \quad \mu_{10} = 0.0698, \quad \text{and } \mu_{11} = -0.0088.$$

This is significantly greater than the 20° observable in the figure. Examination of Figure 14.39a shows the clear presence of at least seventh-order tangential astigmatism, which is used to keep the tangential field reasonably flat and then begins to become strongly undercorrected at about 20°, which is why the estimate by Eq. (13-8) is excessive. The sagittal curve appears to be primarily third- and fifth-order astigmatism. This is a good example of a productive use of the higher-order aberrations to achieve design objectives.

ENDNOTES

[1] M. von Rohr, *Theorie und Geschichte des photographischen Objektivs*, p. 250. Springer, Berlin (1899).

[2] J. H. Dallmeyer, British Patent 2502/1866; U.S. Patent 65,729 (1867).

[3] F. von Voigtländer, British Patent 4756/1878.

[4] E. Glatzel and R. Wilson, "Adaptive automatic correction in optical design," *Appl. Opt.*, 7: 265–276 (1968).

[5] Juan L. Rayces, "Ten years of lens design with Glatzel's adaptive method," in International Lens Design Conference, *Proc. SPIE*, 237:75–84 (1980).

[6] Erhard Glatzer, "New lenses for microlithography," in International Lens Design Conference, R. Fischer (Ed.), *Proc. SPIE*, 237:310–320 (1980).

[7] David Shafer, "Optical design and the relaxation response," *Proc. SPIE*, 766:2–9 (1987).

[8] R. Barry Johnson, "Lenses," in *Handbook of Optics, Third Edition*, Vol. 1, Chapter 17, pp. 17.10–17.11, M. Bass (Ed.), McGraw-Hill, New York (2009).

[9] Bravais, *Annales de Chimie et de Physique*, 33:494 (1851).

[10] James E. Stewart, *Optical Principles and Technology for Engineers*, pp. 68–70, Marcel Dekker, New York (1996).

[11] James R. Johnson, James E. Harvey, and Rudolf Kingslake, U.S. Patent 3,441,338 (1969).

[12] John D. Griffith, U.S. Patent 5,966,252 (1999).

[13] M. von Rohr, *Theorie und Geschichte des photographischen Objektivs*, p. 364. Springer, Berlin (1899). Also see Paul Rudolph, U.S. Patent 444,714 (1891).

[14] P. Rudolph, U.S. Patent 721,240, filed in July (1902).

[15] H. D. Taylor, "Optical designing as an art," *Trans. Opt. Soc.*, 24:143 (1923); also British Patents 22607/1893, 15107/1895.

[16] The lens is called the Cooke triplet after the name of the company Taylor worked for as its optical manager. The company was Cooke & Sons of York. Cooke licensed the design to Taylor, Taylor, and Hobson, and they sold the lens under the brand name of Cooke as they did with some designs of their own.

[17] Robert F. Hopkins, "Third-order and fifth-order analysis of the triplet," *JOSA*, 52(4): 389–394 (1962).

[18] Carl Baur and Christian Otzen, U.S. Patent 2,966,825 (1961).

[19] Rudolf Kingslake, U.S. Patent 3,418,039 (1968).

[20] Muriel D. Ackroyd and William H. Price, U.S. Patent 3,418,040 (1968).

[21] Hiroyuki Hirano, U.S. Patent 5,084,784 (1992).

Chapter 15

Mirror and Catadioptric Systems

Curved mirrors, concave or convex, have often been used as image-forming systems, either alone or in combination with lens elements. Historically, most large astronomical telescopes have used a concave mirror to form the primary image, which is then relayed and magnified by either a second concave mirror (Gregorian) or a convex mirror (Cassegrain), although the objectives of small telescopes are generally achromatic lenses. Very small aspheric or spherical mirror systems have occasionally been used as microscope objectives. A single mirror used alone must generally be aspheric to correct the spherical aberration, but by combining two or more mirrors, with perhaps some lens elements also, it is possible to secure good aberration correction using only spherical surfaces. Summaries of reflecting and catadioptric systems have been given by Villa[1] and Gavrilov.[2]

15.1 COMPARISON OF MIRRORS AND LENSES

Mirrors have many advantages over lenses, principally as follows:

1. A mirror can be made of any size and of any material, even metal, provided it is capable of a high polish. Since good optical glass blanks cannot generally be made in diameters greater than about 20 in., all optical systems larger than that must be mirror systems. Often a mirror is used in conjunction with lens elements for aberration correction; such systems are called *catadioptric*.

2. Mirrors have no chromatic aberrations of any kind; hence a mirror can be focused in the visible and used in any wavelength region in the x-ray, UV or IR if desired. Also, mirrors exhibit no selective absorption through the spectrum as lenses do, but it must be noted that it is difficult to form mirror coatings that reflect well in the extreme ultraviolet.

3. A mirror has only one-quarter the curvature of a lens having the same power; hence mirrors can have a high relative aperture without the introduction of excessive aberration residuals. The Petzval sum of a concave mirror is actually negative.

4. By the use of several mirrors in succession, it is often possible to fold up a system into a very compact space.

On the other hand, mirrors have many features that are disadvantageous by comparison with lenses:

1. There will generally be an obstruction in the entering beam, causing a loss of light and a worsening of the diffraction image. This obstruction may be a secondary mirror or an image receiver, and if the angular field is wide the obstruction may block off nearly all of the incident light.

2. Since all the power is in one mirror surface, that surface must conform extremely closely to the desired shape, because even a slight distortion of the surface by the action of gravity or by temperature variations may cause a severe loss of definition. Flexure of a lens causes merely a trivial change in the aberrations, but flexure in a mirror changes the image position and alters the image quality drastically. The problem of mounting a large mirror without any flexural distortion is a very difficult one.

3. The angular field of a mirror system is generally quite small. It can be increased by the addition of one or more lens elements, but then many of the advantages of a mirror are lost.

4. In most reflective systems it is unfortunately possible for light from an object to proceed directly to the image without striking the mirrors. This must be prevented by the use of suitable baffles if the system is to be used in daylight. No baffles are needed in astronomical instruments since the overall sky brightness is very low at night.

15.2 RAY TRACING A MIRROR SYSTEM

If an optical system contains spherical mirrors, the standard ray-tracing procedure can be readily modified. The surfaces are listed in the order in which they are encountered by the light, with the usual sign convention that radii are regarded as positive if the center of curvature lies to the right of the surface. The separations d, the refractive indices n, and the dispersions Δn are entered as positive quantities if the light is traveling from left to right, but negative if the light is proceeding from right to left. The system should be oriented in such a way that the final imaging rays are moving from left to right so that the image-space index is positive. It may, therefore, be necessary in some cases to regard the object-space index as negative; if this presents difficulties a fictitious plane mirror can be inserted in front of the system to reverse the direction of the incident light.

As an example we will trace a paraxial ray and an $f/1$ marginal ray through a Gabor system (see Table 15.1). This system has a negative corrector lens in

Table 15.1
Ray Trace through a Catadioptric System

			(mirror)			
c		0.20	0.143	0.1	0.6079	0
d	−1.0	−0.35	−4.0	5.286	0.6	
n	−1.0	−1.545	−1.0	1.0	1.545	1.545

Paraxial

ϕ	−0.1090000	0.0779350	0.2	0.3313066
d/n	0.2265372			0.177515
y	2.0	2.049385	2.282510	0.3883495
mu	0.2180000	−0.0582812	−0.3982208	0.177515

y	0.000026
mu	−0.4570327

Marginal ($f/1$)

Q	2.188	2.299947	2.587155	0.1870214	0.0004700
Q'	2.241196	2.289509	2.463413	0.1945020	0.0004302
I	25.9509	9.4244	10.6206	−18.4788	−18.8679
I'	16.4535	14.6544	−10.6206	−11.8382	−29.9757
$\sin U'$	0	−0.1650029	−0.0744115	−0.4306454	−0.3233869	−0.4996328
U'	0	−9.4974	−4.2674	−25.5085	−18.8679	−29.9757

Paraxial $l' = 0.000057$ $f' = 4.376054$

Marginal $L' = 0.000861$ $F' = 4.379217$

$LA' = 0.000804$

front, a concave mirror and a positive field flattener, the light entering from infinity in a right-to-left direction. It should be noted that in the paraxial trace, the sign of the product nu depends on both the sign of n and the sign of u.

15.3 SINGLE-MIRROR SYSTEMS

15.3.1 A Spherical Mirror

A spherical mirror with an object point at its center of curvature is a perfect optical system having no aberrations of any kind. If the object point is displaced from the center of curvature, the paraxial image point moves in the opposite direction along a straight line joining the object point to the center of curvature (Figure 15.1). Because the aperture stop is at the mirror, the system is symmetrical, and for small object displacements there will be no coma. However, some astigmatism will be introduced, the sagittal image coinciding with the paraxial image at the Lagrangian image point while the tangential image is somewhat backward-curving.

It should be remarked that the focal length of a single spherical mirror is exactly half the radius of curvature; the principal points coincide at the vertex of the mirror, while the nodal points coincide at the center of curvature. Because the refractive indices of the overlapping object and image spaces are equal and opposite, the two focal lengths have the same sign, and the distance from the principal point to the nodal point is equal to twice the focal length. This applies to all reflective and catadioptric systems having an odd number of mirrors. With an even number of mirrors, the outside refractive indices have the same sign, and the ordinary rules for a lens system apply.

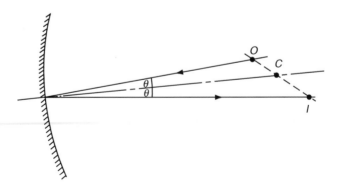

Figure 15.1   The line joining the object and the image passes through the center of curvature of a spherical mirror.

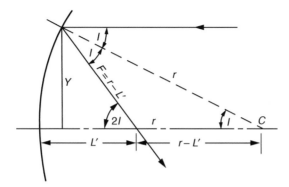

Figure 15.2 The spherical aberration of a spherical mirror.

If a spherical mirror is used with a distant object, undercorrected spherical aberration and overcorrected OSC appear at the focus. The magnitude of these aberrations is seen from the ray diagram in Figure 15.2. Here

$$\sin 2I = Y/(r - L'), \quad \sin I = Y/r$$

Hence

$$\frac{Y}{r - L'} = 2\left(\frac{Y}{r}\right)\left[1 - \left(\frac{Y}{r}\right)^2\right]^{1/2}$$

From which we find

$$L' = r - r^2/2(r^2 - Y^2)^{1/2}, \quad F' = r - L'$$

A few points calculated for a mirror with radius 20 and focal length $f' = 10$ are given in Table 15.2.

In this case the standard OSC formula becomes simplified to $(F'/L - 1)$ because $l'_{pr} = 0$ and $l' = f'$. It should be noted that the spherical aberration is purely primary for apertures less than about $f/6$, and that by $f/5$ the OSC has

Table 15.2

Spherical Aberration and Coma for a Mirror with Radius = 20

Y	L'	$LA' = L' - l'$	$F' = r - L'$	OSC	Aperture
0.1	9.999875	−0.000125	10.000125	0.000025	$f/50$
0.2	9.999500	−0.000500	10.000500	0.000100	$f/25$
0.5	9.996874	−0.003126	10.003126	0.000625	$f/10$
1.0	9.987477	−0.012523	10.012523	0.002508	$f/5$
2.0	9.949622	−0.050378	10.050378	0.010127	$f/2.5$

already reached Conrady's tolerance of 0.0025 for telescope objectives. For apertures under $f/10$ a single spherical mirror is often as good as a parabolic mirror and it is, of course, very much less expensive to manufacture.

15.3.2 A Parabolic Mirror

To determine the correct form for a concave mirror to be free from spherical aberration, we consider a plane wave front reaching the mirror from an axial object point at infinity (Figure 15.3). In this diagram the entering plane wave is PP, and while the axial portion of the wave is traveling a distance $Z + f'$, the marginal part of the wave travels a distance F'. Hence

$$F' = X + f' = [Y^2 + (f' - Z)^2]^{1/2}$$

where

$$Y^2 = 4f'Z$$

This is clearly the equation of a parabola with vertex radius equal to $2f'$.

This property of a parabolic mirror has been known for centuries, and it is the form given to the primary mirror in most reflecting telescopes. However, this mirror suffers from high OSC. The focal length F' of a marginal ray is equal to $[Y^2 + (f' - Z)^2]^{1/2}$ and it increases as Y increases in the same manner as in a spherical mirror with a distant object. The coma corresponding to an object subtending an angle U_{pr} is given by

$$(F' - f') \tan U_{pr} = \{[4f'Z + (f' - X)^2]^{1/2} - f'\} \tan U_{pr} = Z \tan U_{pr}.$$

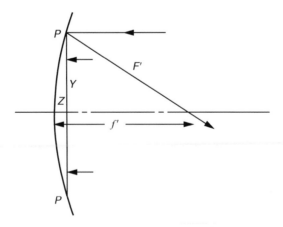

Figure 15.3 Reflection of a plane wave PP by a parabolic mirror.

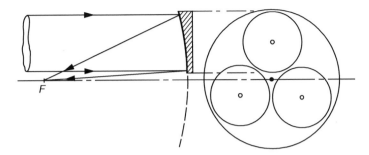

Figure 15.4 Cutting three off-axis parabolic mirrors from one large paraboloid.

If the aperture of the mirror is small, we can write $Z = Y^2/2r$,[3] and the sagittal coma becomes simply

$$\text{coma}_s = h'/16(f\text{-number})^2 \quad \text{or} \quad OSC = 1/16(f\text{-number})^2$$

The same result can be derived from the primary or third-order coma expression in Section 11.7.2. Thus at the prime focus of the Palomar telescope, for example, where the f-number is 3.3, the sagittal coma at a point only 20 mm off-axis has reached a magnitude of 0.115 mm. It will be found that this OSC is the same whether the mirror is a sphere or a parabola, but of course the spherical aberration is quite different in the two cases.

If the obstruction caused by the image receiver is undesirable, a so-called off-axis parabola may be used (Figure 15.4). The only practical way to construct such a mirror is to make a large on-axis mirror and cut as many off-axis mirrors from it as are needed. Such mirrors are used in mirror monochromators of the Wadsworth type, and as Schlieren mirrors for wind tunnel applications.

15.3.3 An Elliptical Mirror

As mentioned in Section 2.7, the equation of a conic section is

$$Z = cY^2/\{1 + [1 - c^2 Y^2(1 - e^2)]^{1/2}\}$$

where c is the vertex curvature and e the eccentricity. For an ellipse, e lies between 0 for a circle and 1 for a parabola. If a and b are the major and minor semiaxes of the ellipse, respectively, then

$$e = [(a^2 - b^2)/a^2]^{1/2}, \quad a = 1/c(1 - e^2), \quad b = 1/c(1 - e^2)^{1/2} = (a/c)^{1/2}$$

In terms of the two semiaxes, the vertex curvature is $c = a/b^2$.

A concave elliptical mirror has the interesting optical property of two "foci," which are such that an object point located at one is imaged at the other without aberration. The two "focal lengths," that is, the distances from the mirror vertex to the two foci, are

$$f_1 = a(1-e), \quad f_2 = a(1+e)$$

Hence

$$e = (f_2 - f_1)/(f_2 + f_1), \quad a = \tfrac{1}{2}(f_1 + f_2), \quad b = (f_1 f_2)^{1/2}$$

All optical paths from one focus to the other via a point on the ellipse are equal, but the magnification along each path is given by the ratio of the two sections of the path, and hence it varies greatly from point to point along the curve. This leads to heavy coma for an off-axis object point.

If the ellipse is turned so that the vertex is at the middle of the long side, we have an oblate spheroid, and then the *conic constant* $1 - e^2$ is greater than 1.0. This situation seldom arises, however, since an oblate spheroid is stronger than a sphere at the margin, and so it has worse spherical aberration.

To manually draw an ellipse, we first construct the two auxiliary circles on the major and minor axes as shown in Figure 15.5a, and we draw any transversal through the midpoint. If this crosses the two circles at A and B, respectively, then the point of intersection of a vertical line through A and a horizontal line through B is a point on the ellipse. By running several such transversals, enough points can be plotted to enable the ellipse to be filled in by use of a French curve. A simpler but less accurate procedure is to calculate the two vertex radii b^2/a and a^2/b and draw arcs with these radii through the ends of the semiaxes as in Figure 15.5b. These arcs almost meet, and the small gaps can be readily filled in with a French curve. A combination of both methods is probably the best procedure. However, a CAD program can make drawing an ellipse quite easy and accurate.

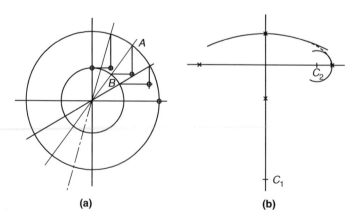

(a)   (b)

Figure 15.5   How to draw an ellipse.

15.3.4 A Hyperbolic Mirror

The eccentricity of a hyperbola is greater than unity, so that the conic constant $1 - e^2$ is negative. A hyperbola has two branches, and a hyperbolic mirror is formed usually by rotating the hyperbola about its longitudinal axis, only one branch being utilized. This may be either a convex or a concave mirror. If convex, then any ray directed toward the inside "focus" will be reflected through the outside "focus," the two focal lengths being

$$f_1 = a(1-e), \quad f_2 = a(1+e)$$

where a is the distance along the axis from the mirror vertex to the midpoint of the complete hyperbola (Figure 15.6). The separation of the vertices of the two hyperbolic branches is, of course, $2a$. The vertex radius is $a(1-e^2)$, so that f_1 and f_2 satisfy the ordinary mirror conjugate relation

$$1/f_1 + 1/f_2 = 2/r$$

A convex hyperbolic mirror is used in the Cassegrain telescope, and a concave hyperbola is used in the Ritchey-Chrétien arrangement.

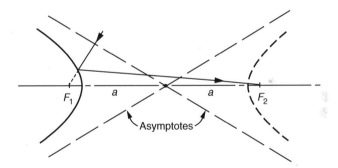

Figure 15.6   A convex hyperbolic mirror ($a = 38$, $r = -12.8$, $e = 1.156$, $f_1 = -5.93$, $f_2 = 81.93$).

15.4 SINGLE-MIRROR CATADIOPTRIC SYSTEMS

It was suggested by F. E. Ross in 1935[4] that it might be possible to remove the coma from a parabolic mirror by inserting an air-spaced doublet lens of approximately zero power into the imaging light beam at a position fairly close to the image to keep the lens small. Since the lens was to be a thin achromat of zero power, the same glass could be used for both elements. Ross found that it is impossible to simultaneously correct all three aberrations, spherical, coma, and field curvature, so he worked to control coma and field, letting the spherical aberration fall where it would. Alternatively, by greatly increasing the lens powers, it is possible to design an aplanat corrected for spherical and coma, but the field is then decidedly inward-curving. Examples of both systems will be given here.

15.4.1 A Flat-Field Ross Corrector

Assuming a parabolic mirror of vertex radius 200 and focal length 100, the spherical aberration is, of course, zero, and the marginal focal length at $f/3.33$ is found to be 100.5625. The OSC of the mirror is therefore 0.005625 when the stop is at the mirror so that $l'_{pr} = 0$. Tracing a principal ray entering the mirror vertex at 0.5°, we find that $Z'_s = 0$ and $Z'_t = -0.00762$. The Petzval sum is -0.01, giving $Z'_{Ptz} = +0.00381$. The tangential astigmatism is exactly three times the sagittal astigmatism at this small obliquity—that is, the distance from the Petzval surface to the sagittal surface is 0.00381 and the distance to the tangential surface is 0.01143, the ratio of these distances being 3.

We will follow through the design of a Ross corrector to be inserted at a distance of 90° from this parabolic mirror. To avoid vignetting at a field of 0.5° the diameter of the corrector must be about 5.0. The entering data for the three rays are

Marginal: $U = -8.57831°$, $Q = 1.49161$
Paraxial: $u = -0.15$, $y = 1.50$
Principal: $U_{pr} = 0.5°$, $Q_{pr} = 0.7853882$

We will start with the following setup. The glass is K-3 with $n_e = 1.52031$ and $V_e = 59.2$:

	c	d	n
	0		
		0.3	1.52031
	0.1		
		0.089228	
	0.07		
		0.65	1.52031
$(D-d)$	-0.036683		

with $f' = 97.5837$, $l' = 9.17044$, $LA' = -0.06648$, $OSC = 0.00221$, $Ptz = -0.00771$; for 0.5°: $Z'_s = -0.00024$, $Z'_t = -0.00636$, distortion $= +0.09\%$. The curvature of the first surface ($c_1 = 0$) is arbitrary and was set to zero (plane surface) and will be retained throughout. The central air space has been calculated to permit the two lenses to be in edge contact at a diameter of 4.8, and the last radius is calculated by the $D - d$ method for perfect achromatism. The dispersion of the glass need not be known since both elements are made of the same material.

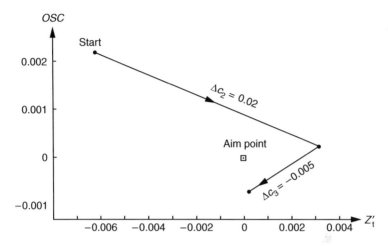

Figure 15.7 Double graph for a flat-field Ross corrector.

The two variables that will be used to achieve coma correction and a flat tangential field are, of course, c_2 and c_3. Making small changes in these variables permits us to plot the double graph shown in Figure 15.7. After a few trials, the final system prescription was as follows:

	c	d	n
	0		
		0.3	1.52031
	0.1169		
		0.149348	
	0.0670		
		0.65	1.52031
$(D - d)$	-0.0576113		

with $f' = 97.4760$, $l' = 9.18666$, $LA' = -0.11509$, $OSC = -0.00001$, Ptz $= -0.00736$; for $0.5°$: $Z'_s = +0.00153$, $Z'_t = -0.00056$, distortion $= +0.19\%$. The passage of axial and oblique rays through this system is shown in Figure 15.8.

The slightly backward-curving sagittal field could probably be corrected by the use of a somewhat higher refractive index for the negative element, but this possibility was not explored. The major problem is, of course, the large residual of spherical aberration, which could be removed only by the use of an aspheric surface. Some recent workers have managed to correct all three aberrations by means of three or more elements.

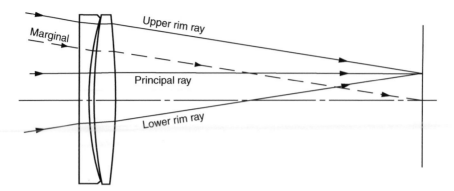

Figure 15.8 Path of rays through Ross corrector.

15.4.2 An Aplanatic Parabola Corrector

By making both elements considerably stronger, it is possible to correct the spherical aberration and *OSC*, and thus design an aplanatic corrector, but only at the expense of a considerable inward field curvature. The thickness of the positive element must be increased, and the central air space must be held at some fixed value because the adjacent surfaces are almost identical.

For a starter we may consider the following setup:

	c	d	n
	-0.1		
		0.3	1.52031
	0.1		
		0.1	
	0.1		
		1.1	1.52031
$(D-d)$	-0.1095215		

with $f' = 97.5847$, $l' = 9.33076$, $LA' = -0.02649$, $OSC = 0.00037$, Ptz = -0.00674; for $0.5°$: $Z'_s = -0.0139$, $Z'_t = -0.0469$, distortion $= -0.12\%$. We will now hold the second surface curvature arbitrarily at 0.1, and vary the other curvatures c_1 and c_3 to plot a double graph (Figure 15.9). The graph for changes in c_1 is found to be decidedly curved, which is not surprising since changes in c_1 represent both a bending and a power change, whereas changes in c_3 are a pure bending. A few trials give us the following final setup:

	c	d	n
	−0.13		
		0.3	1.52031
	0.1		
		0.1	
	0.10387		
		1.1	1.52031
$(D-d)$	−0.1322352		

with $f' = 98.7691$, $l' = 9.58664$, $LA' = 0.00001$, $OSC = 0.00002$, Ptz = −0.00791; for 0.5°: $Z'_s = -0.0196$, $Z'_t = -0.0654$, distortion = −0.21%. This system would be extremely heavy if made in a large size, and the inward tangential field curvature would be obviously undesirable, being about nine times as great as for the mirror alone. Other values of c_2 could, of course, be tried, but the result is likely to be similar to this in performance.

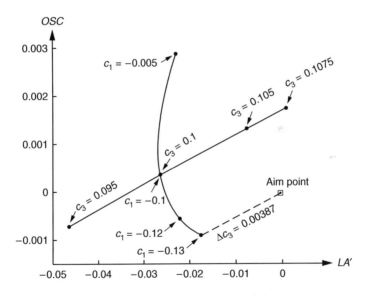

Figure 15.9 Double graph for aplanatic parabola corrector.

15.4.3 The Mangin Mirror

The French engineer Mangin[5] in 1876 proposed replacing the parabolic mirror in a searchlight by a more easily manufactured spherical mirror, with a thin meniscus-shaped negative lens in contact with the mirror to correct the

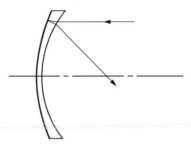

Figure 15.10   A typical Mangin mirror.

spherical aberration (Figure 15.10). The design procedure is simple since there is only one degree of freedom, namely, the outside radius of the lens, because the mirror radius determines the focal length of the system. Using K-4 glass ($n_e = 1.52111$, $V = 57.64$), a few trials give the following setup:

	c	d	n
	0.0981		
		0.3	1.52111
(mirror)	0.06544		

with $f' = 10.0155$, $l' = 9.82028$, LA' ($f/3$) $= 0.00001$, LZA ($f/4.2$) $= -0.00008$, OSC ($f/3$) $= 0.00307$. The OSC is less than half that of a parabolic mirror of the same focal length and aperture, but the chromatic aberration from F to C is found to be 0.0564, while the zonal spherical aberration is negligible. The next step, therefore, is to achromatize the system.

If we replace the simple negative lens with an achromat using the following glasses:

1. F-4: $n_c = 1.61164$, $n_e = 1.62058$, $n_F = 1.62848$
2. K-4: $n_c = 1.51620$, $n_e = 1.52111$, $n_F = 1.52524$

with the flint element adjacent to the mirror, we may start with a plano interface:

	c	d	n
	0.1		
		0.2	1.52111
	0		
		0.3	1.62058
(mirror)	0.062		

with $f' = 9.8332$, $l' = 9.52178$, LA' ($f/3$) $= -0.02094$, zonal chromatic aberration $F - C = -0.03326$.

To plot a double graph for the simultaneous correction of spherical and zonal chromatic aberrations, we make trial changes of $\Delta c_1 = 0.01$ and $\Delta c_2 = 0.01$, respectively. The graph so obtained indicates that we should try $c_1 = 0.1021$ and $c_2 = 0.01625$. This is a great improvement, since $LA' = -0.00705$ and the zonal chromatic aberration $L'_{ch} = -0.00560$. A few further small adjustments gave the following final system:

	c	d	n
	0.10636		
		0.21	1.52111
	0.01489		
		0.3	1.62058
(mirror)	0.062		

with $f' = 10.8324$, $l' = 10.52127$, $LA' = -0.00007$, $OSC = 0.00215$, zonal chromatic aberration $= 0.00002$. By tracing other rays the spherochromatism curves can be plotted as in Figure 15.11. It can be seen that the aberration residuals are very small, the chief residual being the ordinary secondary spectrum typical of a negative achromat. This system is practical if made in small sizes, but for large systems a parabolic mirror would be preferable.

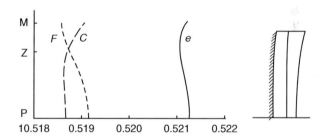

Figure 15.11 Spherochromatism of an achromatic Mangin mirror.

15.4.4 The Bouwers–Maksutov System

During World War II, Bouwers[6] and Maksutov[7] independently proposed the use of a monocentric catadioptric system to cover a wide angular field. This system consisted of a spherical mirror and a thick corrector plate, all three surfaces having a common center C located at the middle of the stop. Such a system has no coma or astigmatism and the image lies on a spherical surface, also concentric about C. The corrector lens can be located either in front of or behind the stop, and it may be thin and strongly curved, or thick and less strongly curved (Figure 15.12). For any given front radius the thickness can be adjusted to eliminate the marginal spherical aberration, but the zonal residual abberation will vary with the thickness.

Although the angular field of this monocentric system is theoretically unlimited, the obstruction caused by the receiving surface increases as the field is widened to the point where eventually no light at all will enter the system. To reduce this effect the relative aperture must be increased as the field is widened, unless, of course, the receiver is a narrow strip crossing the middle of the aperture.

Figure 15.13a shows the zonal spherical aberration of four examples of Maksutov correctors used with a mirror of radius 10.0, the marginal aberration at $f/2.5$ being corrected in each case by using a suitable thickness for the corrector. The four cases are as shown in Table 15.3. For the third case, the chromatic

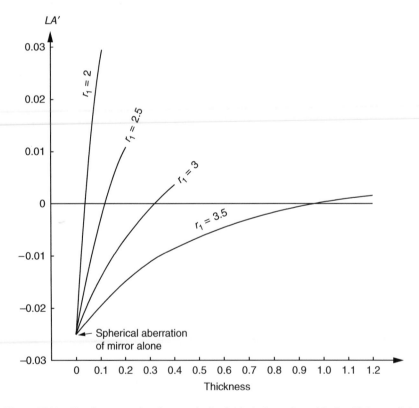

Figure 15.12 Graphs connecting the marginal spherical aberration with the thickness of the corrector plate for various values of r_1 (mirror radius = 10).

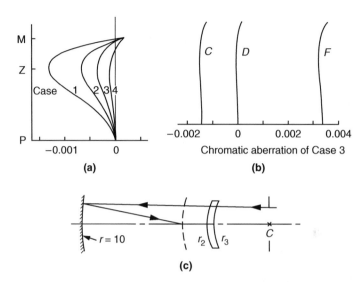

Figure 15.13 Bouwers–Maksutov systems: (a) spherical aberration for four values of r_1; (b) chromatic aberration of case 3; (c) ray diagram of case 3.

Table 15.3

Corrector Thickness for Maksutov System

Case	r_1	Thickness	r_2	Focal length	Back focus
1	2.0	0.040	2.04	4.9172	5.0828
2	2.5	0.121	2.621	4.8463	5.1537
3	3.0	0.320	3.32	4.7386	5.2614
4	3.5	0.950	4.45	4.5260	5.4740

aberration is shown graphically in Figure 15.13b and a scale drawing of the system in Figure 15.13c.

The chromatic aberration of the Bouwers–Maksutov system is decidedly large and could be serious. It can be removed by achromatizing the corrector lens, but then the system is no longer monocentric, and the angular field immediately becomes limited. However, if only a narrow field is desired, then achromatizing the corrector is quite a satisfactory procedure.

15.4.5 The Gabor Lens

In 1941 Dennis Gabor,[8] the inventor of the hologram, patented a catadioptric system that resembled the Bouwers–Maksutov except that it was not monocentric; it was much more compact and covered a narrow field at a high relative aperture. Actually, the example shown by Gabor was not achromatic, but the example to be given here is.

In the absence of a field flattener, if the negative front collector lens has a zero $D - d$ sum, the system will obviously be achromatic. Thus the only requirement for achromatism is that the length D measured along the marginal ray inside the corrector lens should be equal to the axial thickness of that lens. To secure this condition it is helpful to make the lens as thick as practical, and to use a glass of moderately high refractive index such as a barium crown. The front radius is then chosen for spherical aberration correction when used with a spherical mirror, and the second radius is found by the ordinary $D - d$ method. Placing the stop at the front surface, the field is backward-curving, and the Petzval sum is negative. The following $f/1.6$ system was the result of a few easy trials:

	c	d	n_e
	0.25		
		0.4	1.61282
	0.2347439		
		8.0	(air)
(mirror)	0.06		

with $f' = 8.0383$, $l' = 8.59345$, LA' ($f/1.6$) $= 0.00925$, LZA ($f/2.3$) $= -0.00420$, OSC ($f/1.6$) $= 0.00327$, Petzval sum $= -0.1258$. The fields at an obliquity of $1°$ were

$$Z_s' = 0.00104, \quad Z_t' = 0.00064, \quad \text{distortion} = -0.012\%$$

As Gabor indicated in his patent, the negative Petzval sum can be easily eliminated by the addition of a positive field flattener close to the image plane. This lens may conveniently be plano-convex, although it may require a slight bending to flatten the tangential field. A possible starting system with such a field flattener is as follows:

	c	d	n_e	Glass
(as before)	0.25	0.4	1.61282	SK-1
	0.2347439			
		8.0		
(mirror)	0.06			
		8.0		
(field flattener)	0.37162	0.1	1.51173	K-1
	0			

with $f' = 7.2231$, $l' = 0.46712$, LA' $(f/1.6) = -0.00651$, OSC $(f/1.6) = -0.00348$; for $1°$: $Z'_s = 0.00008$, $Z'_t = 0.00024$, distortion $= 0.025\%$. This field flattener has introduced a small amount of negative $D - d$ sum, which is easily removed by a small change in the radius of the second surface of the correcting lens. The front surface was also strengthened slightly to remove the small residual of spherical undercorrection caused by the field flattener. The final system is as follows:

	c	d	n_e
	0.251		
		0.4	1.61282
	0.2348373		
		8.0	
(mirror)	0.06		
		8.0	
	0.37264		
		0.1	1.51173
	0		

with $f' = 7.1775$, $l' = 0.49554$, LA' $(f/1.6) = 0.00501$, LZA $(f/2.3) = -0.00219$, OSC $(f/1.6) = -0.00468$, Ptz $= 0$; for $1°$: $Z'_s = 0.00009$, $Z'_t = 0.00028$, distortion $= 0.025\%$.

To investigate the coma, it is necessary to make a meridional ray plot for the $1°$ beam. This is shown in Figure 15.14 above the corresponding plot for the axial beam. It is clear that there is an excess of negative coma present, which can be removed by shifting the corrector lens along the axis. As this has very little effect on the aberrations, it is advisable to make a large shift, say from 8.0 to 6.0. This causes a slight overcorrection of the spherical aberration,

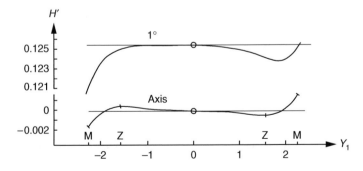

Figure 15.14 Meridional ray plot for Gabor lens with space equal to 8.0.

requiring a weakening of the front surface and a recalculation of the $D - d$ sum. These changes lead to the following results:

	c	d	n_e
	0.246		
		0.4	1.61282
	0.2303761		
		6.0	
(mirror)	0.06		
		8.0	
	0.37264		
		0.1	1.51173
	0		

with $f' = 7.2492$, $l' = 0.49128$, LA' $(f/1.6) = 0.00314$, LZA $(f/2.3) = -0.00299$, OSC $(f/1.6) = -0.00163$. Ptz $= 0.000204$; for $1°$: $Z_s' = 0.00003$, $Z_t' = 0.00009$, distortion $= 0.026\%$. The $(D - d)\Delta n$ in the two lenses is ± 0.0000343. To complete the study, the $1°$ meridional ray plot was drawn (Figure 15.15). The improvement over the previous setup is obvious.

Although this system is well corrected, mechanically something must be done to keep the imaging rays clear of the corrector lens. A possible arrangement is shown in Figure 15.16, using a hole in the middle of the corrector lens, but this hole must be quite large for such a high aperture as $f/1.6$. A plane mirror could be employed to reflect the beam out sideways, or back through the middle of the concave mirror to somewhat mitigate this problem.

This Gabor system is unusual in that each of the six degrees of freedom (five radii and one air space) is almost specific for one particular aberration. The front surface controls the spherical aberration and the second the chromatic aberration; the power of the field lens determines the Petzval sum, while its bending controls

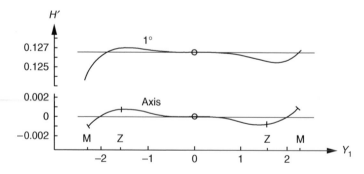

Figure 15.15 Meridional ray plot for Gabor lens with space equal to 6.0.

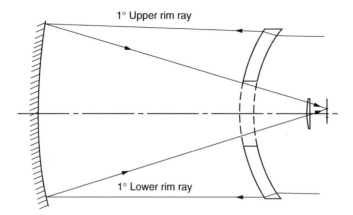

Figure 15.16   Final design of $f/1.6$ Gabor system covering $\pm 1°$.

the field curvature; and finally the central air space is used to vary the coma. The mirror radius, of course, determines the focal length. The two remaining aberrations, lateral color and distortion, are usually negligible at such a narrow field, but if the lateral color should be significant, it may be necessary to achromatize the field lens, which would require a further adjustment of the chromatic correction by c_2. The aperture of the Gabor system can be high, but the angular field is small. This is another good example of why a lens designer should study the lens being optimized to learn how the various parameters affect the several aberrations. Blindly making adjustments can lead to poor outcomes.

15.4.6 The Schmidt Camera

The Schmidt camera[9] consists of a concave spherical mirror with a thin aspheric corrector plate located at the center of curvature of the mirror. By placing the stop at the corrector plate we automatically eliminate coma and astigmatism, although at high obliquities some higher-order aberrations appear, but the useful field of several degrees is much larger than that of most catadioptric systems. The remaining aberration is spherical, which is corrected by a suitable aspheric surface on the corrector plate. The chromatic aberration is ignored.

The simplest way to derive an expression for the shape of the aspheric surface is to select a neutral zone to represent the minimum point on the aspheric surface, where the plate is momentarily parallel, and let the ray through this neutral zone define the focal point of the system. Tracing a paraxial ray backward from this focus and performing an angle solve enables us to determine the vertex radius of the aspheric surface. To determine the thickness of the plate at the neutral zone we must equalize the optical paths along the paraxial ray and the neutral-zone ray. We now have three relationships by which three terms of the aspheric polynomial can be found, namely, the vertex curvature, the sag of the neutral zone, and the slope of the surface at the neutral zone, which is zero. If we need greater precision or if we desire more than three terms in the polynomial, we can trace several other rays backward from the focus and make a least-squares solution for as many terms as we need.

The path of the neutral-zone ray is shown in Figure 15.17a. The point C is the center of curvature of the concave mirror of radius r. The point F is the focus defined by the intersection of the neutral-zone ray with the axis. The focal length of the system is FN', and the back focal distance is a. If θ_0 is the slope of the neutral-zone ray at the image and Y_0 the incidence height of this ray, then

$$\sin\tfrac{1}{2}\theta_0 = Y_0/r, \quad a = r - r/2\cos\tfrac{1}{2}\theta_0$$

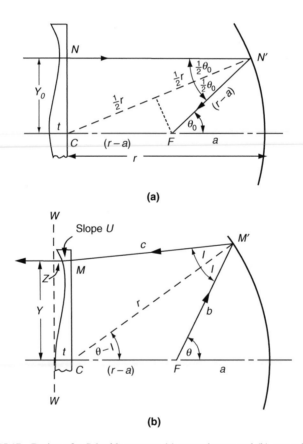

(a)

(b)

Figure 15.17 Design of a Schmidt camera: (a) neutral zone and (b) any other zone.

The path of any other ray traced backward from the focus at a starting angle θ is shown in Figure 15.17b. The angle of incidence I of this ray at the mirror is given by

$$\sin I = (r - a) \sin \theta / r$$

Assuming for simplicity that the plane side of the corrector plate is at the exact center of curvature of the mirror, the ray path in the air can be calculated by

$$b = FM' = r \sin(\theta - I)/\sin \theta$$
$$c = M'M = r \cos(\theta - I)/\cos(\theta - 2I)$$

The slope of the ray inside the corrector plate is found by

$$\sin U = (1/n) \sin(\theta - 2I)$$

The line WW in Figure 15.17b represents a plane wave in the object space, and the optical paths from this wave front to the focus F must be equal along all rays. Along the axis this optical path is evidently $(nt + r + a)$, and along a general ray it is $[b + c + Z + n(t - Z)/\cos U]$. Equating these paths gives the z coordinate of a point on the asphere as

$$Z = \frac{a + r - b - c + nt(1 - \sec U)}{1 - n \sec U}$$

To determine the corresponding height of incidence Y of this ray, we have

$$Y = MC - (t - Z) \tan U, \quad \text{where } MC = c \sin I / \cos(\theta - I)$$

We can apply these formulas to the neutral zone, for which we find

$$b = r - a, \quad c = r \cos \tfrac{1}{2} \theta_0, \quad U = 0$$

Example

For an $f/1$ Schmidt with $r = 4.0$, the focal length is about 2 and the marginal ray enters at a height $Y = 1.0$. We may set the neutral zone at an incidence height of 0.85, where $\sin \tfrac{1}{2} \theta_0 = 0.2125$ and $\theta_0 = 24.5378°$. The focal length of the neutral zone is $F' = 0.85/\sin \theta_0 = 2.046745$ and the back focus is $a = 1.953255$. For the neutral-zone ray we have $b = 2.046745$ and $c = 3.908644$; hence $Z_0 = 0.004080$. The refractive index is 1.523.

We next set the axial thickness of the plate at 0.01, and tracing a paraxial ray backward from the focus, we solve the vertex curvature of the plate to make the paraxial ray emerge parallel to the axis. In this way we find that the vertex radius should be $R = 45.7416$. The paraxial focal length is 2.046899.

Assuming a three-term polynomial of the form

$$Z = AY^2 + BY^4 + CY^6$$

we see that $A = 1/2R = 0.0109310$. For the height of the neutral zone we have

$$Z_0 = A(0.85)^2 + B(0.85)^4 + C(0.85)^6 = 0.004080$$

and for the slope of the surface at the neutral zone we have

$$(dZ/dY)_0 = 2AY + 4BY^3 + 6CY^5$$
$$= 2A(0.85) + 4B(0.85)^3 + 6C(0.85)^5 = 0$$

Solving these three equations simultaneously gives the three coefficients as

$$A = 0.010931, \quad B = -0.00681084, \quad C = -0.00069561$$

Calculating Z for several values of Y gives the data needed to plot the shape of the asphere as shown in Table 15.4. If this curve is plotted, it will be seen that the central bulge is much larger than the curl-up at the rim, so that it might have been better to set the neutral zone a little lower, say at 0.80 instead of 0.85 of the marginal height.

The trivial difference between the zonal and paraxial focal lengths represents an *OSC* of only -0.000075, which is obviously negligible. It could be removed completely by a slight shift of the corrector plate along the axis.

Table 15.4

Shape of Aspheric Corrector Plate for the Schmidt Camera

Y	Z	Y	Z	Y	Z
0.1	0.000109	0.4	0.001572	0.7	0.003639
0.2	0.000426	0.5	0.002296	0.8	0.004077
0.3	0.000928	0.6	0.003020	0.9	0.004016
				1.0	0.003425

15.4.7 Variable Focal-Range Infrared Telescope

Sometimes a telescope is needed that can be used as part of a wide field-of-view scanning system, particularly in the infrared spectrum. Figure 15.18 shows such a $f/1.1$ telescope that can operate in the 8 to 14-μm spectrum in conjunction

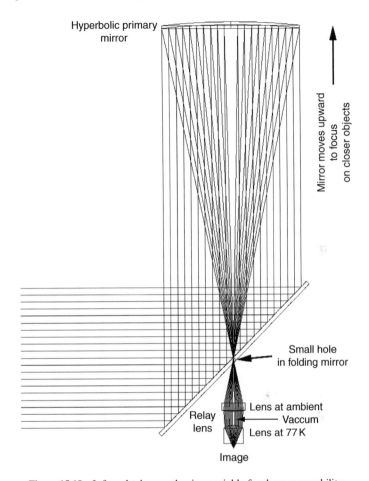

Figure 15.18 Infrared telescope having variable focal range capability.

with an object-space scanner with a single infrared detector having dimensions of about 75 μm by 75 μm.[10,11] Since object-space scanning is used, the telescope only requires a field-of-view needed for the detector. This catadioptric optical system comprises a hyperbolic primary mirror, a folding mirror, and a relay lens. The relay lens has two lenses made of germanium and images the detector at the location of the folding mirror. The folding mirror has a hole in it to allow the passage of the infrared flux. In this case, the hole is about 4-mm diameter for a 105-mm focal length and a 91-mm diameter primary mirror. The detector image at the hole is magnified by a factor of 2.3. This telescope can focus at object distances from about a meter from the primary mirror to infinity; however, it was optimized to have diffraction-limited performance at about 2 meters but still has near diffraction-limited performance over the entire range.

To accomplish this, the primary mirror was made mildly hyperbolic so that longitudinal spherical aberration is zero when the object is about 3 m from the primary mirror. As the object distance increases, the spherical aberration becomes undercorrected and for closer object distances, it becomes overcorrected. Since just a single detector is used, it is prudent to design the useful field-of-view of the telescope to accommodate three to five detectors in order to ease alignment during manufacture. Since this telescope was intended for service in the thermal infrared, the refractive elements were made of germanium since it has very low dispersion in the 8 to 14-μm spectrum. This spectral band was selected rather than the 8 to 12-μm spectrum because it was going to be used typically at short ranges of a few meters, so atmospheric absorption was not an issue until 14 μm is exceeded.

The two lenses comprising the relay lens are elements of a cryogenic dewar. The first lens serves as the dewar window and is at ambient temperature. In contrast, the second lens is mounted on the dewar cold finger along with the infrared detector. It is cooled to the temperature of liquid nitrogen. Consequently, the refractive index of germanium when cooled must be used. Also, the spaces between the two lenses and before the detector are both vacuum and its refractive index should be used rather than air. In this particular case. Omitting the use of vacuum can be compensated by simply adjusting the primary mirror to relay lens distance; however, lack of using the cryogenic refractive index for germanium can result in disaster. The structure of this telescope, excluding the folding mirror, is as follows:

	r	d	n	Conic constant
Primary	−434.884			−1.49534
		See note	Mirror	
Relay lens A	115.949			
		−5.2578	Germanium (ambient)	
	47.14596			
		−9.17448	Vacuum	
Relay lens B	−11.96679			
		−11.2268	Germanium (77K)	
	−4.388507			
		−3.8100	Vacuum	

Note: This thickness depends on object distance Z and is given by

$$d = -246.888 - 167924 \cdot Z^{-1.1448}.$$

DESIGNER NOTE

It may appear that a concave hyperbolic mirror is more difficult to make than a spherical or a parabolic mirror, so the lens designer might be hesitant to use this type of mirror. In this case, the mirror was actually fabricated in production in about 75% the time it took to make a similar parabolic mirror. As with most aspheric lens or mirror components, the lens designer should consider how the element will be tested and the fabrication method. In the case of this hyperbolic mirror, a simple biconvex lens was constructed and used as a null lens so that the optician could use the common knife-edge test method.

The lens was made of BK-7 glass having radii of 102.354 in. and −98.232 in., and thickness of 1.000 in. Separation between the lens and the mirror was 17.000 in. With collimated monochromatic light input to the lens, the image has a peak-to-peak wavefront error of 0.16 lambda or a Strehl ratio greater than 0.9. Since the mirror was used in the 8 to 14-μm spectrum, the wavefront error scales by the ratio of the wavelengths. Consequently, the mirror as used had a Strehl ratio of essentially unity or diffraction-limited performance.

Also, it is often helpful in designing an optical system to view it from the opposite direction. In this case, the lens designer can view the object space from the detector position rather than from object space toward the detector. In a more complicated system, it is often useful to view in the reverse direction to ensure that unexpected vignetting or any other problem does not occur.

15.4.8 Broad-Spectrum Afocal Catadioptric Telescope

Lens designers have found it challenging to design high-magnification afocal telescopes that can operate simultaneously over very broad portions of the visible and infrared spectrums.[12] Such telescopes are frequently used with some type of scanning sensor and require the pupil of the telescope to be external to it so that the sensor and telescope pupils can mate. A possible telescope configuration is shown in Figure 15.19 and comprises a concave mirror as the objective and one-glass-type Schupmann lens (see Section 5.7.2) as the eyepiece or secondary optics. By locating the focus of the primary mirror (objective element) coincident with the internal focal point of the Schupmann lens (eyepiece), the afocal condition is obtained. The aperture stop of the system is located at the primary mirror while the exit pupil position is established by the secondary optics, having a Schupmann configuration, imaging the aperture stop. Since the Schupmann lens has positive optical power, the exit pupil is a real image of the stop and is located external to the telescope. Generally, a folding mirror would be used to allow the image-space beam to be accessible and the resulting obscuration would likely be relatively small.

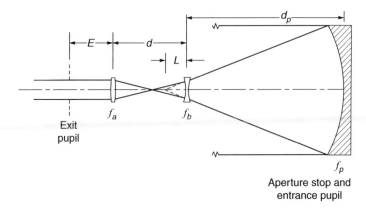

Figure 15.19 Basic afocal catadioptric telescope configuration.

Ludwig Schupmann explored designing and building telescopes incorporating a dialyte objective lens (and other variants) for the purpose of achieving minimal chromatic aberration over a broad spectrum by constructing an objective lens comprising only a single glass type.[13,14,15] It should be noted that the large air space between the Schupmann lens elements allows rays of different colors to become somewhat separated at the negative element, resulting in slightly undercorrected chromatic aberration.[16]

The magnification of the telescope is $M = -f_s/f_p$ where f_s is the effective focal length of the secondary optics. As given in Section 5.7.2, the separation factor is $k = d/f_a$ and the back focal length is $L = -(k-1)f_s$. The distance between the primary mirror and the negative lens of the secondary optics is $d_p = f_p \pm L$. The separation of the elements comprising the Schupmann lens is

$$d = \frac{f_s k^2}{k-1}.$$

Particular attention should be given to the sign of each parameter. After some algebraic manipulations, the pupil relief or distance between the exit pupil and lens a is found to be given by

$$E = \frac{1 - M + k}{(k-1)^2} f_s.$$

Using these equations and those in Section 5.7.2, the afocal optical system is now corrected for axial chromatic aberration. Now we impose the condition that the Petzval sum equals zero (or whatever value may be desired). Since the Petzval contribution of the mirror is opposite that of the secondary optics, we can write

$$-\frac{\phi_a + \phi_b}{n} + \phi_p = 0$$

where n is the refractive index of the two lenses. By combining the preceding equations, we find that

$$M = -\frac{1}{n}\left(\frac{k-2}{k-1}\right) \text{ or } k = \frac{nM+2}{nM+1}.$$

With the Petzval equal to zero, the exit pupil relief becomes

$$E = [1 - M + (nM + 2)(nM + 1)] f_s.$$

Although the telescope is now corrected for primary and secondary axial chromatic aberrations and Petzval, the system magnification is a function of the spectral variation of refractive index. It is simple to show that

$$\frac{\partial M}{\partial n} = -\frac{\phi_p}{\phi_s}\frac{\partial \phi_s}{\partial n}$$

and

$$\frac{\partial \phi_s}{\partial n} = \frac{k\phi_s}{(n-1)(k-1)}.$$

Therefore, the fractional variation in magnification is given by

$$\frac{\Delta M}{M} = \frac{k}{V(k-1)}.$$

This equation shows that the telescope will suffer lateral chromatic aberration and expresses the amount of aberration in object space for a unity principal ray angle ($n \tan u_{prin}$) at the exit pupil. This could set the useful field-of-view of the telescope; however, if a lens is placed proximate to the positive lens in the secondary optics, then the lateral chromatic aberration can be reasonably mitigated.[17,18,19]

Johnson has presented a detailed design procedure[20] that begins with the specification of the aperture stop diameter, magnification, and exit pupil location. The magnification, optical powers, element separations, and refractive index are parametrically related by the preceding first-order equations such that the primary and secondary axial color and Petzval are corrected while potentially realizing the desired specifications.

As we mentioned previously, the insertion of finite thicknesses into the thin lenses often upsets the correction of the system. A commonly used technique to maintain the first-order behavior is the measure all of the distances from the principal points of the lens elements. As we have seen, the principal points will move about spatially as a lens' bending is changed. One design procedure that the lens designer could follow is to select a curvature c_{1j} for the surface nearest the exit pupil and then select the remaining curvatures. The formula is

$$c_{2j} = \frac{c'_{2j}}{1 + c_{1j}(1-n)(t_j/n)}$$

where $j = [a, b]$, t_j is the thickness of the jth element, and c'_{2j} is the curvature of c_{2j} when $t_j = 0$. The optical power of each thin-lens element, which we already know from the initial design, is

$$\phi_j = (c_{1j} - c'_{2j})(n - 1).$$

Now that the curvatures are known, a thick-lens layout can be determined using the following equations to locate distances from curvature vertices rather than principal points. In general, this is preferred when inputting data into a lens design program or manually ray tracing the system. Using paraxial ray trace methods, the equations can be formulated. The distance from the exit pupil to the vertex of c_{1a} is

$$t_E = E - c_{2a}(1-n)\left(\frac{t_a}{n}\right)f_a$$

where t_a is the thickness of lens a. The distance between the vertex of c_{1b} and the vertex of c_{2a} is

$$t_d = d - c_{1a}(1-n)\left(\frac{t_a}{n}\right)f_a - c_{2b}(1-n)\left(\frac{t_b}{n}\right)f_b$$

while the distance from the vertex of c_{2b} to the primary mirror vertex is

$$t_p = d_p - c_{1b}(1-n)\left(\frac{t_b}{n}\right)f_b$$

where t_b is the thickness of lens b.

Final correction of the telescope requires perhaps making the primary mirror slightly conic and bending the lenses to minimize the spherical aberration and the *OSC*, and perhaps deviating the spacings slightly to correct any residual axial chromatic aberration. It should also be recognized from Section 6.1.6 that either or both of the lenses in the secondary optics can be split to provide additional aberration correction. The example telescope discussed in the Endnote 20 reference is a 0.5° total object-space field-of-view afocal telescope for the 3 to 12-μm spectrum having $M = -0.05$ (20X), germanium lenses, $f_p = 100$, aperture stop diameter of 1.0, $k = 2.25$, and $f_s = 5$. For a thick-lens configuration with the units in inches, the primary mirror is hyperbolic, the lens bendings are used to correct the aberrations, and a third lens is added for lateral color correction as previously discussed. Over the entire spectrum, the Strehl ratios near the axis are near unity and about 0.85 at the edge of the field-of-view.

15.4.9 Self-Corrected Unit-Magnification Systems

Two very interesting systems have been proposed for 1:1 imagery, which are automatically corrected for all the primary aberrations. Altman also invented a unit magnification catadioptric system comprising a concave spherical mirror and a system of lenses for correction of the mirror aberrations.[21]

The Dyson Catadioptric System

This is a monocentric system (Figure 15.20), the object and image lying in the same plane on opposite sides of the center of curvature C.[22] A marginal ray from C returns along its own path, thus automatically removing spherical and chromatic aberrations. The radius of curvature of the lens is set at $(n-1)/n$ times the radius of curvature of the mirror, to give a zero Petzval sum. The aperture stop is at the mirror, making a symmetrical system that is automatically corrected for the three transverse aberrations. The seventh aberration, astigmatism, is zero near the middle of the field and the sagittal field is flat, but the tangential field bends somewhat backward at increasing distances out from the axis. A typical system is the following:

c	d	n
0		
	3.434012	1.523
0.2912046		
	6.565988	(air)
(mirror) 0.1		

with

$$l = l' = 0, \quad m = -1$$
$$H' = 1: \ Z'_s = 0, \quad Z'_t = 0.01460$$
$$H' = 1.5: Z'_s = 0, \quad Z'_t = 0.08776$$

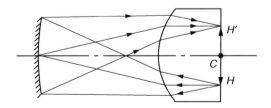

Figure 15.20   The Dyson autocollimating system.

As can be seen, the system would be telecentric except for the spherical aberration of the principal ray at the lens surface: the principal ray for $H' = 1.5$ enters at a slope angle of almost 4° in air (2.58° in the glass).

Caldwell designed a catadioptric relay system, similar in many respects to the Dyson system, for use with very compact projection lenses with dual DMD projectors.[23,24] This system is more compact than the Dyson system and provides near diffraction-limited performance. The references mentioned in endnotes 23 and 24 contain the optical prescriptions.

The Offner Catadioptric System

This monocentric system[25] is similar to the Dyson arrangement, except that a small convex mirror is placed midway between the concave mirror and the object to give a zero Petzval sum, and the beam is reflected twice at the concave mirror (Figure 15.21). The aperture stop is at the small convex mirror and the system is virtually telecentric.

Because the two mirrors are concentric about C, an object point placed there would be imaged on itself without aberration. However, this is academic because the entire axial beam is blocked out by the secondary mirror. For object points lying off-axis, the vignetting becomes progressively less and finally disappears for object points with H and H' equal to or greater than the diameter of the convex mirror. The symmetry about the stop ensures that coma and distortion are absent. There are, of course, no chromatic aberrations of any kind.

The remaining aberration, astigmatism, is zero for object points near the axis, and the sagittal field is flat, as for the Dyson case. However, the tangential field bends slightly backward for extraaxial object points.

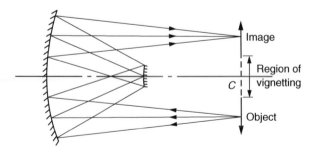

Figure 15.21   The Offner autocollimating system.

As an example of this system we may consider the following:

	c	d
Concave	0.1	
		5
Convex	0.2	
		5
Concave	0.1	

with

$$l = l' = 10.0, \quad m = -1$$

$$H' = 1: \quad Z'_s = 0, \quad Z'_t = 0.00205$$

$$H' = 2: \quad Z'_s = 0, \quad Z'_t = 0.03519$$

It is observed that the astigmatism is much smaller than in the Dyson system, and moreover the long air space between the mirrors and the object plane permits the insertion of plane mirrors to deflect the beam if desired.

15.5 TWO-MIRROR SYSTEMS

The classical two-mirror systems used in telescopes date from the seventeenth century. They were either of the Gregorian form, with a concave parabolic primary mirror and a concave elliptical secondary, or of the Cassegrain form, with the same parabolic primary but a convex hyperbolic secondary.[26] The Gregorian form was popular for a hundred years as a small erecting telescope for terrestrial observation. Because of the near impossibility of making an accurate convex hyperboloid, the Cassegrain form only gradually came into use as grinding and polishing techniques were improved. Today Cassegrain telescopes are found in most astronomical observatory.

15.5.1 Two-Mirror Systems with Aspheric Surfaces

Suppose we lay out a simple Cassegrain system as shown in Figure 15.22. The primary mirror has a radius of curvature equal to 8.0, a focal length of 4.0, and a clear aperture of 2.0 ($f/2$). The secondary mirror has a radius of 3.0 with conjugate distances of -1 and $+3$, forming a final image at the middle of the primary mirror at a magnification of three times. Thus the overall system has a focal length of 12.0 and a relative aperture of $f/6$.

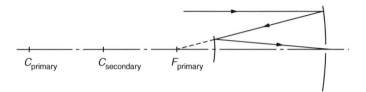

Figure 15.22   A simple Cassegrain system.

Starting with two spherical mirrors, we find a large residual of undercorrected spherical aberration. This was eliminated in the classical Cassegrain system by making the primary mirror parabolic, of eccentricity equal to 1.0, and the secondary mirror hyperbolic, with an eccentricity of 2.0 in our case. This served to remove the spherical aberration perfectly, leaving an *OSC* residual of 0.001736. The paths of a marginal ray through each of these systems are shown in Table 15.5. For the *OSC* calculation, the stop was assumed to be at the primary mirror, its image being at a distance of $l'_{\text{pr}} = -1$ from the secondary mirror.

In the late 1920s Ritchey and Chrétien recognized that the cause of the coma in the classical Cassegrain is that the final U' of the marginal ray is too small, making the marginal focal length F' too long. They therefore suggested departing from the conventional forms of the two mirrors and using shapes that are somewhat flattened at the edge. A few trials show that in our example the eccentricity of the primary mirror should be raised from 1.0 to 1.0368 (a weak hyperbola) and that of the secondary from 2.0 to 2.2389. These changes completely remove both the spherical aberration and the *OSC*, as can be seen in the fourth ray trace in Table 15.5.

The amateur telescope maker finds it almost impossible to make the mirrors required for these well-corrected systems, especially the convex hyperboloid of the classical Cassegrain. He is therefore tempted to use the Dall-Kirkham design, in which the secondary is a convex sphere while the primary is a concave ellipse. A few trials reveal the desired eccentricity of this ellipse in any particular case. For our example the primary ellipse should have an eccentricity of 0.839926, as shown in the fifth ray trace in Table 15.5. It is clear that the real problem here is coma, which is five times as large as in the classical Cassegrain. Obviously it is wrong to strengthen the rim of the primary, as in the Dall-Kirkham, when it should be weakened, as in the Ritchey-Chrétien form. However, the Dall-Kirkham does have the additional advantage that the elliptical primary can be tested in the workshop before assembly by the use of a pinhole source at one focus and a knife-edge at the other. In our example the two focal lengths are 4.35 and 50.0, respectively.

Table 15.5

Two-Mirror Telescope Systems

c		−0.125		−0.3333333		
d			−3		1	
n	1		1			
Paraxial ray						
φ		0.25		−0.6666666		f′ = 12.0
d/n			3			
y		1		0.25		l′ = 3.0
nu	0			−0.25	−0.083333	
Spherical surfaces						
Q		1.0		0.2401960		F′ = 11.522999
Q′		0.9843135		0.2435727		L′ = 2.806690
U	0		14.36151		−4.97856	
Y		1.0		0.2453707		LA′ = −0.19331
Z		−0.0627461		−0.0100513		OSC = 0.009013
Classical Cassegrain						
e		1.0		2.0		
Q		1.0		0.2461538		F′ = 12.020833
Q′		0.9846154		0.2495667		L′ = 3.0
U	0		14.25003		−4.77189	
Y		1.0		0.251309		LA′ = 0
Z		−0.0625		−0.0104710		OSC = 0.001736
Ritchey–Chrétien						
e		1.0368		2.2389		
Q		1.0		0.2465948		F′ = 12.0000
Q′		0.9846376		0.2500017		L′ = 3.0000
U			14.24178		−4.78019	
Y		1.0		0.2517515		LA′ = 0
Z		−0.062482		−0.0104896		OSC = 0
Dall–Kirkham						
e		0.839926		0		
Q′		1.0		0.2444135		F′ = 12.104064
Q		0.9845275		0.2478498		L′ = 3.0000
U	0		14.28260		−4.73900	
Y		1.0		0.2495620		LA′ = 0
Z		−0.0625721		−0.0103982		OSC = 0.008672

15.5.2 A Maksutov Cassegrain System

Many Cassegrain systems have been constructed using only spherical mirrors, the spherical aberration being corrected by means of a meniscus corrector lens placed in the entering beam. The secondary mirror can be conveniently formed by depositing an aluminized reflecting disk on the rear surface of the corrector.

Table 15.6

A Classical Cassegrain System

		Concave		Convex		
c		-0.1		-0.4		
d			-4		1	
n	1		-1			
ϕ		0.2		-0.8		
d/n			4			
y		1		0.2		$l' = 5.0$
nu	0		-0.2		-0.04	$f' = 25.0$

As an example, suppose that before adding the corrector lens we have two mirrors separated by a distance of 4.0, the concave primary placing its image at 1.0 units behind the secondary mirror, which in turn projects the final image to a point 1.0 behind the primary mirror, through a hole. The focal length of the primary is 5.0, and the secondary mirror magnifies this by five times, thereby giving an overall focal length of 25. The paraxial ray trace is shown in Table 15.6. It will be seen that the l' after the concave mirror is -5.0, and this value must be maintained after adding a corrector lens in order for the final image to remain at a distance of 1.0 behind the primary mirror. This is achieved by recalculating the curvature of the primary mirror each time a change is made in the system.

To design the correcting lens, we start with some guessed value of c_1, retain the radius of the secondary mirror $c_2 = c_4 = -0.4$, trace a paraxial ray to solve for c_3 (primary mirror) to give the desired back focus, and then add a marginal ray at $f/10$. This gives us the spherical aberration and also the $(D - d)\,\Delta n$ value arising at the lens. After tracing a paraxial principal ray through the front lens vertex, we find l'_{pr} and so determine the *OSC*. Our trials yielded the following:

c_1	c_3	LA'	f'
-0.5	-0.117763	$+14.7$	20.5666
-0.42	-0.101396	-0.4475	23.8626
-0.43	-0.103723	$+0.5673$	23.3938
(Setup A) -0.425	-0.102571	$+0.0397$	23.6259

Taking this last case for further study, we find that the zonal spherical aberration is -0.0170, the $(D - d)\,\Delta n$ value is -0.0000009 (insignificant), and $l'_{pr} = -1.204$, giving $OSC = -0.00226$.

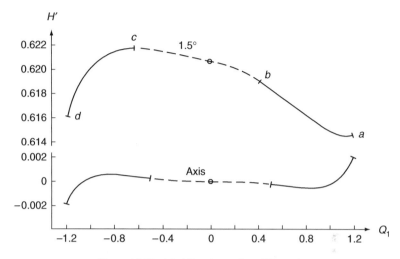

Figure 15.23 Meridional ray plot of Setup A.

To investigate the seriousness of this *OSC* residual we next make a meridional ray plot at an angular field of, say, $-1.5°$ (Figure 15.23). This is a very small field angle, but it serves to determine the size of the hole in the primary mirror, and if the field is too wide there will be very little mirror left for image formation. The paths of the upper and lower limiting rays are shown in Figure 15.24, where it will be seen that the hole in the primary mirror is the factor that determines which rays get through and which do not. A front view of the system, looking upwards along the 1.5° beam is shown in Figure 15.25.

There are two branches to the meridional ray plot, the left-hand branch containing those rays that strike the primary mirror below the hole and the right-hand branch containing those above the hole. There is obviously a large amount of negative coma in this system and there is some degree of inward-curving field, although the Coddington fields are meaningless here since the principal ray is blocked out by the secondary mirror. The Petzval sum, arising mainly at the secondary mirror, is very large (0.5863).

The most effective way to improve this system is to increase the central air space. We will therefore increase this to 5.0, and to maintain the focal length at 25.0 we repeat the paraxial layout (Table 15.7).

After adding the corrector lens, we must determine the curvature of the primary mirror, at such a value that the $l'_3 = -6.978947$ in order to place the image once more at 1.0 behind the hole in the primary mirror. Utilizing the previous procedure we end up with the following system (Setup B):

	c	d	n_e	Glass
	−0.249			
		0.25	1.52111	K-4
	−0.233333			
		5.0		
(concave)	−0.0786549			
		5.0		
(convex)	−0.233333			

with $f' = 23.82816$, $l' = 6.0000$, $LA' = 0.00272$, $LZA = -0.00383$, $OSC = 0.00014$, Ptz = 0.3046.

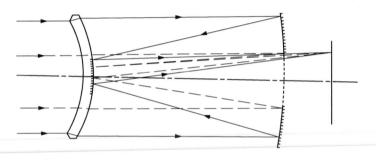

Figure 15.24   Ray diagram of Setup A.

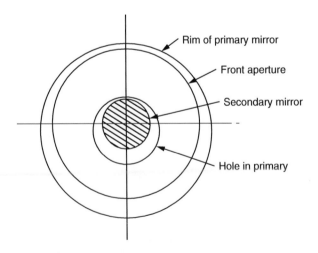

Figure 15.25   Front view of system, looking upward at 1.5° obliquity.

Table 15.7
A Cassegrain with Increased Separation

c	−0.076		−0.233333		
d		−5			
n	1	−1		1	
ϕ	0.152		−0.466666		
d/n		5			
y	1		0.24		$l' = 6.0$
nu	0	−0.152		−0.04	$f' = 25.0$

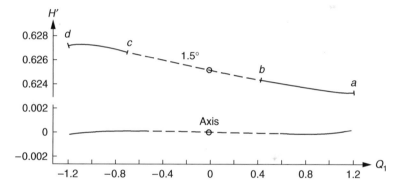

Figure 15.26 Meridional ray plot for the longer Setup B.

The meridional ray plot is shown in Figure 15.26, and the lens will be seen to be almost perfect except for a strongly inward-curving field. To remove the Petzval sum entirely requires that the two mirrors have the same radius; this occurs when the central air space is about 9.0. At an even longer space, at 12.0, the secondary mirror becomes a plane and all the power is in the primary. At these increased lengths the coma correction becomes a problem.

In an effort to reduce the Petzval sum with a reasonably short system, we can insert a negative field flattener in the hole in the primary mirror. We must then redetermine the radius of the primary to restore the back focus at 1.0 beyond the field flattener. The addition of this negative lens increases the focal length (the system is now an extreme telephoto), and it makes both the spherical aberration and the *OSC* more positive. The spherical aberration can be corrected by an adjustment of c_1, giving Setup C, as shown in the following table:

	c	d	n
Corrector lens	$\left\{\begin{array}{l} -0.24865 \\ -0.233333 \end{array}\right.$	0.25	1.52111
		5.0	
Primary mirror	-0.0786009		
		5.0	
Secondary mirror	-0.233333		
		5.13	
Field flattener	$\left\{\begin{array}{l} -0.5 \\ 0 \end{array}\right.$	0.1	1.52111

with $f' = 30.325$, $l' = 0.099991$, $LA' = 0.01809$, $LZA = -0.01401$, $OSC = 0.00090$, Ptz $= 0.1329$.

The 1.5° meridional ray plot of this lens is shown in Figure 15.27 alongside that of the axial image. The zonal aberration has become much larger and the positive coma is decidedly serious. The coma can be reduced by slightly reducing the central air space, as illustrated in the following system (Setup D):

c	d	n
$\left.\begin{array}{l} -0.28192 \\ -0.265446 \end{array}\right\}$	0.25	1.52111
	4.74	
-0.0837446		
	4.74	
-0.265446		
	4.84	
$\left.\begin{array}{l} -0.5 \\ 0 \end{array}\right\}$	0.1	1.52111

with $f' = 30.0756$, $l' = 1.00001$, $LA' = 0.03100$, $LZA = -0.02070$, $OSC = 0.00047$, Ptz $= 0.1865$. We have evidently not gone quite far enough since the OSC is still positive. This final system is illustrated in Figure 15.28 and its meridional ray plot at 1.5° is shown in Figure 15.29. It will be seen that shortening the system has indeed reduced the coma, but it has greatly increased the zonal spherical aberration, which causes the ends of the 1.5° graph to depart quickly from the desired form.

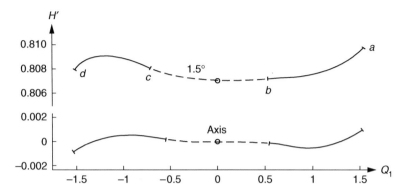

Figure 15.27 Meridional ray plot of Setup C.

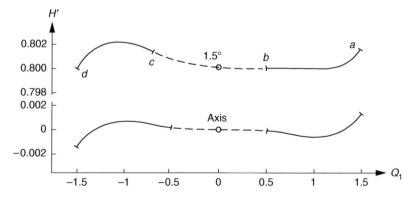

Figure 15.28 Ray diagram of final Setup D. [For the sake of clarity the limiting rays (*b*) and (*c*) have not been drawn in this figure.]

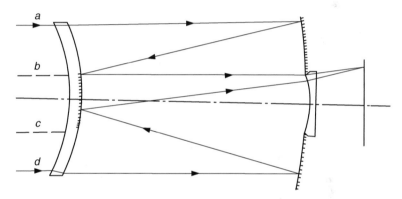

Figure 15.29 Meridional ray plot of the final Setup D.

Systems of this general type have been used frequently in the longer focal lengths for 35-mm SLR cameras. The field angles can be readily found since the picture diagonal is 43 mm:

Focal length (mm)	Semifield (deg)
500	2.5
750	1.6
1000	1.2

It must be remembered that these reflecting systems need very careful and complete internal baffling to prevent light from going straight to the film without being reflected by the mirrors. Also, it is almost impossible to introduce an iris diaphragm to vary the lens aperture, so all the exposure control must be made by varying the shutter speed.

15.5.3 A Schwarzschild Microscope Objective

It was discovered by Karl Schwarzschild in about 1904 that a two-mirror system of the *reversed telephoto type*—that is, one in which the entering parallel light first encounters the convex mirror from which it is reflected over to the large concave secondary mirror—has the remarkable property that if both mirrors are spherical and have a common center C, then the primary spherical aberration, coma, and astigmatism are all automatically zero provided the ratio of the mirror radii is equal to $(\sqrt{5} + 1)/(\sqrt{5} - 1) = 2.618034$.[27] This conclusion can be easily verified for primary aberrations by use of the stop-shift formulas given in Section 11.7.2.

At finite aperture this system suffers from a very small spherical overcorrection, an example being as follows:

	c	d
(convex)	1.0	
		1.618034
(concave)	0.381966	

with $f' = 0.809017$, $l' = 3.427051$, LA' $(f/1) = 0.00137$, OSC $(f/1) = 0.00129$, LA' $(f/2) = 0.00008$, OSC $(f/2) = 0.00007$. For the OSC calculation it was assumed that the stop is at the concave mirror, making $l'_{pr} = 0$, hence $OSC = (F'l'/f'L') - 1$.

However, when this system is intended for use as a microscope objective, the object must be at such a finite distance as to give the desired magnification.

To correct the spherical aberration and coma it is then necessary to weaken the concave mirror appropriately, the two mirrors remaining concentric about the common center C. The separation is, of course, equal to $r_1 - r_2$. A few trials tell us quickly what separation should be used. The following is an example of a $10\times$ objective of this type:

	c	d
(convex)	1.0	
		2.07787
(concave)	0.3249	

with

$$L = l = -7.14694, \quad \sin U = u = -0.05$$

$$l' = 3.89256, \quad m = -0.1$$

$$NA = 0.5: \quad LA' = -0.000002, \quad OSC = -0.000003$$

$$NA = 0.35: \quad LA' = -0.000394, \quad OSC = -0.000382$$

There is a small zonal residual of spherical aberration, decidedly less than the zonal tolerance of $6\lambda/\sin^2 U'_m$ given in Section 6.5.2, which in this case amounts to 0.00052, assuming that the unit of length is the inch. A scale drawing of the system is shown in Figure 15.30. It will be seen that the diameter of the convex mirror must be 0.72 to catch the marginal ray at $NA = 0.5$, and this blocks out the middle of the beam so that the lowest ray has an NA of 0.193. The diameter of the hole in the concave mirror must be about 0.56, but this is not a limiting aperture and it can be made somewhat larger. However, if it is too large it will

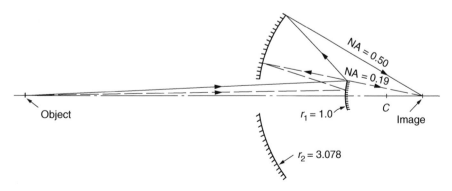

Figure 15.30 A Schwarzschild microscope objective.

pass unwanted light, which is undesirable. The obstruction is not large enough to cause a serious degradation of the definition.

15.5.4 Three-Mirror System

An interesting modification of the classical Gregorian reflecting system has been suggested by Shafer,[28] in which the marginal ray is reflected twice at the primary mirror, at equal distances on opposite sides of the axis, and once at the concave secondary mirror, the final image being formed at the center of the secondary mirror. The paraxial layout is shown in Figure 15.31. If the mirror radii are respectively 10 and 7.5, the separation is 7.5 and the focal length is -7.5. A few trials indicate that for the simultaneous correction of spherical aberration and *OSC* with a semiaperture of 1.0 (i.e., an *f*/3.75 system), the primary must be a concave ellipse with eccentricity 0.63782 and the secondary a concave hyperboloid with eccentricity 2.44. The aperture stop and both the pupils are at the primary mirror. Since the middle half of the entering beam is blocked out by the secondary mirror, the image receiver can cover the middle half of the secondary mirror without introducing any further obstruction. The angular field of our *f*/3.75 system is therefore $\pm 1.9°$. If the aperture is doubled, the angular field will also be doubled to $\pm 3.8°$.

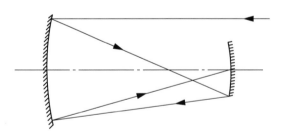

Figure 15.31 A three-reflection aplanatic system.

15.6 MULTIPLE-MIRROR ZOOM SYSTEMS

In the past several decades, there has been some interesting work on multiple mirror systems that often have a zooming capability. There are two general types, namely, obscured and unobscured pupils. In this section, both types will be discussed with some being fixed focused and other being zoom capable.

15.6.1 Aberrations of Off-Centered Entrance Pupil Optical Systems

When the entrance pupil is decentered with respect to the optical axis of the optical system of an otherwise rotationally symmetric system, it breaks the normal symmetry and the system becomes known as a plane-symmetric system. The aberrations no longer appear as they do for a rotationally symmetric system and other new aberration coefficients appear in the aberration expansion equation. We will not discuss this expansion, but will now look at the general behavior of the aberrations. Both the terms *decentered* and *off-centered* are used interchangeably in the literature.

Figure 15.32 shows the distortion behavior for an off-centered optical system having zero, positive and negative aberration values for distortion, coma, astigmatism, and Petzval.[29] When the pupil is centered, the coma, astigmatism, and Petzval do not affect the distortion of the image. Spherical aberration for an *f*/2 optical system with a centered pupil having a diameter of 5 is shown in the center of Figure 15.33. A close-up view of the focal region is shown at the top of the figure and five focus positions are included at the bottom. These are what we are accustomed to viewing. Now consider an off-centered entrance pupil system having a focal length of 10 and operating at *f*/5.

Figure 15.34 presents the spherical aberration for three pupil offset displacements of 2, 3, and 4 as well as for four defocus positions. In the zero defocus position, the spot diagram rather appears to have a comatic shape while the defocused positions appear to be a combination of astigmatism and coma;

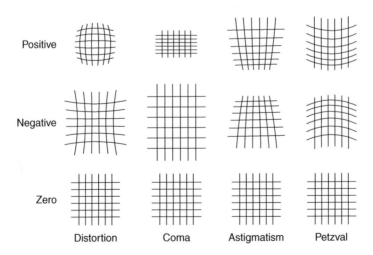

Figure 15.32   Distortion for optical systems having an off-centered entrance pupil.

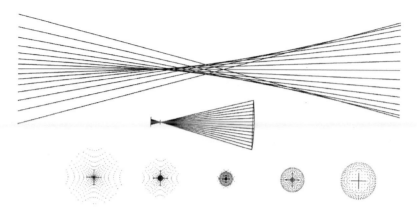

Figure 15.33 Spherical aberration for a centered-pupil optical system showing spot diagrams for several defocused positions.

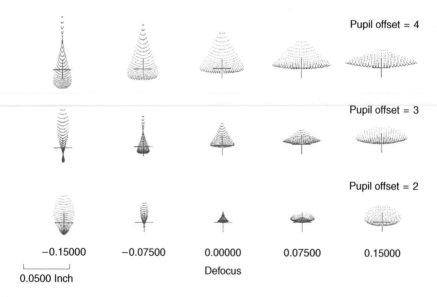

Pupil offset = 4

Pupil offset = 3

Pupil offset = 2

−0.15000   −0.07500   0.00000   0.07500   0.15000

0.0500 Inch   Defocus

Figure 15.34 Spherical aberration for an off-centered entrance pupil optical system showing spot diagrams for several defocused and entrance pupil offset positions.

however, they have their own unique shapes. Coma for the decentered pupil condition appears much like the centered pupil situation. Figure 15.35 illustrates the behavior for the $f = 10$, $f/5$ optical system with the source 1° off-axis and the pupil decentered by 2. In the lower right corner of the figure, the coma for the same system with a centered pupil is shown and is scaled up by a factor of 2.

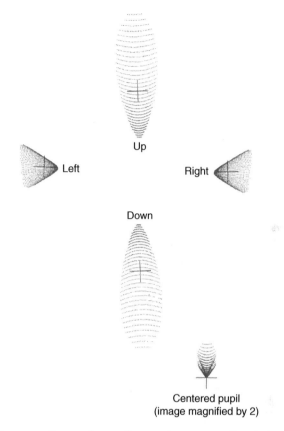

Figure 15.35 Coma spot diagrams for an off-centered entrance pupil optical system compared with the spot diagram for the same system having a centered entrance pupil.

In the following examples, the spot diagrams presented will be seen to contain what will appear as perhaps oddly shaped distributions but are consistent with the introduction just presented.

15.6.2 All-Reflective Zoom Optical Systems

Woehl published the first paper on all-reflective zoom systems.[30] The principal purpose was to provide a means for beam shaping and image manipulation. The unobscured, off-axis six-mirror configuration had two fixed mirrors for input focusing and output reimaging, and two pairs of dual moving mirrors to affect zooming. The zoom range was 30:1 with the field-of-view limited by aberrations.

The first patent for an all-reflective zoom optical system was filed by Pinson in 1986.[31] This system has a centered and obscured entrance pupil with no field-of-view offset. The patent covers a variety of configurations having two or more mirrors with examples for two to six mirrors. Practical designs require more than two mirrors due to aberration control, with four being a compromise. The generic form of the optics is back-to-back Cassegrain-type optics. The pupil shape is unusual and varies as a function of field angle and zoom ratio. Reasonably good resolution can be realized for f-numbers less than four. The system is uncompensated; that is, the images move as the optical system zooms. Two proof-of-concept four-mirror prototypes have been built.[32]

The first-generation unit was made from spherical and conic surfaces with a primary mirror diameter of about four inches. The zoom range was 4:1 with a field-of-view range of 1.5° to 6° and $f/3.5$. Remarkably, the overall length of the telescope varied from 13.5–15.5 inches while the focal length varied from 2.5 to 10 inches. The second-generation unit's mirrors were all conics with higher-order aspherics on the secondary and tertiary mirrors. With an $f/3.3$ and a zoom ratio of 4:1, the primary mirror diameter was 4.9 inches and the overall length of the unit was about one inch less than the first generation. On-axis performance was diffraction limited and suffered some coma off-axis.

In 1989, Rah and Lee published a description of a four-mirror zoom telescope that maintained the aplanatic condition throughout the zoom range.[33] This obscured-pupil, uncompensated design used spherical mirrors in a cascaded Cassegrain–Cassegrain configuration.[34] By this is meant the order of surfaces is first primary mirror, first secondary mirror, second primary mirror, and second secondary mirror, in contrast to the Pinson configuration of first primary mirror, first secondary mirror, second secondary mirror, and second primary mirror. With the 2:1 zoom range, the f-number varied from 4 to 8 and the field of view (FOV) was maintained at a constant one degree throughout the zoom range. The FOV is limited by astigmatism; however, Rah and Lee observed that conic mirrors could be used to correct this aberration. It should be noted that the overall length of this structure changes quite dramatically with zoom.

In early 1991, Cook was awarded a patent for an all-reflective continuous zoom optical system.[35] This obscured entrance pupil configuration comprises three mirrors arranged to form an anastigmat. The primary and secondary mirrors form a Cassegrain, with the tertiary mirror moving to affect the zooming function, basically serving as a relay mirror. The image is uncompensated and the line-of-sight changes with zoom in some realizations. This telescope was designed for a scanning system and has a narrow along-scan field-of-view and a wide cross-scan field-of-view. The image surface is flat and has a constant offset from the optical axis while the intermediate image has a varying offset. The basic design shown was a 2:1 zoom with the f-number varying from $f/5.14$ to $f/10.2$, focal length of 154.2 to 305.5, entrance pupil diameter of 30, and FOV offset of 3° to 1.5°. The structure of this system is as shown in

Table 15.8

All-Reflective Continuous 2:1 Optical System Patented by Cook

Surface	Radius	Conic const.	Thickness	Zoom position
Primary	−104.067	−0.92298	−39.4831	
Secondary	−32.8640	−1.9639	100.005	A: 3°
			91.329	B: 2.25°
			86.351	C: 1.5°
Tertiary	−38.6032	1.0489	−32.847	A: 3°
		$A6 = 0.32497 \cdot 10^{-5}$	−39.065	B: 2.25°
		$A8 = 0.36639 \cdot 10^{-8}$	−46.062AA	C: 1.5°
Image	Flat			

Table 15.8, where $A6$ and $A8$ are the sixth- and eigth-order aspheric deformation coefficients.

The following year, Kebo received a patent for another all-reflective zoom optical system which taught afocal telescopes with object-space FOVs of up to a couple of degrees and zoom ranges of 2:1 (1.7×–3.6×) and 4:1 (0.125×–0.5×).[36] The 2:1 design is a reasonably compact four-mirror unobscured-pupil configuration, having a common optical axis, that uses off-axis sections of rotationally symmetric mirrors for a FOV coverage of 0.95° to 2°. Unlike the prior systems by Cook, the FOV is centered on the optical axis in angular coordinates (but spatially translated). The basic mirror shapes are parabolic primary, hyperbolic secondary, and spherical tertiary and quaternary mirrors. To achieve the zoom function, it is required to move the final three mirrors. An intermediate image is formed by the primary and secondary mirrors at the location of the tertiary mirror. An interesting aspect of this design is that the exit pupil remains fixed with respect to the primary mirror and optical axis while the entrance pupil, lying prior to the primary mirror, utilizes different portions of the primary mirror with zoom. The on-axis 80% geometric blur diameters are 0.36 mrad (1.7×) and 0.09 mrad (3.6×).

The second of Kebo's designs uses a three-mirror unobscured-pupil configuration with the primary mirror being the stationary mirror. It has a 4:1 (0.25°–1° object space) zoom range, and the remotely located entrance pupil in front of the primary mirror is fixed with respect to the tertiary mirror and the optical axis. The primary and secondary mirrors are ellipsoidal, and the tertiary mirror is hyperbolic, all having a common optical axis and being off-axis sections. In this case the output beam moves over the tertiary mirror with zoom. Also, this design is not compact and the tertiary mirror is much greater in size with respect to the primary mirror. The on-axis 80% geometric blur diameters are 0.084 mrad (0.125×) and 0.15 mrad (0.5×).

Also in 1992, Korsch investigated a dynamic three-mirror obscured-pupil zoom telescope for potential use in planetary observations.[37] This design had a 4:1 (0.125°–0.5°) zoom range operating at f/3.3 at a FOV of 0.5°. The image and the tertiary mirror remain fixed with respect to one another during zoom. Both the primary and secondary mirrors move during zoom. An unusual aspect of Korsch's design was the incorporation of a dynamically deformable primary mirror to correct for both aberrations and focus during zoom. The telescope has a flat image field, essentially no distortion, and excellent resolution.

An unobscured-pupil five-mirror all-spherical mirrors zoom telescope was designed by Shafer in 1993.[38] The stop is located remotely in front of the primary mirror and remains fixed with respect to the primary mirror while the other mirrors and image move about during zoom. The FOV range is 8° to 3.2° (diameter) with the corresponding *f*-number range being 3.5 to 8.75. The geometric resolution over the entire FOV and zoom range is 100 μrad. Maintaining alignment during zoom can be challenging.

Many of these mirror systems are designed to have an accessible entrance pupil location so that they can mate with another optical system such as a camera or other sensor. The afocal type can be used to change the FOV of a sensor while the focal type may be used as a collimator or projector for coupling efficiently with a sensor under test or calibration.

15.6.3 Off-Centered Entrance Pupil Reflective Optical Systems

In 1994, Johnson investigated a variety of three-mirror unobscured-pupil zoom telescopes for planetary science missions.[39] Since the 1970s, a variety of fixed-focused, three-mirror, unobscured-pupil, anastigmatic optical systems have been developed and are often used in specialized space-based sensors and custom collimators for testing infrared sensors.[40,41] Although the image is uncompensated, the configuration shown in Figure 15.36 required as small a volume as possible while providing a zoom range of 1.5° to 3° (square) and operating at *f*/3 at FOV = 3°, and needed a flat image field. The aperture stop (entrance pupil) of diameter 152 mm is located at about a constant 1.5 m from the primary mirror. A decentered entrance pupil is used with all of the mirrors having a common optical axis. The mirrors are all segments of rotational symmetric forms, which are conic with aspheric deformations up to tenth order. The FOV center is offset from the optical axis by 5°, which means that all of the useful FOV is located at an off-axis portion of the image field of the telescope—that is, the actual image area is located no closer than 2° to the optical axis.

This technique allows the use of different portions of aspherized secondary and tertiary mirrors to be used in aberration control during zoom as shown in the Figure 15.36. The image is flat field, less than 1% distortion, and has little anamorphic error. This telescope also forms an intermediate image and a real

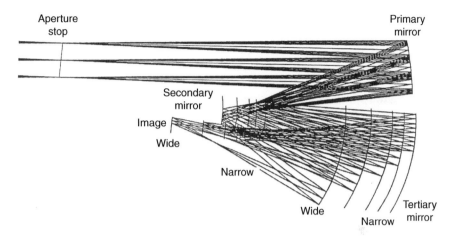

Figure 15.36 Three-mirror unobscured-pupil zoom telescope with 2:1 zoom range.

exit pupil. It was found that the angular offset of the FOV is critical in achieving good optical performance over wide zoom ranges of up to 6:1 at relatively low f-numbers. Also, care must be given in selecting parameters since the size of the tertiary mirror and motion of the mirrors can become unacceptably large.

An example of an actual system used as an infrared collimator is shown in Figure 15.37. This configuration has a well-formed intermediate image formed

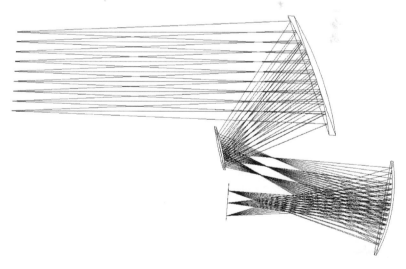

Figure 15.37 Three-mirror unobscured-pupil $f/4$ telescope having a common optical axis and remote entrance pupil.

by the primary and secondary mirrors which is relayed to the image plane by the tertiary mirror. A real and accessible exit pupil is also formed by the system. The total field-of-view of this system is 4.4° by 4.4° with a focal length of 600 mm operating at $f/4$, and the image field is flat. The central beam in the input FOV is +4° with respect to the optical axis of the telescope.

In Figure 15.37, this beam is horizontal and it can be seen that the optical system is tilted by 4 clockwise (notice the image plane). The structure of the system is given by

Surface	r	t	κ	CA radius	Decenter
Stop		630.428		75	−176.350
Primary	−487.148		−0.818	110	210
		−206.191			
Secondary	−153.290		−4.676	40	60
		326.819			
Tertiary	−283.223		−0.233	90	−30
		−319.240			
Image	Infinite			35	−40

and the sixth- through the twelfth-order aspheric deformation coefficients are shown in Table 15.9. The active object-space FOV is located at 1.8° to 6.2° in elevation (y-axis) and ±2.2° in azimuth (x-axis). Defining the structure of such a non-rotationally symmetric optical system is more complicated than the typical rotationally symmetric optical system. This is also true of optimizing such systems.

Figure 15.38 presents the geometric spot diagram for this telescope and also shows the Airy disk diameter for flux at 10 μm. As should be expected, the shape of the images over the FOV varies significantly yet has symmetry about the meridional plane. Achieving the relatively large FOV was accomplished by using different portions of the secondary and tertiary mirrors in the image formation as the FOV angles change. The beam footprint on the secondary mirror is about 30% of the total active area of this mirror and, in a like manner, is about 40% for the tertiary mirror.

Table 15.9

Aspheric Coefficients for Optical System Shown in Figure 15.37

Aspheric coefficient	Primary	Secondary	Tertiary
A6	$3.647 \cdot 10^{-16}$	$8.273 \cdot 10^{-13}$	$-6.127 \cdot 10^{-14}$
A8	$-3.812 \cdot 10^{-21}$	$-1.206 \cdot 10^{-16}$	$6.422 \cdot 10^{-18}$
A10	$2.452 \cdot 10^{-26}$	$5.437 \cdot 10^{-21}$	$3.647 \cdot 10^{-16}$
A12	$-2.657 \cdot 10^{-32}$	0.000	0.000

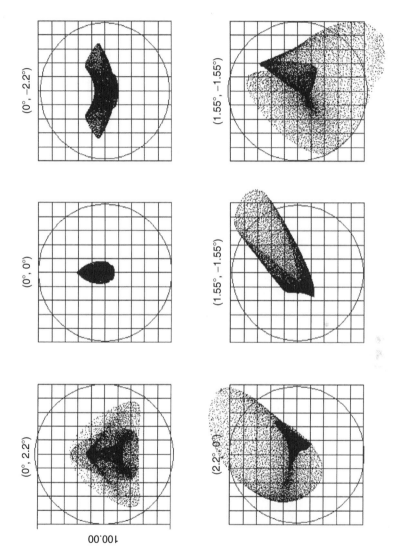

Figure 15.38 Spot diagram for collimation shown in Figure 15.37. Scale is 100 μm on a side. Circles indicate the Airy disk diameter for a wavelength of 10 μm.

This design is actually for a 2:1 zoom telescope, but only the fixed-focus design for the wide FOV is given. For the infrared spectrum, this type of telescope can be manufactured using diamond turning technology with excellent results.

Another example of a compact three-mirror unobscured-pupil telescope with an accessible entrance pupil is illustrated in Figure 15.39. This configuration has an accessible intermediate image formed by the primary and secondary mirrors and folded upward by a flat mirror. The field stop is located at this image, which is relayed to the image plane by the tertiary mirror. A real exit pupil is also formed by the system, but is not accessible. Although very compact, this configuration can provide excellent stray-light suppression by the inclusion of baffles and the field stop at the intermediate image, which is tilted at about 20° with respect to the optical axis. The total field-of-view of this system is 1° by 1° with a focal length of 1000 mm operating at $f/4$. The central beam in the input FOV is +1.5° with respect to the optical axis of the telescope, that is, the meridional FOV is +1° to +2° (positive slope angle).

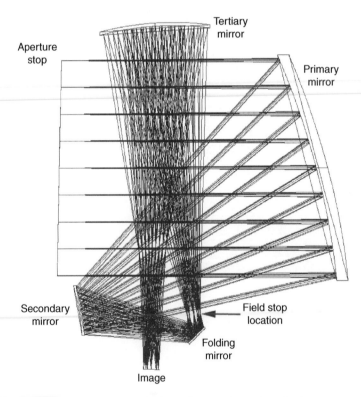

Figure 15.39   Compact $f/4$ three-mirror unobscured-pupil telescope.

In Figure 15.39, this central beam is horizontal and the optical system is tilted by 1.5 degrees clockwise. The structure of the system is given by

Surface	r	d	κ	CA radius	Decenter
Stop	Infinite	630.428	0.000	75	−201.6
Primary	−705.888		−0.801	116.9	209.206
		−289.364			
Secondary	−221.581		−3.704	26.9	48.15
		101.448			
Fold Mirror with 45° tilt	Infinite			$10.3 \times 12.9\ y$	34.83
		−359.333			
Tertiary	340.844		−0.176		
		373.439			
Image	Infinite			9.35	−25.7

and the sixth- through the fourteenth-order aspheric deformation coefficients are shown in Table 15.10.

The active object-space FOV is located at 1.0° to 2.0° in elevation (*y*-axis) and ±0.5° in azimuth (*x*-axis). Defining the structure of such a folded non-rotationally symmetric optical system is more complicated than the typical rotationally symmetric optical system. As mentioned before, this is also true of optimizing such systems.

Figure 15.40 presents the geometric spot diagram for this telescope and also shows the Airy disk diameter for flux at 10 µm. As seen in the prior example, the shape of the images over the FOV varies significantly yet has symmetry about the meridional plane. Unlike the prior example, most of the area of each of the three powered mirrors is utilized in image formation as the FOV angles change.

Rodgers developed a folded four-mirror zoom collimator that can be either focal or afocal.[42,43,44] Notice that this telescope utilizes the fold mirror in the prior two optical systems as a powered element; however, these optical systems were independently developed. Two examples are presented in this discussion. The first is illustrated in Figure 15.41 on page 496 and provides a 2:1 zoom

Table 15.10

Aspheric Coefficients for Optical System Shown in Figure 15.39

Aspheric coefficient	Primary	Secondary	Tertiary
A6	$6.880 \cdot 10^{-16}$	$-6.019 \cdot 10^{-13}$	$-8.347 \cdot 10^{-15}$
A8	$-1.065 \cdot 10^{-20}$	$4.016 \cdot 10^{-16}$	$2.154 \cdot 10^{-18}$
A10	$8.875 \cdot 10^{-26}$	$-1.079 \cdot 10^{-19}$	$-1.187 \cdot 10^{-21}$
A12	$-3.389 \cdot 10^{-31}$	$1.369 \cdot 10^{-23}$	$9.898 \cdot 10^{-26}$
A14	$3.895 \cdot 10^{-37}$	$-6.771 \cdot 10^{-28}$	$2.363 \cdot 10^{-31}$

capability with an FOV range of 1.5° to 3° (square) and a corresponding f-number range of $f/4.3$ to $f/8.6$. The internal negative power mirror is used to control field curvature of the extended FOV. The specific novelty of this design is that the "folding mirror" has a weakly powered, highly aspheric, and highly tilted surface located near the internally formed image. This mirror serves as a field element and provides control of field aberrations. The tilt of this mirror is about 30° with respect to the beam incident upon it, which allows the fourth mirror to be located above the optical system and the image formed below it. The image is uncompensated with zoom.

The second optical system is presented in Figure 15.42. It follows the design methodology of the preceding telescope. Since this system is being used as a collimator, the exit pupil is located in front of the primary mirror, as illustrated in the figure. The zoom ratio is 2:1 with an FOV range of 1.5° to 3° (square) and a corresponding f-number range of $f/4.3$ to $f/8.6$. The exit pupil diameter is fixed at 100 mm with a pupil relief of more that 1200 mm. The folding mirror is

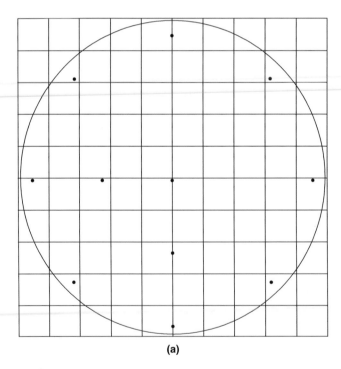

(a)

Figure 15.40   Eleven point images are shown in (a) with the blowup of each image shown in (b). The scale on each side of the grid in (b) is 200 μm. Circles indicate the Airy disk diameter for a wavelength of 10 μm.

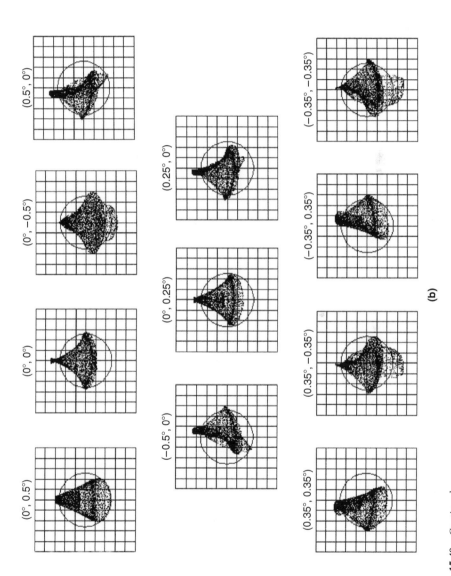

Figure 15.40 *Continued*

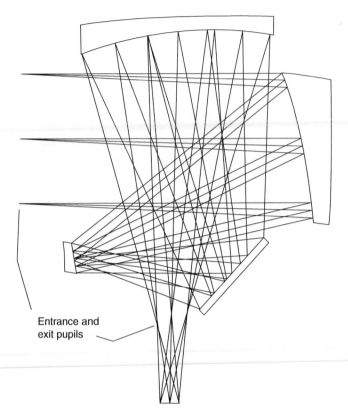

Figure 15.41   Compact four-mirror unobscured-pupil zoom telescope.

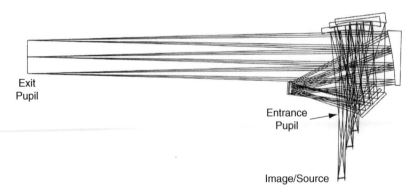

Figure 15.42   Compact four-mirror unobscured-pupil zoom telescope that has a very remote entrance pupil and accessible exit pupil.

similar to that shown in Figure 15.41. The rms wavefront error is $< 0.37\ \lambda$ for $\lambda = 1\ \mu$m and the distortion is less than 1.5%. Although the entrance pupil is accessible (located near the source/image plane), the internally formed image is not. The source/image plane is uncompensated and is observed to move a significant distance with zoom.

15.7 SUMMARY

In this chapter we presented a broad range of mirror and catadioptric systems having rotational or plane-symmetric symmetry, with obscured and unobscured pupils. In many systems, the field-of-view is centered on the optical axis while in others the entire field-of-view is remote from the optical axis. This seemingly peculiar location of the FOV is actually necessary to achieve the desired optical performance. Although this chapter was rather comprehensive, there are many more such systems that have been conceived and often built. Such systems include the afocal telescope designs by Marin Mersenne in about 1636 that uses coaxial parabolas,[45] tilted component telescopes,[46] and complex multiple-mirror telescopes such as those used at the six-mirror MMT Observatory from 1979 through 1998.[47] Wilson describes the MMT and many other telescopes in his excellent books.[48]

There are many other examples in the literature and patent files, such as a compact four-mirror afocal telescope with dual exit pupils as described by Rodgers.[49] The use of mirror and catadioptric systems is common in infrared systems[50] and astronomical and scientific instrumentation.[51,52] With the ability to make highly aspherized reflective surfaces that can be rotationally symmetric or not (off-axis section), or even be free-form, the lens designer has the opportunity to invent additional mirror and catadioptric systems.

ENDNOTES

[1] J. Villa, "Catadioptric lenses," *Opt. Spectra.*, 1:57(Mar.–Apr.), 49(May–June) (1968).

[2] D. V. Gavrilov, "Optical systems using meniscus lens mirrors," *Sov. J. Opt. Technol.* (Engl. transl.), 34(May–June):392 (1967).

[3] This same formula is a useful approximation for a spherical mirror near its optical axis; see Eq. (2-2) in Section 2.4.1.

[4] F. E. Ross, "Lens systems for correcting coma of mirrors," *Astrophys. J.,* 81:156 (1935).

[5] A. Mangin, "Mémorial de l'officier du génie" (Paris), 25(2):10, 211 (1876).

[6] A. Bouwers, *Achievements in Optics*, Elsevier, New York (1946).

[7] D. D. Maksutov, "New catadioptric meniscus systems," *J. Opt. Soc. Am.,* 34:270 (1944).

[8] D. Gabor, British Patent 544,694, filed January (1941).

[9] B. Schmidt, *Mitt. Hamburg Sternw. Bergedorf,* 7:15 (1932).

[10] R. Barry Johnson, "A high spatial and thermal resolution infrared camera for the 8–14 micrometer spectrum," *Proc. EO Sys. Des. Conf.,* 221+ pages (1970).

[11] Ralph B. Johnson, "Target-Scanning Camera Comprising a Constant Temperature Source for Providing a Calibration Signal," U.S. Patent 3,631,248 (1971).

[12] T. H. Jamieson, "Ultrawide waveband optics," *Opt. Engr.,* 23(2):111–116 (1984).

[13] L. Schupmann, *Die Medial-Fernrohre,* Druck and Verlag von B. G. Teubner, Leipzig (1899).

[14] L. Schupmann, U.S. Patent 620,978 (1899).

[15] J. A. Daley, *Amateur Construction of Schupmann Medial Telescopes,* Daley, New Ipswich (1984).

[16] The power of the negative element generally requires a slight adjustment from the power determined by the first-order equations in order to optimize the chromatic aberration performance.

[17] R. D. Sigler, "All-spherical catadioptric telescope with small corrector lenses," *Appl. Opt.,* 21(2):2804–2808 (1982).

[18] T. L. Clarke, "A new flat field eyepiece," *Telescope Making,* 21:14–19 (1983).

[19] T. L. Clarke, "Simple flat-field eyepiece," *Appl. Opt.,* 22(12):1807–1811 (1983).

[20] R. Barry Johnson, "Very-broad spectrum afocal telescope," *International Optical Design Conference, SPIE,* 3482:711–717 (1998).

[21] Fred E. Altman, U.S. Patent 2,742,817 (1956).

[22] J. Dyson, "Unit magnification optical system without Seidel aberrations," *J. Opt. Soc. Am.,* 49:713 (1959).

[23] J. Brain Caldwell, "Catadioptric relay for dual DMD projectors," in *International Optical Design Conference 1998,* Leo. R. Gardner and Kevin P. Thompson (Eds.), *SPIE,* 3482:278–281 (1998).

[24] J. Brain Caldwell, "Compact, wide angle LCD projection lens," in *International Optical Design Conference 1998,* Leo. R. Gardner and Kevin P. Thompson (Eds.), *SPIE,* 3482:269–273 (1998).

[25] A. Offner, "New concepts in projection mask aligners," *Opt. Eng.,* 14:131 (1975).

[26] H. P. Brueggemann, *Conic Mirrors,* Focal Press, London (1968).

[27] P. Erdös, "Mirror anastigmat with two concentric spherical surfaces," *J. Opt. Soc. Am.,* 49:877 (1959).

[28] D. R. Shafer, "New types of anastigmatic two-mirror telescopes," *J. Opt. Soc. Am.,* 66:1114, Abs. ThE-17 (1976).

[29] Dietrich Korsch, *Reflective Optics,* Chap. 6, Academic Press, New York (1991).

[30] Walter E. Woehl, "An all-reflective zoom optical system for the infrared," *Opt. Engr.,* 20(3): 450–459 (1981).

[31] George T. Pinson, U.S. Patent 4,812,030 (1989).

[32] R. Barry Johnson, James B. Hadaway, Tom Burleson, Bob Watts, and Ernest D. Parks, "All-reflective four-element zoom telescope: design and analysis," *Proc. SPIE,* 1354:669–675 (1990).

[33] Seung Yu Rah and Sang Soo Lee, "Four-spherical-mirror zoom telescope continuously satisfying the aplanatic condition," *Opt. Engr.,* 28(9):1014–1018 (1989).

[34] Perhaps Schwarzschild should be used rather than Cassegrain since spherical mirrors are used.

[35] Lacy. G. Cook, U.S. Patent 4,993,818 (1991).

36 Reynold S. Kebo, U.S. Patent 5,144,476 (1992).

37 D. Korsch, "Study of new wide-field, medium resolution telescope designs with zoom capability for planetary observations," *SBIR Phase I Final Report*, Korsch Optics, Inc., NASA Contract NAS7-1188 (December 1992).

38 Allen Mann, "Infrared zoom lenses in the 1990s," *Opt. Engr.*, 33(1):109-115, Figure 9 (1994); private communication with David R. Shafer.

39 R. Barry Johnson, "Unobscured three-mirror zoom telescopes for planetary sciences missions," *NASA SBIR Phase I Final Report*, Optical E.T.C., Inc., NAS7-1268 (July 1994).

40 Lacy G. Cook, "The last three-mirror anastigmat (TMA)?" SPIE CR41, *Lens Design*, pp. 310–323 (1992). This paper includes a chronology of reflective optical forms and an excellent summary of TMA studies and patents.

41 Allen Mann and R. Barry Johnson, "Design and analysis of a compact, wide field, unobscured zoom mirror system," *Proc. SPIE*, 3129:97–107 (1997).

42 J. Michael Rodgers, U.S. Patent 5,309,276 (1994).

43 J. Michael Rodgers, "Design of a compact four-mirror system," *ORA News Supplement*, Winter (1995).

44 Private communications. Drawing and data provided courtesy of J. Michael Rodgers and included with permission.

45 Henry C. King, *The History of the Telescope*, pp. 48–49, Dover, New York (1979).

46 Richard A. Buchroeder, "Tilted component telescopes. Part I: Theory," *Appl. Opt.*, 9: 2169–2171 (1970).

47 Daniel J. Schroeder, *Astronomical Optics*, Chapter 17, Academic Press, San Diego (1987).

48 R. N. Wilson, *Reflective Telescope Optics, Second Edition*, Vols. I and II, Springler-Verlag, Berlin (2004).

49 J. M. Rodgers, "Four-mirror compact afocal telescope with dual exit pupil," *Proc. SPIE*, 6342:63421J (2006).

50 R. Barry Johnson and Chen Feng, "A history of infrared optics," *SPIE Critical Reviews of Optical Science and Technology*, Vol. CR38, pp. 3–18, Bellingham (1991).

51 Lloyd Jones, "Reflective and catadioptric objectives," in *Handbook of Optics, Second Edition*, Vol. II, Chapter 18, Michael Bass (Ed.), McGraw-Hill, New York (1995).

52 Richard A. Buchroeder, "Application of aspherics for weight reduction in selected catadioptric lenses," NTIC, AD-750 758 (1971).

Chapter 16

Eyepiece Design

An eyepiece differs fundamentally from a photographic objective in that the entrance and exit pupils are outside the system. The lens itself must therefore have a large diameter, which is determined far more by the angular field to be covered than by the relative aperture. The latter is set by the objective lens and has little relation to the structure of the eyepiece itself.

So far as aberration correction is concerned, the axial spherical and chromatic are usually unimportant, and they can be corrected in the objective if necessary. On the other hand, lateral color and coma must be corrected as well as possible. Most eyepieces have a large Petzval sum, which leads to a large amount of astigmatism at the edge of the field. Because the observer naturally prefers to relax his accommodation on axis and accommodate as much as necessary when viewing at the edges of the field, it is customary to aim at a flat sagittal field and a backward-curving tangential field, including the objective, relay (if any), prisms, and so on, in the computation. An attempt to reduce the astigmatism by making the tangential field less backward-curving generally leads to an inward-curving sagittal field, which is unpleasant to the observer. Of course, the situation is much improved if some way can be found to reduce the Petzval sum of the entire system, but this is difficult because the eyepiece has a short focal length and therefore a large Petzval sum, while the objective has a longer focal length and a smaller Petzval sum.

For ease of use, the eyepoint, where the emerging principal ray crosses the axis, should be at least 20 mm from the last lens surface. This is difficult to achieve in a high-power eyepiece, and often requires a deep concavity close to the internal image plane. There may also be a serious amount of spherical aberration of the exit pupil. This causes the principal rays of oblique beams to cross the axis at points that become progressively closer to the rear lens surface at increasing obliquities, so that the eye must be moved forward to view the edges of the field. The eye is then not in the best position to view the intermediate parts of the field, resulting in a "kidney-bean" shadow, which moves about as the eye is moved. One way to correct this is to include a parabolic surface somewhere in the eyepiece, or to insert an aspheric plate in the focal plane to make the extreme principal rays diverge before entering the eyepiece.

All eyepieces suffer from some degree of distortion typically of the pincushion type as seen by the eye. This is often of such an amount that the oblique magnifying power is given more by the ratio of the emerging and entering angles themselves than by the ratio of their tangents. When this is the case the distortion amounts to about 6% at an apparent field of 24° and perhaps 10% at 35°. However, because of the circular shape of the eyepiece field these large distortions are seldom bothersome to the observer.

The design procedures for a number of the simpler types of eyepiece have been described by Conrady. (The following page numbers refer to Conrady's book.) These include the Huygenian (p. 484), the Ramsden (p. 497), the Kellner, or achromatized Ramsden (p. 503), the simple achromatic (p. 761), and the various cemented or air-spaced triplet types (p. 768). Other more complicated eyepieces have been described by Rosin.[1] In this chapter we will discuss the design of an eyepiece of the excellent so-called military type, consisting of two cemented doublets mounted close together, and also one of the Erfle type commonly used in wide-angle binoculars. In the preceding chapter (Section 15.4.8), a Schupmann eyepiece was discussed as the secondary or collimating optics for an afocal telescope having a remote exit pupil, which also can be viewed as the eyepoint for a visible version of the telescope. Clarke has presented significant information on using this single glass-type eyepiece with astronomical telescopes (Chapter 15, refs. 18 and 19).

16.1 DESIGN OF A MILITARY-TYPE EYEPIECE

As an example of the design of an eyepiece of this type, we will assume a focal length of 1 in. and a clear aperture of just over 1 in., for use with a 10-in. telescope doublet objective having a clear aperture of 2 in. ($f/5$). The true field will be 2.4° at the objective, giving an apparent field at the eye of about 25°. It should be noted that in the absence of distortion the apparent field would be given by the tangent ratio being equal to the focal-length ratio, or $\tan U'_{pr} = 10 \tan 2.4°$, where $U'_{pr} = 22.7°$. The actual emerging ray slope is more likely to be about 25°, with about 10% distortion.

16.1.1 The Objective Lens

For the objective lens, we will take the $f/5$ aplanatic doublet described in Section 10.3, scaled down to $f' = 10.0$ in e light:

	c	d	n_e	
	0.1554			
		0.32	1.56606	SK-11
	−0.2313			
		0.15	1.67158	SF-19
$(D - d)$	−0.0549164			

with $f' = 9.99963$, $l' = 9.76247$, $LA'(f/5) = 0.00048$, $LZA(f/7) = -0.00168$, $OSC(f/5) = 0.00011$. The upper and lower rim rays at 2.4° are next traced with $Y_1 = \pm 1.0$ at the front vertex, and also the principal ray midway between them. By Coddington's equations traced along the principal ray we find

$$H'_{pr} = 0.419107, \quad X'_s = -0.01455, \quad X'_t = -0.03154$$

16.1.2 Eyepiece Layout

To lay out the eyepiece, we may decide to use the same glasses as for the objective, and keep the outside surfaces plane. As a start we can make all the other surfaces of the same curvature, which for the prescribed focal length of 1.0 is 1.0337. Tentative thicknesses are set at 0.4 for the crowns and 0.1 for the flints with a separation of 0.05. As a check on these thicknesses we trace the lower rim ray entering the objective at 2.4°, and we find that it intersects the six surfaces of the eyepiece at these heights:

Field lens	Eye lens
0.5175	0.4877
0.5400	0.4362
0.5490	0.3854

A scale drawing (Figure 16.1) indicates that the thicknesses of the crown elements should be changed to 0.5 and 0.35, respectively. Having done this, we restore the focal length by changing c_4 to 1.0237.

Lateral Color

Our first task is to calculate the angular lateral color $U'_F - U'_C$ at the eye by tracing principal rays in C and F light through the entire system including the objective lens. It is best to do this at two obliquities so that a nice balance can be obtained. This is shown in Table 16.1.

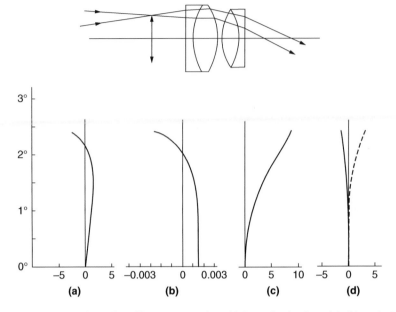

Figure 16.1 Aberrations of a military-type eyepiece. (a) Lateral color (arcmin), (b) equivalent *OSC*, (c) distortion (percent), and (d) astigmatism (diopters).

Table 16.1

Lateral Color of Military Eyepiece in Figure 16.1

Field angle (deg)	$U'_F - U'_C$ (deg)	Minutes of arc
2.4	−0.0505	−3.03
1.5	+0.0241	+1.44

This might well be considered an excellent balance since it favors the intermediate parts of the field and lets the extreme edge go. However, we can slightly reduce the lateral color at 2.4° by weakening c_5 to −1.0 and holding the focal length of the eyepiece by changing c_4 to 1.0226. This brings the lateral color at 2.4° to −2.30 minutes of arc, which we will accept.

Coma

We must now direct attention to the coma. This is found by tracing the upper and lower rim rays through the whole system, and then calculating their point of intersection by the formulas given in Eq. (8-3a); see also Section 4.3.4.

Table 16.2

Coma of the Military Eyepiece

Field angle (deg)	L'_{ab}	H'_{ab}	H'_{pr}	Coma$_t$	Equivalent OSC
2.4	−9.6221	4.6920	4.7415	−0.0496	−0.00348
1.5	−147.927	40.2579	40.3045	−0.0466	−0.00039

The vertical distance of this point above or below the principal ray is a direct measure of the tangential coma; we get the "equivalent OSC" by dividing the coma$_t$ by 3 and by H'_{pr}. Once again it is best to calculate this at two obliquities and try to secure the best balance between them, letting the extreme value go somewhat in order to favor the intermediate fields. Our present system has the characteristics shown in Table 16.2.

We must obviously try to make the coma more positive. We can do this by weakening the field lens, say by 5%, and then repeating the correction of the lateral color and focal length. These changes yield the following system:

		c	d	n_e
Field lens	⎧	0		
			0.1	1.67158
	⎨	0.982		
			0.5	1.56606
	⎩	−0.982		
			0.05	
Eye lens	⎧	1.07227		
			0.35	1.56606
	⎨	−1.03		
			0.1	1.67158
	⎩	0		

with $f' = 1.0000$, $l = -0.59779$; lateral color: $2.4° = -2.43$ arcmin, $1.5° = +1.79$ arcmin; distortion: $2.4° = 8.23\%$, $1.5° = 3.13\%$. For the coma, Table 16.3 shows what we find.

Table 16.3

Coma of Modified Military Eyepiece

Field angle (deg)	U'_{pr} (deg)	L'_{ab}	H'_{ab}	H'_{pr}	Coma$_t$	Equivalent OSC
2.4	24.4	−9.7363	4.6896	4.7282	−0.0385	−0.00271
1.5	15.1	−410.79	111.275	111.1386	+0.1263	+0.00114
					Paraxial:	+0.00156

The paraxial *OSC* is assumed to be equal and opposite to the *OSC* at the internal image, found by tracing a marginal ray back into the eyepiece from the exit pupil. Since these corrections appear to be reasonable, we next turn to the astigmatism.

Astigmatism

The astigmatism of the system is found by calculating Coddington's equations along the traced principal rays, including the objective lens as well as the eyepiece. The closing formulas give the oblique distances s' and t' from the eyepoint, which is here assumed to be at a distance of 0.7 beyond the rear surface. It is more meaningful to convert the final s' and t' values to diopters of accommodation at the eye; this is done by dividing the calculated values into 39.37, the number of inches in a meter, and reversing the sign. Table 16.4 shows what we have for our last system.

Table 16.4

Astigmatism of Modified Eyepiece in Figure 16.1

Field angle (deg)	s'	t'	Diopters at eye s'	t'	Eye relief L'_{pr} (in.)
2.4	34.054	−12.076	−1.16	+3.26	0.69
1.5	62.94	−136.19	−0.63	+0.29	0.76
				Paraxial:	0.81

In this table, a positive diopter value represents a backward-curving field that the observer can readily accommodate; a negative sign indicates an inward field, which requires the observer to accommodate beyond infinity, an almost impossible requirement for most people. Thus the negative values should be kept as small as possible, and certainly less than one diopter. The various aberrations of this final system are shown graphically in Figure 16.1.

16.2 DESIGN OF AN ERFLE EYEPIECE

When it is desired to provide an apparent angular field approaching ±35°, it is necessary to weaken the inner convex surfaces of the two-doublet "military" eyepiece and insert a biconvex element between them. This type of eyepiece was patented in 1921 by H. Erfle.[2]

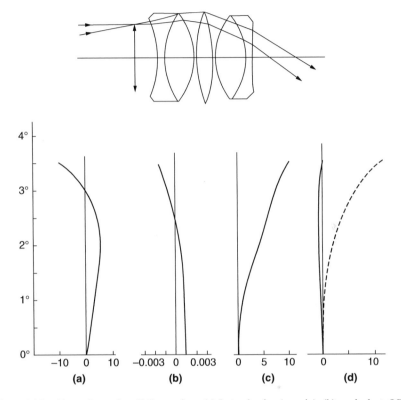

Figure 16.2 Aberrations of an Erfle eyepiece. (a) Lateral color (arcmin), (b) equivalent *OSC*, (c) distortion (percent), and (d) astigmatism (diopters).

Because of the great length of the eyepiece, and because the clear aperture must be considerably greater than the focal length, it is usual to weaken the field lens and provide a deep concave surface close to the internal image plane, so as to keep the eye relief as long as possible. The concave surface near the image also helps reduce the Petzval sum (Figure 16.2).

In view of these considerations, we will assign a power of 0.1 to the field lens and 0.4 to the middle lens; the eyelens will then come out to have a power of about 0.36 for an overall focal length of 1.0. This is an entirely arbitrary division of power and some other distribution might be better. We will use the same glasses as for the military eyepiece, with BK-7 for the middle lens. Since we have more degrees of freedom than we need to correct three aberrations, we can make some of the positive elements equiconvex for economy in manufacture. The starting system, to be used with the same objective lens as before, will be as follows:

	c	d	n_e
	-0.6		
		0.1	1.67158
Field lens $\phi = 0.1$	0.6		
		0.6	1.56606
	-0.833563		
		0.05	
	0.3949		
Middle lens $\phi = 0.4$		0.35	1.51871
	-0.3949		
		0.05	
	0.8175		
		0.6	1.56606
Eye lens	-0.8175		
		0.1	1.67158
	0.05		

with $f' = 1.0$, $l = -0.34460$, 2.5° lateral color = 8.38 arcmin.

Clearly, our first task must be to reduce the lateral color; to do this we strengthen c_7 and solve for the overall focal length by c_6. The chosen thicknesses are just sufficient to clear the 3.5° beam from the objective. Our second setup is as follows:

c	d	n_e
-0.6		
	0.1	1.67158
0.6		
	0.6	1.56606
-0.833563		
	0.05	
0.3949		
	0.35	1.51871
-0.3949		
	0.05	
0.83321		
	0.6	1.56606
-0.96		
	0.1	1.67158
0.05		

with $f' = 1.0$, $l = -0.34987$; lateral color: 3.5° $= -5.67$ arcmin, 2.5° $=$ $+5.58$ arcmin; equivalent OSC: 3.5° $= -0.00301$, 2.5° $= -0.00049$, 1.5° $= 0.00057$, axis $= -0.00096$. This lateral color is probably satisfactory, although an increase in the negative value at 3.5° would be advantageous since it would tend to reduce the lateral color at the intermediate fields. As before, the so-called equivalent *OSC* was found by tracing upper, principal, and lower rays at each obliquity and finding the intersection of the upper and lower rays in relation to the principal-ray height. The coma$_t$ found was divided by $3H'$ as before to give the equivalent *OSC*. For the axial *OSC*, a marginal ray was traced backwards, entering the eye-lens parallel to the axis at a height of $Y_1 = 0.1$, and finding the ordinary *OSC* at the internal image. The equivalent *OSC* at the eye was then taken as being equal and opposite to the true *OSC* at the internal image.

It is clear that we must reduce the negative *OSC* at the 3.5° obliquity. The simplest way to do this is to strengthen the interface c_2 in the field lens and re-adjust the interface in the eye lens to restore the lateral color correction, always holding the focal length by c_6. It is also advantageous to deepen c_8 slightly and to reduce the two air spaces between the elements. With all these changes we get the following:

	c	d	n_e
	$\begin{cases} -0.6 \end{cases}$		
		0.1	1.67158
Field lens $\phi = 0.1$	0.7		
		0.6	1.56606
	-0.846516		
		0.03	
Middle lens $\phi = 0.4$	0.3949		
		0.35	1.51871
	-0.3949		
		0.03	
	0.83941		
		0.6	1.56606
Eye lens	-0.85		
		0.1	1.67158
	0.1		

with $f' = 1.0$, $l = -0.37806$; at the internal image: $LA = +0.00612$ (under-correction), $OSC = -0.00099$ (overcorrection). The results are shown in Table 16.5.

The properties of this eyepiece were shown graphically in Figure 16.2. There is a good balance in the lateral color and also in the equivalent *OSC*. The tangential field is decidedly backward-curving, which is desirable, especially since the sagittal field is flat. The only sure way to change the field curvature is to redesign the entire eyepiece with other glasses, chosen to have a smaller index difference across the internal surfaces, but keeping a large V difference for the sake of lateral color correction.

Table 16.5

Performance of Erfle Eyepiece

Field (deg)	U'_{pr} (deg)	Lateral color (arcmin)	L'_{ab}	Equivalent OSC	Distortion (%)	Diopters s'	t'	L'_{pl}
3.5	33.9	−9.57	−2.32	−0.00150	9.50	−0.09	+11.04	0.57
2.5	24.7	+4.90	−8.93	−0.00012	5.75	−0.51	+3.88	0.64
1.5	14.9	+5.78	−54.20	+0.00068	2.17	−0.27	+0.91	0.69
			Paraxial:	+0.00099			Paraxial:	0.72

16.3 DESIGN OF A GALILEAN VIEWFINDER

The common eye-level viewfinder used on many cameras is a reversed Galilean telescope, with a large negative lens in front and a small positive lens near the eye. The rim of the front lens serves as a mask to delimit the viewfinder field, but of course since it is not in the plane of the internal image, there will be some mask parallax and the mask will appear to shift relative to the image if the observer should happen to move his eye sideways.

To design such a viewfinder, it is necessary to specify the size of the negative lens, the length of the finder, and the angular field to be covered in the object space. It is usual to assume that the eye will be located about 20 mm behind the eye lens. The magnifying power of the system follows from the given dimensions. The axial magnifying power is given by the ratio of the focal length of the negative lens to the focal length of the eyelens, which is the ratio y_1/y_4 for a paraxial ray entering and leaving parallel to the lens axis. The oblique magnifying power is given by $\tan U'_{pr}/\tan U_{pr}$ and generally varies across the field. It can be made equal to the axial magnifying power, to eliminate distortion (see Section 4.3.5), by the use of an aspheric surface on the rear of the front lens; a concave ellipsoid is a useful form for this aspheric.

As an example, we will design a Galilean viewfinder having a front negative lens about 30 mm diagonal to cover a ±24° field, a central lens separation of 40 mm, and an eyepoint distance of about 20 mm. We start by guessing at a possible front negative element. A paraxial ray is traced through it, entering parallel to the axis, and by a few trials we ascertain the radii of a small equiconvex eye lens to make the system afocal. A 24° principal ray is then traced with a starting Q_1 equal to 15 mm, and the oblique magnifying power and L'_{pr} are found. The distortion is also calculated by $MP_{\text{oblique}} - MP_{\text{axial}}$.

A concave ellipse is then substituted for the second spherical surface, of course with the same vertex curvature so as not to upset the paraxial ray, and by experimentally varying its eccentricity the distortion can be eliminated. If the L'_{pr} is then about 20 mm the problem is solved. If not, then it is necessary to change c_2 and repeat the whole process.

The following design resulted from the procedure just outlined (all dimensions in centimeters):

	c	d	n
Ellipse with $e = 0.5916$	0.1		
		0.30	1.523
$(1 - e^2) = 0.65$	0.38		
		4.00	(air)
	0.089698		
		0.25	1.523
	−0.089698		

with $L'_{pr} = 2.043$; magnifying power: $24° = 0.6250$, $15.8° = 0.6247$, axis $= 0.6249$; focal length: front lens $= -6.686$, rear lens $= 10.699$. After tracing the corner-principal ray at 24° to locate the eyepoint, other principal rays can be traced right-to-left through this eyepoint out into the object space. It will be seen that this particular elliptical surface has completely eliminated the distortion. A diagram of the system is given in Figure 16.3. In practice, of course, the front lens is cut into a square or rectangular shape to match the format of the camera, and to match its vertical and horizontal angular fields. For safety, the viewfinder is often constructed to indicate a field slightly narrower than that of the camera itself.

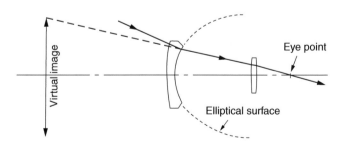

Figure 16.3 A Galilean eye-level viewfinder.

ENDNOTES

[1] S. Rosin, "Eyepieces and magnifiers," in *Applied Optics and Optical Engineering*, R. Kingslake (Ed.), Vol. III, p. 331. Academic Press, New York (1965).

[2] H. Erfle, U.S. Patent 1,478,704, filed in August 1921.

Chapter 17

Automatic Lens Improvement Programs

Many of the methods of lens design outlined in this book were the only procedures available up to about 1956, when electronic computers that had sufficient speed to be used for lens design became available. Many people in several countries then began work on the problem of how to use a high-speed computer, not only to trace sufficient rays to evaluate a system but to make changes in the system so as to improve the image quality. A brief history of this evolution is presented in Section 17.7. It is our purpose in this chapter to indicate how such a computer program is organized and how some "boundary conditions" are handled.[1,2]

When using this type of program, a starting system is entered into the program, and the computer then proceeds to make changes that will reduce a calculated "merit function" to its lowest possible value. The starting system need not be a particularly good lens, and often a very rough approximation to the desired system can be used. Indeed, some designers have even submitted a set of parallel glass plates to the computer, leaving it up to the program to introduce curved surfaces where necessary. Lenses designed in this way are not likely to be as good as those in which the initial starting system is already fairly well corrected. To gain further knowledge in the use of any of the automatic lens design programs, we suggest that the reader consult the user manual for the program of interest and consult the books cited in Section 17.8.2.

Mastery of the material contained in this treatise can serve the lens designer well by providing a solid foundation of the fundamentals of lens design. Blind use of a lens design program can and has at times provided useful results; however, the resulting design may be difficult to manufacture or align, or it may have marginal performance. Application of lens design fundamentals will almost always result in a preferable design and also provide guidance for the lens designer to control/redirect the optimization path being taken by the lens design program. For example, in Chapter 7 we showed there can be four solutions for a spherically corrected achromat. Which of the solutions is best for a particular

513

optical system design project will be difficult for almost any lens design program to select because it doesn't "know" there are multiple solutions. The designer can interject his knowledge and assist the program to follow a better path.

Perhaps in the future, knowledge engineering and artificial intelligence will achieve adequate capability that can be integrated with a lens design program to produce acceptable designs from the engineer/designer providing just the desired detailed requirements.[3] Even with methodologies to search merit-function space to find the global minimum, the resulting design achieved by one designer may be quite different from that of another designer if they should have, as is often the case, different merit functions. Arguably, the skill, experience, and creativity of the lens designer will be important in lens design for the foreseeable future.

17.1 FINDING A LENS DESIGN SOLUTION

The basic lens design optimization program includes modules for ray tracing, aberration generation, constraints, merit function, and optimization. Programs also include a variety of analysis modules to aid the lens designer in assessing progress other than by the merit function. In this section, we will present a basic understanding of optimization methods, generation of a merit function, and constraints. The lens designer should carefully study the user manual for the lens design and evaluation software being used. A certain commonality in structure, terms, parameters, optimization, and so forth, exists between the various programs, but often subtle and significant differences are present and must be understood by the lens designer for successful utilization.

17.1.1 The Case of as Many Aberrations as There Are Degrees of Freedom

We will consider first the simple case of a lens having the same number of degrees of freedom, N, as there are aberrations to be corrected. By degrees of freedom, or variables, in a lens we refer to the surface curvatures, air spaces, and sometimes lens thicknesses, although thickness changes do not generally help very much.

We first evaluate all the aberrations of our starting system. Next, small experimental changes in each of the N variables are made in turn, and we evaluate the change in each aberration resulting from this small change in each variable. (This procedure was followed in the design of a telephoto lens in Section 14.2.)

To remove all the aberrations we must now solve N equations of the form

$$\Delta ab_1 = \left(\frac{\partial ab_1}{\partial v_1}\right)\Delta v_1 + \left(\frac{\partial ab_1}{\partial v_2}\right)\Delta v_2 + \left(\frac{\partial ab_1}{\partial v_3}\right)\Delta v_3 + \dots$$

$$\Delta ab_2 = \left(\frac{\partial ab_2}{\partial v_1}\right)\Delta v_1 + \left(\frac{\partial ab_2}{\partial v_2}\right)\Delta v_2 + \left(\frac{\partial ab_2}{\partial v_3}\right)\Delta v_3 + \dots$$

where ab represents an aberration and v represents a variable, or degree of freedom, in the lens. There are, of course, N equations in N unknowns, with N^2 coefficients that we must evaluate by making small experimental changes.

Provided the variables have been chosen to be effective in changing the particular aberrations so that the equations are well conditioned, then the N equations can be solved simultaneously. If everything were linear, the solution would tell us how much each variable should be changed to yield the desired changes in the aberrations. Unfortunately, a lens is about as nonlinear as anything in physics, and it will probably happen that at least some of the calculated changes are far too large to be used, and well out of the linear range. Consequently, we take a fraction of the changes, say 20% to 40%, and apply these to the lens parameters. This should yield an improved system, but nowhere near the desired solution. Then we repeat the process, and we must now reevaluate the N^2 coefficients because the changes that we have introduced will alter the path of all the traced rays and hence all the subsequent coefficients. In the next iteration we shall be closer to the solution and the changes will be smaller, so we can take a much larger fraction, say 50% to 80%. After a third iteration we should be so close to the solution that the whole of the calculated changes can be applied. This process can be manually applied, but becomes challenging when N is very large.

17.1.2 The Case of More Aberrations Than Free Variables

Suppose we have M aberrations and only N variables, where M is greater than N. Then our procedure will yield M equations in N unknowns, and a unique solution is impossible. The equations to be solved can be written in simple form:

$$y_1 = a_1 x_1 + a_2 x_2 + a_3 x_3 + \dots$$
$$y_2 = b_1 x_1 + b_2 x_2 + b_3 x_3 + \dots$$
$$\vdots$$

where the y are the desired changes in the aberrations and the x are the changes in the variables. The quantities a, b, ... are the coefficients determined by making small experimental changes in the variables.

Although an exact solution is now impossible, we can ascertain a set of changes x that will minimize the sum of the squares of the aberration residuals R, where

$$R_1 = a_1 x_1 + a_2 x_2 + a_3 x_3 + \ldots - y_1$$
$$R_2 = b_1 x_1 + b_2 x_2 + b_3 x_3 + \ldots - y_2$$

Obviously the Rs are in the nature of aberrations. Our problem is to find the set of x values that will minimize the sum

$$\phi = R_1^2 + R_2^2 + R_3^2 + \ldots$$

there being as many R as there are aberrations and as many x as there are variables. The sum ϕ is called a *merit function*, and our aim is to reduce its value as far as possible.

There are two reasons that we sum the squares of the residuals instead of the residuals themselves. One is that all squares are positive, and of course we do not wish to have one negative aberration compensating some other positive aberration. Another reason is that any large residual will be greatly increased on squaring so that it will receive most of the correcting effort of the program, while a small residual when squared becomes smaller still and is ignored by the program. Eventually all the residuals end up at about the same value and the image of a point source will then be as small as it can become. However, the values of the quantities $a, b, \ldots$ can vary many orders of magnitude, which can cause computation problems, and the solution obtained may not actually yield the best image performance achievable for that lens configuration.

17.1.3 What Is an Aberration?

"What is an aberration?" may seem like an odd question to ask, but actually it is rather important to an understanding of the optimization problem. Perhaps a better term for aberration would be *defect*. Throughout this treatise we have discussed many image aberrations and measures of image quality. We could use the conventional aberrations, provided they are all expressed in some comparable terms such as their transverse measure, but this will often be found to be inadequate in achieving an acceptable solution.

Almost always, it is desirable to have significantly more defects than parametric variables, as will be explained in the following section. It has been found useful to trace a number of rays and regard as an aberration the departure of each ray from its desired location in the image plane. In a like manner, the optical path difference (OPD) for each ray can be computed and used, but it should be recalled that the OPD and transverse ray error are related. The OPD states the departures of the wavefront from the ideal spherical form while the transverse ray error uses the slope of the wavefront, the second being the derivative of the first. One can also use various forms of chromatic errors, differential ray traces,[4] aberration coefficients, Strehl ratio, *MTF*, encircled energy, and so on, for image defects. Combining the image defects into the merit function must be done with care since the magnitude of the errors can be dramatically different. For example, the Strehl ratio and *MTF* for a well-corrected system are somewhat less than unity while the wavefront error will be a fraction of a wavelength.[5] To compensate for the numerical disparities of the constituents of the merit function and relative importance to the lens designer, an appropriate weight is assigned to each defect.

17.1.4 Solution of the Equations

For the merit function ϕ to be a mathematical minimum, we must solve a set of equations of the form

$$\frac{\partial \phi}{\partial x_1} = 0, \quad \frac{\partial \phi}{\partial x_2} = 0, \quad \frac{\partial \phi}{\partial x_3} = 0, \dots$$

with there being as many of these equations as there are variables in the lens. Differentiating our expression for ϕ, we get the appropriate set of equations

$$\frac{\partial \phi}{\partial x_i} = 2R_1 \left(\frac{\partial R_1}{\partial x_i} \right) + 2R_2 \left(\frac{\partial R_2}{\partial x_i} \right) + 2R_3 \left(\frac{\partial R_3}{\partial x_i} \right) + \dots = 0$$

for $i = 1, 2, \dots, N$.

Entering the successive derivatives of the R with respect to x_1 for the first equation gives

$$\frac{1}{2}\frac{\partial \phi}{\partial x_1} = (a_1 x_1 + a_2 x_2 + a_3 x_3 + \dots)a_1 + (b_1 x_1 + b_2 x_2 + \dots)b_1 + \dots = 0$$

or

$$x_1(a_1^2 + b_1^2 + \dots) + x_2(a_1 a_2 + b_1 b_2 + \dots) + \dots - (a_1 y_1 + b_1 y_2 + \dots) = 0$$

Carrying out this differentiation in turn for each of the N variables, we obtain the so-called normal equations. They are simultaneous linear equations and have a unique solution. This is the well-known least-squares procedure invented by Legendre in 1805.

17.2 OPTIMIZATION PRINCIPLES

In the early stages of the optimization process it is common to find the program demanding large changes in some of the variables, which are then reversed at the next iteration. To prevent this kind of oscillation it was suggested by Levenberg[6] and others[7,8] that the merit function should be modified to include the sum of the squares of the changes in the variables x so that

$$\phi = \sum R^2 + p \sum x^2$$

The "damping factor" p is made large at first to control the oscillations, but of course when it is large the improvement in the lens is very slow.

For each iteration thereafter, the value of p is gradually reduced until the procedure finally becomes an almost perfect least-squares solution with no damping. This process replaces the use of fractions of the calculated changes suggested in Section 17.1.1. Typically, the damping factor is reduced until the merit function begins to increase again. The last three values are used to estimate the best value of p generally by a parabolic interpolation.

Lens design is an extremely nonlinear optimization problem which is linearized to the best degree practicable to allow rational constructional changes. A number of schemes have been explored by researchers over the past decades to provide the mathematical method for lens design optimization.[9] The results of these efforts indicated, most strongly, that a least-squares or minimum-variance formulation is preferable.

The overall quality of a lens system has, for the purpose of design, been found to be best described by a single-value merit function. The typical merit function used in practice includes not only image quality factors but numerous constructional parametrics. If f_i denotes the ith defect of the lens system, then potential merit functions (ϕ) include the following:

(i) $\phi = \sum_{i=1}^{M} f_i$

(ii) $\phi = \sum_{i=1}^{M} |f_i|$

(iii) $\phi = \sum_{i=1}^{M} f_i^2$

In general, defects can be expressed as

$$f_i = w_i(e_i - t_i)$$

where w_i is a weighting factor, e_i is the actual value of the ith defect, and t_i is the target value of the ith defect. The functions, f_i, have as design variables x_j, which are the constructional parameters of the system. In the following discussion, M defects and N variables are assumed. For best results, it is desired that the number of defects exceed the number of variables.

The most common merit function has the form $\phi = \sum_{i=1}^{M} f_i^2$. In matrix notation,

$$\phi = F^T F$$

where F is a column vector. Expanding each f_i in a Taylor series and ignoring terms higher than the first derivative terms yields

$$\phi = \sum_{i=1}^{M} \left[f_{0i} + \sum_{j=1}^{N} \frac{\delta f_i}{\delta x_j}(x_j - x_{0j}) \right]^2$$

$$= \sum_{i=1}^{M} f_{0i}^2 + 2\sum_{i=1}^{M} \left[f_{0i} \sum_{j=1}^{N} A_{ij}(x_j - x_{0j}) \right] + \sum_{i=1}^{M}\sum_{j=1}^{N}\sum_{k=1}^{N} A_{ij}A_{ik}(x_j - x_{0j})(x_k - x_{0k})$$

where $A_{ij} = \frac{\delta f_i}{\delta x_j}$. The term $\sum f_{0i}^2$ is a constant and is therefore ignored. Combining the remaining terms in matrix form yields

$$\phi = (X - X_0)^T A^T A (X - X_0).$$

Now,

$$f_i = f_{0i} + \sum_{j=1}^{N} A_{ij}(x_j - x_{0j})$$

or

$$F = F_0 + A(X - X_0).$$

The change vector, assuming a linear system, that would yield $F = 0$ is given by

$$(X - X_0) = -A^{-1}F_0.$$

Since the lens design problem is highly nonlinear, this solution is very unlikely to be acceptable.

Rather than requiring each f_i to equal zero, the nonlinear nature of the problem implies that it is more realistic to minimize the residuals of the f_i's. Hence,

$$\frac{\delta\phi}{\delta x_k} = \sum_{i=1}^{M} 2f_{ik}A_{ik} = 0$$

and then

$$\sum_{i=1}^{M} \left[f_{0i} + \sum_{j=1}^{N} A_{ij}(x_j - x_{0j}) \right] A_{ik} = 0$$

or

$$A^{T}A(X - X_0) + A^{T}F_0 = 0.$$

Therefore, the appropriate predicted change vector is

$$(X - X_0) = -(A^{T}A)^{-1}A^{T}F_0.$$

This result typically provides improved prediction, but changes are undamped. Without some form of damping, ill-conditioning ($A^{T}A$ close to singularity) and nonlinearities in F will cause the new value of the merit function to be worse, in general, than the starting system.

To overcome these problems, a number of damping schemes have been tried. The basic formulation is to add the damping term to the merit function to form a new merit function. Thus,

$$\phi_{NEW} = \phi_{OLD} + p^2 \sum_{j=1}^{N} (x_j - x_{0j})^2.$$

If $\frac{\delta\phi}{\delta x_k} = 0$ as before, then the change vector for *additive damped least squares* becomes

$$(X - X_0) = -(A^{T}A + p^2I)^{-1}A^{T}F_0.$$

It is evident that the change vector components are attenuated as the value of p increases. Furthermore, p affects each element of the change vector in a like manner.

An improved damping method is known as *multiplicative damping* and is given by

$$(X - X_0) = -(A^{T}A + p^2Q)^{-1}A^{T}F_0$$

where Q is a diagonal matrix with elements

$$q_j^2 = \sum_{i=1}^{M} A_{ij}^2$$

This has the effect of damping variables that cause ϕ to change rapidly. Although this often seems to work very well, it is not justifiable on theoretical grounds since the q_j values should be based on the second derivatives.[10]

Buchele[11] discussed an improved method of damping the least-squares process. Basically, it is much the same as multiplicative damping except that the damping uses a damping matrix:

$$d_{ij} = \frac{\partial^2 f_i}{\partial x_j^2}$$

which means the diagonal terms are the second derivatives and the off-diagonal terms are the partial derivatives. Although this method should be rather robust and maintain control over the merit function oscillations, the amount of effort to compute all N^2 second derivative terms can be unreasonable. An alternative pseudo–second-derivative matrix approach by Dilworth has demonstrated in actual practice both a reasonable level of computation and very impressive performance.[12]

The problems of ill-conditioning and nonlinearity mentioned above can actually limit the ability of the optimization routine to find the "optimum" solution. Ill-conditioning shows up in damped least squares as a short solution vector which limits the size of the parametric changes. To overcome these difficulties, Grey[13,14] developed a methodology that orthogonalizes the solution vectors by creating a set of orthogonal parameters (curvatures, thicknesses, refractive indices, etc.) from the original set of parameters. These orthogonal parameters can be considered as a linear combination of the original set of parameters. When the solution is found, ill-conditioning still shows up as short vectors. Since these solution vectors are orthogonal, unlike the highly correlated solution vectors in the conventional damped least-squares approaches, they are simply set to zero. Each of the remaining vectors is then scaled until nonlinearities are observed. The Grey orthonormalization process is very powerful particularly when used with the Grey merit function; however, it has been observed that use of the conventional damped least-squares method is best when "roughing-in" the lens and then switching over to the Grey method once the design is in its final stages. An interpretation of Grey's merit function was made by Seppala and clearly explains the process of aberration balancing.[15]

A variety of other techniques have been applied to the lens optimization problem including the so-called direct search, steepest descent, and conjugate gradients. None of these have been shown to be generally superior to damped least squares or orthonormalization. Glatzel and Wilson[16] developed an adaptive approach for aberration correction. Basically the weights and targets of the various aberrations are dynamically adjusted during the optimization process while attempting to keep the solution vector within the linear region. As was discussed in Chapter 4 and elsewhere, higher-order aberrations are more stable with respect to changes in constructional parameters than are the lower-order aberrations. The Glatzel and Wilson process attempted to gain control of the higher-order aberrations first and then correct the next lower order and so on. They and others have realized many successful designs using this adaptive method.[17]

It should be evident that these methods all are looking for a minimum value of the merit function in a local region of the solution space rather than the absolute minimum value in all of the solution space, that is, a global minimum. Most likely the first such effort to find the global minimum was by Brixner.[18,19] His lens design program essentially started with a series of flat plates that the program manipulated to achieve the lowest merit function value.[20] By running the program multiple times with the program trying different potential regions of the solution space, it was thought that the global minimum could be found. In the 1990s as great computing power became readily available at low cost, methods for allowing the computer program to search for a global solution became a seriously investigated topic involving simulated annealing, genetic and evolutionary algorithms, artificial intelligence, and expert systems.[21,22,23,24,25,26,27,28]

Although impressive results are often obtained, the lens designer still needs to be involved to guide and select alternative paths for the program to follow. It is noted that, at times, these solution space search methods have found new configurations that were totally unexpected. New manufacturing methods have allowed the fabrication of diffractive optics, highly aspherized surfaces, and free-formed surfaces. Each of these advances adds to the complexity and capability of the programs. Only in recent years have polarization issues been addressed in some lens design programs.[29]

One may ponder the question "Will a lens design program ultimately be able to design, without human intervention, an optical system meeting a user's requirements?" Perhaps so, but it will be at a far distant time. The lens designer provides an insight and system overview that is difficult to imagine a computer achieving. One should remember that designing the lens is only a part of the engineering activities necessary to manufacture an optical system. Tolerancing, manufacturing methods to be used, mechanical and thermal considerations, antireflective coatings, and so on, are complex factors to be included in the total design of an optical system.

17.3 WEIGHTS AND BALANCING ABERRATIONS

The optimization program has no way to know which aberration is more important than another; it only can tell the contribution the aberration makes to the merit function. The lens designer needs to assign weights to the aberrations/defects such that the contribution of each is appropriately balanced to achieve the desired correction of the lens system. For example, consider an axial monochromatic image and that a sharp image core with some flare is acceptable, as was discussed in an earlier chapter. In this case, the weighting of each meridional ray should decrease toward the marginal ray. The relative weighting can influence the amount of flare.

Many lens design programs have default merit functions that include a variety of image defect terms and associated weights. Often these can take a crude lens design and make significant progress toward an acceptable design. As the design activity progresses, the lens designer most often needs to adjust the weights and mix of aberrations to guide the program to achieve the goal. For example, in the early stages of a design, the use of transverse ray aberrations may be fruitful. As the design progresses, the use of wavefront errors or OPDs may be appropriate.[30] In some cases, final tweaking of the design may be best done using *MTF*, encircled energy, or Strehl ratio. And, of course, some clever combination of these may be necessary.

The lens designer should also be careful not to try to control aberrations that are uncorrectable. Consider, for example, the aplanatic doublets discussed in Chapter 10. We taught that one may correct axial chromatic aberration, spherical aberration, and *OSC* (coma). Attempting to control astigmatism would be imprudent in this case.

In Chapter 4, we discussed balancing aberrations. Recall that in Chapter 6 (Figures 6.3 and 6.18) we discussed how defocus was used to compensate for the residual spherical aberration. It was also demonstrated how third-order and fifth-order spherical aberration and defocus could be balanced to achieve several different outcomes depending on the lens designer's requirements.

17.4 CONTROL OF BOUNDARY CONDITIONS

In addition to the reduction of the merit function to improve the image quality, a computer optimization program must be able to control several so-called boundary conditions, for otherwise the lens may not be producible. The principal boundary conditions that must be controlled are axial thicknesses, edge thicknesses, length of lens, vignetting, focal length, *f*-number, back focal length, and overall length. At times it is important to control pupil locations, stop position, nodal points, internal image locations, and so forth. There are various ways to accomplish control of these boundary conditions.

One approach is the use of Lagrange multipliers, which are a method to constrain the solution of the least-squares optimization in such a way that the constraints are met. This approach has been successfully used and also has met with failure in the hands of an inexperienced lens designer. Should the lens designer specify a set of constraints where two or more are in conflict, the optimization program will generally abort.

Rather than attempting to demand that the program satisfy the specified constraints, it is often preferable to include them in some manner in the merit function in the form of a defect. Consider, for example, controlling the axial thickness of a lens element where the lens designer wants to keep the lens thickness of the jth surface at least 1.5 units, an edge thickness of 0.1 unit, and a maximum axial thickness of 5 units. The defects could be written as follows:

$$f_i = w_i(\text{thickness}_j - 1.5) \quad \text{or} \quad (= 0 \text{ if positive})$$
$$f_{i+1} = w_{i+1}(\text{edge thickness}_j - 0.1) \quad \text{or} \quad (= 0 \text{ if positive})$$
$$f_{i+2} = w_{i+2}(\text{thickness}_j - 5) \quad \text{or} \quad (= 0 \text{ if negative})$$

so that no contribution is made to the merit function when the constraint is satisfied. Although this approach generally works fine, it can create boundary noise that can foul the optimization process somewhat if the derivative isn't handled correctly. By making the transition softer at the boundary, the problem of discontinuity of the first derivative is typically alleviated.

One of the most common and perhaps important constraints is the focal length. As we have explained in this treatise, there are multiple ways to determine the focal length. Perhaps the most obvious way is to define it as a defect. This can work well, but at times this approach degrades the performance of the optimization. Setting the image height of a principal ray (at say 10% of the FOV) as a defect can be used to define the focal length. The reason for using a fractional image height is to avoid distortion which can cause an error in computing the focal length. A third, and often preferred, approach is to use a *curvature solve* (see Sections 2.4.5, 3.1.4, and 3.1.8) on the last surface of the lens to achieve the desired marginal ray slope angle.

The lens designer has the responsibility to adjust the weights on the multitude of defects so that the lens can achieve the desired performance. It is generally a good rule to minimize the number of defects used to those really needed to control the progress of the lens design. The reason is simple; the more defects that are present in the merit function, the less impact any given defect will have on the merit function. If the lens design program you are using offers an option to view the derivatives of the defects with respect to the design parameters, then it can be very instructive to study them as an aid to deciding if more or fewer defects would be helpful, and to provide guidance in changing the defect weights.

17.5 TOLERANCES

Closely related to the design optimization process is determination of the tolerances for the design. Establishing the tolerances for a lens system can be a major portion of the entire lens design project.[31] All of the major lens design programs provide extensive tools for establishing manufacturing tolerances including attempting to utilize existing test plates. Using existing test plates can necessitate tweaking the design to maintain performance. Some programs allow the lens designer to include tolerances in the merit function such that they are desensitized. Even for a rotationally symmetric optical system, aberrations that are caused by lens element decentration, tilt, and wedge must be given consideration.[32,33,34,35,36]

It has been mentioned that Glatzel and Shafer have each written about reducing the strain in an optical design as a means to lower the tolerancing requirements.[37] The principle basically is to minimize the angle of incidence of rays at element surfaces, which aids in not generating high-order aberrations rather than attempting to mitigate these aberrations. (See Section 6.1.6 also.)

17.6 PROGRAM LIMITATIONS

Optimization programs are generally written so that it is impossible to make a change in a lens that will increase the merit function, even though the next iteration will effect a large improvement. Also, in general no program will tell the designer that he should add another element or move the stop into a different air space. However, if an intelligent lens designer stops the program after a small number of iterations to see what is happening, he will quickly realize that an element should be divided into two, that the stop should be shifted, or that he should eliminate a lens element that is becoming so weak as to be insignificant. He may also decide to hold certain radii at values for which test glasses are available, letting the program work on only a few variables to effect the final solution.

It is also essential to remember that a computer optimization program will only improve the system that is given to it, so that if there are two or more solutions, as in a cemented doublet or a Lister-type microscope objective, the program will proceed to the closest solution and ignore the possibility of there being a much better solution elsewhere. It is this limitation that makes it very necessary for the operator to know how many possible solutions exist and which is the best starting point to work from.

17.7 LENS DESIGN COMPUTING DEVELOPMENT

The early computers used for lens design were humans performing manual calculations for a meridional ray at speeds up to perhaps 40 seconds per ray-surface.[38] In 1914, C. W. Frederick was hired by Eastman Kodak to establish a lens-design facility within the company. Although he stated he knew nothing about lens design, he was responsible for developing lens design methods and formulas adequate for lens production.

In 1937, it was recognized that Frederick (age 67) would soon retire. Rudolf Kingslake, an associate professor at The Institute of Optics, was invited to join Frederick's group with the intent that Kingslake would succeed Frederick, which occurred in 1939.[39] Kingslake retained a close relationship with The Institute of Optics for many decades thereafter. During World War II, Kingslake's group designed many lenses important for the war effort using human computers with Marchant calculators. During this same period, Robert Hopkins and Donald Feder were the principal lens designers at The Institute of Optics and also made important contributions. After the war, a few computers became available and Feder moved to the National Bureau of Standards (NBS).[40]

By 1950, ray-tracing programs had been written but the issue of automatic design was found to be quite difficult. Nevertheless, by 1954 work on automatic design programs was progressing at Harvard, University of Manchester, and the National Bureau of Standards. From 1954 to 1956, Feder explored at NBS various methods of optimizing lenses and achieved promising results. He approached Kingslake for a job at that time and soon developed an automatic design program for the new Bendix G-15 digital computer, which evolved into the LEAD (Lens Evaluation and Design) program, beginning use in 1962.[41] Manual skew-ray tracing through a single spherical surface in 1950 required over eight minutes and just one second on the G-15. By 1970, the time dropped to 50 μs on a CDC 6600 computer.

As mentioned earlier, in the 1950s digital computers of very modest capability became available (although costly) and the age of digital computer-aided lens design was born. In 1955, Gordon Black wrote about ultra-high-speed skew-ray tracing in *Nature*, where he stated that several digital computers in Britain and the United States were achieving 1 to 2 seconds per ray surface, with the fastest being about ½ second per ray surface.[42]

During the late 1950s and throughout the 1960s, groups around the world spent significant effort in developing lens design and evaluation software. Some of the activity occurred at universities while others were performed within companies for their own proprietary use. Pioneering work was performed at Imperial College London, The Institute of Optics, Eastman Kodak, Bell & Howell, Texas Instruments, PerkinElmer, and others. In Britain, SLAMS (Successive Linear Approximation at Maximum Steps) was developed by Nunn and Wynne.[43] Donald Feder[44] developed LEAD at Eastman Kodak. At The Institute of Optics, ORDEALS (Optical Routines for Design and Analysis of Optical Systems) was developed under the leadership of Robert Hopkins, and Gordon Spencer wrote the code for ALEC (Automatic Lens Correction) as part of his Ph.D. dissertation, which evolved into FLAIR,[45] POSD (Program for Optical System Design),[46] and ACCOS (Automatic Correction of Centered Optical Systems).[47]

In 1963, after ten years at Bell & Howell, Thomas Harris started Optical Research Associates and was joined by Daryl Gustafson a couple of years later. They developed their own software, which became known as CODE V, to support their consulting business and made it available commercially in the mid-1970s. CODE V rapidly became widely accepted in industry and government and has continued to remain one of the principal programs used today. Donald Dilworth also began development of SYNOPSYS (SYNthesis of OPtical SYStems) in the 1960s and made it available commercially in 1976.

The 1960s were an exciting period in the development of optical design software in part because computers were becoming available to designers and the computing power seemed to be growing exponentially year by year! It should be pointed out that computing time was rather expensive, input was by keypunch into paper cards, and turnaround time when using a mainframe was often days. In 1965, IBM introduced the IBM 1130 Computing System, which was a mini-computer about the size of an office desk. Spencer and his group developed POSD, an extension of ALEC, for the IBM 1130 which became available in 1966.

At Texas Instruments, we had an IBM 1130 dedicated for lens design and the proprietary OPTIK program written by Howard Kennedy for use on the mainframe. Even with the slow skew-ray tracing speed of 10 ray surfaces per second for POSD compared to the seemingly blazing speed of the IBM 360 running OPTIK (about 5000 ray surfaces per second), the humble IBM 1130 frequently allowed design work to proceed in an orderly manner while the use of the mainframe turnaround was often days or longer if a keypunch error had been made.

Around 1970, Control Data Corporation (CDC) had public data terminals called Cybernet which were tied into a network of CDC 7600 computers scattered around the United States, fortunately with one being in Dallas, Texas. The advantage of this was that turnaround was now measured in minutes rather than days. Also, optical design software was available on the CDC computers that could be used for a quite reasonable fee. Programs included ACCOS, GENII,[48] and David Grey's COP programs (FOVLY, MOVLY, and COVLY). Soon thereafter, a local CDC terminal was installed within the work area of the Texas Instruments lens design group and the improvement in productivity was nothing short of dramatic. This CDC capability made some of the best optical design software available to anyone and arguably changed the dynamics of optical design from just the few companies to any company being able to participate in the optics business.

Another important event in the evolution of optical design tools occurred in 1972 when Hewlett Packard introduced the HP 9830A, which looked like a desktop calculator but actually blurred the distinction between calculators and traditional computers. The programming language was BASIC and it had under 8K words of RAM and 31K words of ROM. A critical aspect of its power was the special plug-in Matrix Operations ROM that made possible the development of a lens design program for it.

Teledyne Brown Engineering (TBE) purchased an HP 9830A for our group in early 1973 for about $10,000 (about $40,000 equivalent in 2009 dollars). In short order, I wrote ALDP (A Lens Design Program) for it for the initial purpose of using it as a training tool for those in our group desiring to learn how to design and evaluate lenses. The aberrations were based primarily on those presented in Chapter 4, with tolerance control following the method developed by William Peck for GENII. Optimization choices included additive damped least squares, multiplicative damped least squares, and orthonormalization. Remarkably, many of the designers used the program for actual design work rather than just for training. Again, mainframe turnaround lag time was a consideration in this utilization. TBE considered ALDP a proprietary program and rejected any request even for publication of a technical paper. Soon thereafter, Douglas Sinclair independently began developing lens design software for the HP 9800 series that resulted in the formation of Sinclair Optics in 1976. This program was known as OSLO (Optical System and Lens Optimization) and became quite widely used.

In the 1980s, personal computers (PCs) became more available and affordable, although serious computing power really became available in the late 1990s and thereafter continued impressively increasing. During this period, ACCOS, OSLO,[49] GENII (with option for Grey's programs),[50,51] SYNOPSYS,[52] CODE V,[53] SIGMA,[54] EIKONAL,[55] and others were ported to the PC.

Some others developed code specifically for the PC, most notably Kenneth Moore's ZEMAX (after his Samoyed named Max), which was introduced in the early 1990s and has arguably become the most widely used optical design program. At the writing of this book, a PC system can be purchased for a few thousand dollars that provides ray tracing speed of tens of millions of ray surfaces per second, which is millions of times faster than the humble IBM 1130 of 40 years ago.

Another point often overlooked is that the PC cost per run and turnaround time are insignificant compared to running on a mainframe. Also, the capability of these PC-based programs has rapidly expanded to handle almost any imaginable optical configuration including those containing diffractive surfaces, nonimaging systems, nonsequential systems, free-form surfaces, polarization, birefringent materials, and so on. Also, these codes have evolved over the past 30 years to meet the ever increasing performance demands of microlithographic lenses, which are difficult to design, fabricate, and align.[56] Extraordinary analysis capability is contained in these programs that give the designer the tools often necessary to understand and explain the behavior of a lens and how it may perform in an actual system. As optical technology evolves, it is clear that the code developers will enhance their software to model the technological innovations.

17.8 PROGRAMS AND BOOKS USEFUL FOR AUTOMATIC LENS DESIGN

The following lists of lens design programs and books are intended to provide additional material that may be helpful to the lens designer using any of the various software packages. It should be noted that there are additional software packages that have specialized applications and limited capabilities, and are no longer commercially available which have not been included. No attempt was made to be all-inclusive. No representation of suitability, quality, capability, accuracy, and so on, is made by the author whether or not a software package or a book is included or excluded from the following lists. Some of the books cited are focused on the use of a specific lens design program; however, much can be still be learned by reading the material even if you are using a different program.

17.8.1 Automatic Lens Design Programs

The following are some of the automatic lens design programs available, including information about where to obtain them.

CODE V – Optical Research Associates, 3280 East Foothill Boulevard, Suite 300, Pasadena, CA 91107-3103

OSLO – Lambda Research Corporation, 25 Porter Road, Littleton, MA 01460

SYNOPSYS – Optical Systems Design, Inc., P.O. Box 247, East Boothbay, ME 04544

ZEMAX – ZEMAX Development Corporation, 3001 112th Avenue NE, Suite 202, Bellevue, WA 98004-8017

17.8.2 Lens Design Books

For further information about the subject, refer to the following books as needed.

Michael Bass (Ed.), *Handbook of Optics, Third Edition*, McGraw-Hill, New York (2009) [contains numerous related chapters].

H. P. Brueggemann, *Conic Mirrors*, Focal Press, London (1968).

Arthur Cox, *A System of Optical Design*, Focal Press, London (1964).

Robert E. Fischer, Biljana Tadic-Galeb, and Paul R. Yoder, *Optical System Design, Second Edition*, McGraw-Hill, New York (2008).

Joseph M. Geary, *Introduction to Lens Design*, Willmann-Bell, Richmond, VA (2002).

Herbert Gross (Ed.), *Handbook of Optical Systems: Vol. 3, Aberration Theory and Correction of Optical Systems*, Wiley-VCH, Weinheim (2007).

Michael J. Kidger, *Fundamental Optical Design*, SPIE Press, Bellingham (2002).

Michael J. Kidger, *Intermediate Optical Design*, SPIE Press, Bellingham (2004).

Rudolf Kingslake (Ed.), *Applied Optics and Optical Engineering*, Vol. 3, Academic Press, New York (1965). [Chapters regarding eyepieces, photographic objectives, and lens design.]

Rudolf Kingslake, *Optical System Design*, Academic Press, Orlando (1983).

Rudolf Kingslake, *A History of the Photographic Lens*, Academic Press, San Diego (1989).

Rudolf Kingslake, *Optics in Photography*, SPIE Press, Bellingham (1992).

Dietrich Korsch, *Reflective Optics*, Academic Press, San Diego (1991).

Milton Laikin, *Lens Design, Fourth Edition*, Taylor & Francis, New York (2006).

Daniel Malacara and Zacarias Malacara, *Handbook of Lens Design*, Marcel Dekker, New York (1994).

Virendra N. Mahajan, *Optical Imaging and Aberrations, Part I*, SPIE Press, Bellingham (1998).

Virendra N. Mahajan, *Optical Imaging and Aberrations, Part II*, SPIE Press, Bellingham (2001).

Pantazis Mouroulis and John Macdonald, *Geometrical Optics and Optical Design*, Oxford Press, New York (1997).

Sidney F. Ray, *The Photographic Lens*, Focal Press, Oxford (1979).

Sidney F. Ray, *Applied Photographic Optics, Second Edition*, Focal Press, Oxford (1994).

Harrie Rutten and Martin van Venrooij, *Telescope Optics: Evaluation and Design*, Willmann-Bell, Richmond (1988).

Robert R. Shannon, *The Art and Science of Optical Design*, Cambridge University Press, Cambridge (1997).

Robert R. Shannon and James C. Wyant (Eds.), *Applied Optics and Optical Engineering*, Vol. 8, Academic Press, New York (1980). [Chapters regarding aspheric surfaces, photographic lenses, automated lens design, and image quality.]

Robert R. Shannon and James C. Wyant (Eds.), *Applied Optics and Optical Engineering*, Vol. 10, Academic Press, New York (1987). [Chapters regarding afocal lenses and Zernike polynomials.]

Gregory H. Smith, *Practical Computer-Aided Lens Design*, Willmann-Bell, Richmond, VA (1998).

Gregory H. Smith, *Camera Lenses from Box Camera to Digital*, SPIE Press, Bellingham (2006).

Warren J. Smith, *Modern Lens Design, Second Edition*, McGraw-Hill, New York (2005).

Warren J. Smith, *Modern Optical Engineering, Fourth Edition*, McGraw-Hill, New York (2008).

W. T. Welford, *Aberrations of Optical Systems*, Adam Hilger, Bristol (1986).

R. N. Wilson, *Reflecting Telescope Optics I, Second Edition*, Springer-Verlag, Berlin (2004).

ENDNOTES

[1] D. P. Feder, "Automatic optical design," *Appl. Opt.*, 2:1209 (1963).

[2] William G. Peck, "Automatic lens design," in *Applied Optics and Optical Engineering*, Vol. 8, Chap. 4, Robert R. Shannon and James C. Wyant (Eds.), Academic Press, New York (1980).

[3] R. Barry Johnson, "Knowledge-based environment for optical system design," *1990 Intl Lens Design Conf*, George N. Lawrence (Ed.), *Proc. SPIE*, 1354:346–358 (1990).

[4] D. P. Feder, "Differentiation of raytracing equations with respect to construction parameters of rotationally symmetric optics," *J. Opt. Soc. Am.*, 58:1494 (1968).

[5] The actual defect for these is actually a target value (typically unity) minus the Strehl ratio or *MTF* value.

[6] K. Levenberg, "A method for the solution of certain non-linear problems in least squares," *Quart. Appl. Math.*, 2:164 (1944).

[7] G. Spencer, "A flexible automatic lens correction procedure," *Appl. Opt.*, 2:1257 (1963).

[8] J. Meiron, "Damped least-squares method for automatic lens design," *JOSA*, 55:1105 (1965).

[9] T. H. Jamieson, *Optimization Techniques in Lens Design*, American Elsevier, New York (1971). [This is an excellent monograph on the subject; however, methods concerning global solutions were not broadly investigated until years later.]

[10] H. Brunner, "Automatisches korrigieren unter berucksichtigung der zweiten ableitungen der gutefunktion (Automatic correction taking into consideration the second derivative of the merit function)," *Optica Acta*, 18:743–758 (1971). [Paper is in German.]

[11] Donald R. Buchele, "Damping factor for the least-squares method of optical design," *Appl. Opt.*, 7:2433–2435 (1968).

[12] Donald C. Dilworth, "Pseudo-second-derivative matrix and its application to automatic lens design," *Appl. Opt.*, 17:3372–3375 (1978).

[13] David S. Grey, "Aberration theories for semiautomatic lens design by electronic computers. I. Preliminary Remarks," *J. Opt. Soc. Am.*, 53:672–676 (1963).

[14] David S. Grey, "Aberration theories for semiautomatic lens design by electronic computers. II. A Specific Computer Program," *J. Opt. Soc. Am.*, 53:677–687 (1963).

[15] Lynn G. Seppala, "Optical interpretation of the merit function in Grey's lens design program," *Appl. Opt.*, 13:671–678 (1974).

[16] E. Glatzel and R. Wilson, "Adaptive automatic correction in optical design," *Appl. Opt.*, 7:265–276 (1968).

[17] Juan L. Rayces, "Ten years of lens design with Glatzel's adaptive method," *Proc. SPIE*, 237:75–84 (1980).

[18] Berlyn Brixner, "Automatic lens design for nonexperts," *Appl. Opt.*, 2:1281–1286 (1963).

[19] Berlyn Brixner, "The LASL lens design procedure: Simple, fast, precise, versatile," Los Alamos Scientific Laboratory, LA-7476, UC-37 (1978).

[20] Berlyn Brixner, "Lens design and local minima," *Appl. Opt*, 20:384–387 (1981).

[21] Donald C. Dilworth, "Applications of artificial intelligence to computer-aided lens design," *Proc. SPIE*, 766:91–99 (1987).

[22] G. W. Forbes and Andrew E. Jones, "Towards global optimization and adaptive simulated annealing," *Proc. SPIE*, 1354:144–153 (1990).

[23] Donald C. Dilworth, "Expert systems in lens design," *Proc. SPIE International Optical Design Conf.*, 1354:359–370 (1990).

[24] Thomas G. Kuper and Thomas I. Harris, "Global optimization for lens design: An emerging technology," *Proc. SPIE*, 1781:14 (1993).

[25] Thomas G. Kuper, Thomas I. Harris, and Robert S. Hilbert, "Practical lens design using a global method," *OSA Proc. SPIE International Optical Design Conf.*, 22:46–51 (1994).

[26] Andrew E. Jones and G. W. Forbes, "An adaptive simulated annealing algorithm for global optimization over continuous variables," *J. Global Optimization*, 6:1–37 (1995).

[27] Simon Thibault, Christian Gagné, Julie Beaulieu, and Marc Parizeau, "Evolutionary algorithms applied to lens design," *Proc. SPIE*, 5962–5968 (2005).

[28] C. Gagné, J. Beaulieu, M. Parizeau, and S. Thibault, "Human-competitive lens system design with evolution strategies," Genetic and Evolutionary Computation Conference (GECCO 2007), London (2007).

[29] Russell A. Chipman, "Polarization issues in lens design," *OSA Proc. International Optical Design Conf.*, 22:23–27 (1994).

[30] Joseph Meiron, "The use of merit functions based on wavefront aberrations in automatic lens design," *Appl. Opt.*, 7:667–672 (1968).

[31] Jessica DeGroote Nelson, Richard N. Youngworth, and David M. Aikens, "The cost of tolerancing," *Proc. SPIE*, 7433:74330E-1 (2009).

[32] L. Ivan Epstein, "The aberrations of slightly decentered optical systems," *J. Opt. Soc. Am.*, 39:847–847 (1949).

[33] Paul L. Ruben, "Aberrations arising from decentration and tilts," M.S. Thesis, Institute of Optics, University of Rochester, New York (1963).

[34] Paul L. Ruben, "Aberrations arising from decentrations and tilts," *J. Opt. Soc. Am.*, 54:45–46 (1964).

[35] G. G. Slyusarev, *Aberration and Optical Design Theory, Second Edition*, Chapter 8, Adam Hilger, Bristol (1984).

[36] S. J. Dobson and A. C. Cox, "Automatic desensitization of optical systems to manufacturing errors," *Meas. Sci. Technol.*, 6:1056–1058 (1995).

[37] David Shafer, "Optical design and the relaxation response," *Proc. SPIE*, 766:2–9 (1987).

[38] This section is not intended to be an exhaustive history, but rather a terse history from the perspective of author R. Barry Johnson.

[39] R. Kingslake, D. P. Feder, and C. P. Bray, "Optical design at Kodak," *Appl. Opt.*, 11:50–53 (1972).

[40] The National Bureau of Standards is today named The National Institute of Standards and Technology.

[41] Donald P. Feder, "Optical design with automatic computers," *Appl. Opt.*, 11:53–59 (1972).

[42] Gordon Black, "Ultra high-speed skew-ray tracing," *Nature*, 176:27 (July 1955).

[43] M. Nunn and C. G. Wynne, "Lens designing by electronic digital computer: II," *Proc. Phys. Soc.*, 74:316–329 (1959).

[44] Donald P. Feder, "Automatic lens design with a high-speed computer," *J. Opt. Soc. Am.*, 52:177–183 (1962).

[45] Was available from The Institute of Optics, University of Rochester, until the late 1970s.

[46] Was available from IBM for the IBM 1130 and then later for the IBM 360 until the late 1970s.

[47] Developed by Gordon Spencer and Pat Hennessey at Scientific Calculations, Inc., in the 1960s and became widely used in the industry during the 1970s.

[48] Developed by William Peck at Genesee Computing Center.

[49] Douglas C. Sinclair, "Super-Oslo optical design program," *Proc. SPIE*, 766:246–250 (1987).

[50] William G. Peck, "GENII optical design program," *Proc. SPIE*, 766:271–272 (1987).

[51] David. S. Grey, "Computer aided lens design: program PC FOVLY," *Proc. SPIE*, 766:273–274 (1987).

[52] Donald C. Dilworth, "SYNOPSYS: a state-of-the-art package for lens design, *Proc. SPIE*, 766:264–270 (1987).

[53] Bruce C. Irving, "A technical overview of CODE V version 7," *Proc. SPIE*, 766:285–293 (1987).

[54] Michael J. Kidger, "Developments in optical design software," *Proc. SPIE*, 766:275–276 (1987).

[55] Juan L. Rayces and Lan Lebich, "RAY-CODE: An aberration coefficient oriented lens design and optimization program," *Proc. SPIE*, 766:230–245 (1987). [Development started in 1970 and evolved later into EIKONAL as a commercial program.]

[56] Yasuhiro Ohmura, "The optical design for microlithographic lenses," *Proc. SPIE*, 6342, 63421T:1–10 (2006).

Appendix

A Selected Bibliography of Writings by Rudolf Kingslake

The Early Years

L. C. Martin and R. Kingslake, "The measurement of chromatic aberration on the Hilger lens testing interferometer," *Trans. Opt. Soc.*, XXV(4):213–218 (1923–24).

H. G. Conrady, "Study of the significance of the Foucault knife-edge test when applied to refracting systems," *Trans. Optical Soc.*, XXV(4):219–226 (1923–24).

R. Kingslake, "An experimental study of the minimum wavelength for visual achromatism," *Trans. Opt. Soc.*, XXVIII(4):173–194 (1926–27).

R. Kingslake, "A new type of nephelometer," *Trans. Opt. Soc.*, XXVI(2):53–62 (1924–25).

R. Kingslake, "The interferometer patterns due to the primary aberrations," *Trans. Opt Soc.*, XXVII(2):94–105 (1925–26).

R. Kingslake, "Recent developments of the Hartmann test," *Proc. Opt. Conf.*, Part II: 839–848 (1926).

R. Kingslake, "The analysis of an interferogram," *Trans. Opt. Soc.*, XXVII(1):1–20 (1926–27).

R. Kingslake, "Increased resolving power in the presence of spherical aberration," *Mon. Nat. Roy. Astron. Soc.*, LXXXVII(8):634–638 (1927).

R. Kingslake, "The 'absolute' Hartmann test," *Trans. Opt. Soc.* XXIX(3):133–141 (1927–28).

R. Kingslake, "18-inch coelostat for Canberra Observatory," *Nature* (Feb. 18, 1928).

The Institute Years

R. Kingslake, "A new bench for testing photographic lenses," *J. Opt. Soc. Amer.*, 22(4):207–222 (1932).

R. Kingslake and A. B. Simmons, "A method of projecting star images having coma and astigmatism," *J. Opt. Soc. Amer.*, 23:282–288 (1933).

R. Kingslake, "The development of the photographic objective," *J. Opt. Soc. Amer.*, 24(3):73–84 (1934).

R. Kingslake, "The measurement of the aberrations of a microscope objective," *J. Opt. Soc. Amer.*, 26(6):251–256 (1936).

R. Kingslake, "The knife-edge test for spherical aberration," *Proc. Phys. Soc.*, 49:289–296 (1937).

R. Kingslake and H. G. Conrady, "A refractometer for the new infrared," *J. Opt. Soc. Amer.*, 27(7):257–262 (1937).

R. Kingslake, "An apparatus for testing highway sign reflector units," *J. Opt. Soc. Amer.*, 28(9):323–326 (1938).

DOI: 10.1016/B978-0-12-374301-5.00023-1

The Kodak Years

R. Kingslake, "Recent development in photographic objectives," *J. Phot. Soc. Amer.*, 5(2):22–24 (1939).

R. Kingslake, "The design of wide-aperture photographic objectives," *J. Appl. Phys.*, 11(4):56–69 (1940).

R. Kingslake, "Lenses for amateur motion-picture equipment," *J. Soc. Motion Picture Eng.*, 34:76–87 (1940).

R. Kingslake, "Resolution tests on 16 mm. projection lenses," *J. Soc. Motion Picture Eng.*, 37:70–75 (1941).

R. Kingslake, "Lenses for aerial photography," *J. Opt. Soc. Amer.*, 32(3):129–134 (1942).

R. Kingslake, "Optical glass from the viewpoint of the lens designer," *J. Amer. Ceramic Soc.*, 27(6):189–195 (1944).

R. Kingslake, "The effective aperture of a photographic objective," *J. Opt. Soc. Amer.*, 35(8):5189–5520 (1945).

R. Kingslake, "A classification of photographic lens types," *J. Opt Soc. Amer.*, 36(5):251–255 (1946).

R. Kingslake, "Recent developments in lenses for aerial photography," *J. Opt. Soc. Amer.*, 37(1):1–9 (1947).

R. Kingslake, "The diffraction structure of elementary coma image," *Proc. Phys. Soc.*, 61:147–158 (1948).

R. Kingslake and P. F. DePaolis, "New optical glasses," *Nature*, 163:412–417 (1949).

R. Kingslake, "The development of the zoom lens," *J. Soc. Motion Picture Eng.*, 69:534–544 (1960).

R. Kingslake, "Trends in zoom," *Perspective*, 2:362–373 (1960).

R. Kingslake, "The contribution of optics to modern technology and a buoyant economy," in *Applied Optics at Imperial College* (1917–18 and 1967–68); also *Optica Acta*, 15(5):417–429 (1968).

R. Kingslake, D. P. Feder, and C. P. Bray, "Optical design at Kodak," *Appl. Opt.*, 11(91):50–59 (1972).

R. Kingslake, "Some impasses in applied optics" (Ives Medal Lecture), *J. Opt. Soc. Amer.*, 64(2):123–127 (1974).

R. Kingslake, "My fifty years of lens design," *Forum, Optical Engineering*, 21(2):SR-038–039 (1982).

Books and Edited Volumes

R. Kingslake, *Lenses in Photography*, Garden City Books, New York (1951); Second Edition, A.S. Barnes, New York (1963).

A. E. Conrady, *Applied Optics and Optical Design, Part II*, Dover Publications, New York (1960).

R. Kingslake, "Classes of Lenses" and "Projection" in *SPSE Handbook of Photographic Science and Engineering*, pp. 234–257, 982–998, Woodlief Thomas (Ed.), Wiley Interscience, New York (1973).

R. Kingslake, "Camera Optics" in *Leica Manual*, Fifteenth Edition, pp. 499–521, D. O. Morgan, D. Vestal, and W. Broecker (Eds.), Morgan and Morgan, New York (1973).

R. Kingslake, *Lens Design Fundamentals*, Academic Press, New York (1978).

R. Kingslake, *Optical System Design*, Academic Press, New York (1983).

R. Kingslake, *A History of the Photographic Lens*, Academic Press, New York (1989).

R. Kingslake, *Optics in Photography*, SPIE Press, Bellingham (1992).

R. Kingslake, *The Photographic Manufacturing Companies of Rochester New York*, George Eastman House, Rochester (1997).

R. Kingslake (Ed.), *Applied Optics and Optical Engineering*, Vol. I–III, Academic Press, New York (1965); Vol. IV (1967), Vol. V (1969).

R. Kingslake and B.J. Thompson (Eds.), *Applied Optics and Optical Engineering*, Vol. VI, Academic Press, New York (1980).

Index

Note: Bold numbers indicate the pages on which important information can be found.

Printed and bound by CPI Group (UK) Ltd, Croydon, CR0 4YY

17/10/2024

01775346-0001